JOHN TYNDALL

HEAT

A MODE OF MOTION

Elibron Classics
www.elibron.com

Elibron Classics series.

ISBN 1-4021-6852-7 (paperback)
ISBN 1-4212-8274-7 (hardcover)

This Elibron Classics Replica Edition is an unabridged facsimile of the edition published in 1870 by Longmans, Green, and Co., London.

HEAT

A MODE OF MOTION.

LONDON: PRINTED BY
SPOTTISWOODE AND CO., NEW-STREET SQUARE
AND PARLIAMENT STREET

HEAT

A MODE OF MOTION.

BY

JOHN TYNDALL, LL.D. F.R.S.

ETC.

PROFESSOR OF NATURAL PHILOSOPHY IN THE ROYAL INSTITUTION
OF GREAT BRITAIN.

FOURTH EDITION.

LONDON:
LONGMANS, GREEN, AND CO.
1870.

TO

HIS FRIEND AND TEACHER ;

ROBERT BUNSEN

THIS BOOK IS DEDICATED

BY

JOHN TYNDALL.

1870.

'Again the concretion of ice will not endure a dry attrition without liquation; for if it be rubbed long with a cloth it melteth.'

SIR THOMAS BROWNE, *Pseudodoxia Epidemica*, Book 2, chap. i.
Pickering's Edition, 1835, vol. ii. p. 271.

PREFACE

TO

THE FOURTH EDITION.

THE FOURTH EDITION of '*Heat as a Mode of Motion*' embraces some account of my recent researches on the chemical reactions of light, considered from a dynamical point of view. In former editions the relations of gaseous matter to the longer waves of the spectrum were developed. In this one the relations of the same form of matter to the shorter waves are also submitted to experimental treatment. The two lines of enquiry are united by the conception of molecular motion, which is common to both.

The bearing of some of the results on two of the greatest questions of Meteorology—the Blue of the Sky, and the Polarisation of its Light—is likewise pointed out and explained.

I have also added a hypothesis regarding the constitution of Comets; and a few remarks on the Polarisation of Heat.

J. T

ROYAL INSTITUTION:
March 1870.

PREFACE

TO

THE THIRD EDITION.

SOME YEARS AGO I had the honour of holding a number of Examinerships under the Council for Military Education. It was also my privilege to be Examiner for the University of London.

These, and the examinations connected with my public lectures, gave me an opportunity of making myself acquainted, to some extent, with the knowledge and needs of England as regards the department of Natural Knowledge which it is my vocation to cultivate.

The experience thus obtained was supplemented by that derived from conversation with eminent scholars, who deprecated and deplored the utter want of scientific knowledge, and the utter absence of sympathy with scientific studies, which mark the great bulk of our otherwise cultivated English public.

Though regarding original investigation as the great object of my life, I thought it no unworthy work to attempt to supply the deficiencies here indicated. The idea arose, and gained consistence by reflection, of taking, as far as time permitted, the various parts of Natural

Philosophy, treated in my lectures, in succession, and of describing and illustrating with clearness and simplicity such conceptions regarding them as the best culture I could command enabled me to entertain.

The firstfruit of this idea was the work on Heat, the third edition of which is now before the reader.

The reception of the work proved that it met a general want. Not only has its success in this country been far greater than was ever hoped for, but large editions of it have been published and circulated in France, Russia, and the United States.

Something more, however, than its rapid diffusion among the general public was needed to convince me that the work was such as I desired it to be.

This assurance came to me, both privately and publicly, from scientific sources, and lately in a very striking form from Germany. The beautiful translation of the work by Helmholtz and Wiedemann, issued by Vieweg of Brunswick, and the reception of that translation by the press of Germany, are to me the best guarantee and the most gratifying evidence that I have not entirely missed my aim.

That aim was to combine soundness of matter with a style which should arouse interest and sympathy in persons uncultured in science. I had, also, reason to believe that the more specially scientific student would find in the work help and furtherance, in forming definite conceptions of those molecular processes which underlie both chemical and physical phenomena.

The second instalment of the task contemplated was

the work on Sound recently published by Longmans. The reception of the work in this country has been also far more flattering than I had ventured to anticipate. It has, moreover, been already published in America. In France a translation of it is being prepared by M. Gauthier-Villars, while in Germany the two eminent men already named have taken it under their protection.

All this convinces me that if a scientific man take the trouble, which in my case is immense, of thinking, and writing, with life and clearness, he is sure to gain general attention. It can hardly be doubted, if fostered and strengthened in this way, that the desire for scientific knowledge will ultimately correct the anomalies which beset our present system of education.*

Besides other additions and alterations, a considerable amount of matter, derived mainly from my own recent investigations, is added, in a new chapter, to the present edition. In order to prevent the book from assuming an inconvenient size, I have omitted most of the Supplementary Appendices. Within the coming year I hope to collect and publish, in a single volume, the original memoirs on Experimental Physics, which I have communicated to the 'Philosophical Transactions' and 'Philosophical Magazine' during the last eighteen years. These memoirs will embrace all the supplementary matter referred to, and they may be consulted by those who wish to carry their studies beyond the limits prescribed to an elementary work.

* Huxley's 'Lessons in Elementary Physiology' is a great step in this direction.

It will interest the scientific student to learn that Mayer and Clausius have recently published, in a collected form, their celebrated researches on the Dynamical Theory of Heat, an English translation of the first Part of the memoirs of Clausius having been edited by Professor Hirst. It is to be hoped that the investigations of Joule, Helmholtz, Thomson, and Rankine, on this, the greatest scientific subject hitherto unfolded by the human mind, may ultimately be rendered equally accessible. The memoirs of Sir William Thomson, at once varied and profound, would be of especial interest and importance.

ROYAL INSTITUTION:
January 1868.

PREFACE

TO

THE SECOND EDITION.

IN the First Edition of this work, the language employed was almost verbatim that of the lecture-room. For this, had time permitted, I should willingly, in this edition, have substituted a graver style, though it may be doubted whether the change would have added to the clearness of the exposition.

The word 'Lecture,' formerly used as a heading, has, however, been abandoned, the work being now divided into thirteen chapters. I have sought to embody in it an abstract of my own researches on radiant heat, completed since the publication of the First Edition. That portion of the work which treats of the phenomena of vitality has also, for the most part, been rewritten.

ROYAL INSTITUTION:
February 1865.

PREFACE

TO

THE FIRST EDITION.

IN the following Lectures I have endeavoured to bring the rudiments of a new philosophy within the reach of a person of ordinary intelligence and culture.

The first seven Lectures of the course deal with *thermometric heat*; its generation and consumption in mechanical processes; the determination of the mechanical equivalent of heat; the conception of heat as molecular motion; the application of this conception to the solid, liquid, and gaseous forms of matter; to expansion and combustion; to specific and latent heat; and to calorific conduction.

The remaining five Lectures treat of *radiant heat*; the interstellar medium, and the propagation of motion through this medium; the relations of radiant heat to ordinary matter in its several states of aggregation; terrestrial, lunar, and solar radiation; the constitution of the sun; the possible sources of his energy; the relation of this energy to terrestrial forces, and to vegetable and animal life.

My aim has been to rise to the level of these questions from a basis so elementary, that a person possessing any imaginative faculty and power of concentration, might accompany me.

Wherever additional remarks, or extracts, seemed likely to render the reader's knowledge of the subjects referred to in any Lecture more accurate or complete, I have introduced such extracts, or remarks, as an Appendix to the Lecture.

For the use of the Plate at the end of the volume, I am indebted to the Council of the Royal Society; it was engraved to illustrate some of my own memoirs in the 'Philosophical Transactions.' For some of the Woodcuts I am also indebted to the same learned body.

To the scientific public, the names of the builders of this new philosophy are already familiar. As experimental contributors, Rumford, Davy, Faraday, and Joule, stand prominently forward. As theoretic writers (placing them alphabetically), we have Clausius, Helmholtz, Kirchhoff, Mayer, Rankine, Thomson; and in the memoirs of these eminent men the student who desires it must seek a deeper acquaintance with the subject. MM. Regnault and Séguin also stand in honourable relationship to the Dynamical Theory of Heat, and M. Verdet has recently published two lectures on it, marked by the learning for which he is conspicuous. To the English reader it is superfluous to mention the well-known and highly-prized work of Mr. Grove.*

* The beautiful experiments of M. Favre ought to be referred to here: and also, in connection with a subject treated in Chapter XIII., a most important experiment by M. Foucault, which is described in the 'Philosophical Magazine,' vol. xix. p. 194 (Feb. 1865).

I have called the philosophy of Heat a new philosophy, without, however, restricting the term to the subject of Heat. The fact is, it cannot be so restricted; for the connection of this agent with the general energies of the universe is such, that if we master it perfectly, we master all. Even now we can discern, though but darkly, the greatness of the issues which connect themselves with the progress we have made—issues which were probably beyond the contemplation of those, by whose industry and genius the foundations of our present knowledge were laid.

In a Lecture on the 'Influence of the History of Science on Intellectual Education,' delivered at the Royal Institution, Dr. Whewell has shown 'that every advance in intellectual education has been the effect of some considerable scientific discovery, or group of discoveries.' If the association here indicated be invariable, then, assuredly, the views of the connection and interaction of natural forces—organic as well as inorganic—vital as well as physical—which have grown, and which are to grow, out of the investigation of the laws and relations of Heat, will profoundly affect the intellectual discipline of the coming age.

In the study of Nature two elements come into play, which belong respectively to the world of sense and to the world of thought. We observe a fact and seek to refer it to its laws,—we apprehend the law, and seek to make it good in fact. The one is Theory, the other is Experiment; which, when applied to the ordinary purposes of life, becomes Practical Science. Nothing could illustrate more forcibly the wholesome interaction of these two elements, than the history of our present subject. If the steam-engine had not been invented, we should assuredly stand below the theoretic level which we now occupy. The achievements of Heat through the steam-engine have forced, with augmented emphasis, the question upon thinking minds—'What is this agent, by means of which we can supersede the force of winds and rivers—of horses and of men? Heat can produce mechanical force, and mechanical force can produce Heat; some common quality must therefore unite this agent and the ordinary forms of mechanical power.' This relationship established, the generalising intellect could pass at once to the other energies of the universe, and it now perceives the principle which unites them all. Thus the triumphs of practical skill have promoted the development of philosophy. Thus, by the interaction of thought and fact, of truth conceived and truth executed, we have made our Science what it is,—the noblest growth of modern times, though as yet but partially appealed to as a source of individual and national might.

As a means of intellectual education its claims are still disputed, though, once properly organised, greater and more beneficent revolutions await its employment here, than those which have already marked its applications in the material world. Surely the men whose noble vocation it is to systematise the culture of England, can never allow this giant power to grow up in their midst without endeavouring to turn it to practical account. Science does not need their protection, but it desires their friendship on honourable terms; it wishes to work with them towards the great end of all education,—the bettering of man's state. By continuing to decline the offered hand, they invoke a contest which can have but one result. Science must grow. Its development is as necessary and as irresistible as the motion of the tides, or the flowing of the Gulf-Stream. It is a phase of the energy of Nature, and as such is sure, in due time, to compel the recognition, if not to win the alliance, of those who now decry its influence and discourage its advance.

ROYAL INSTITUTION:
February 1863.

CONTENTS.

CHAPTER I.

Instruments—Generation of Heat by Friction, Compression, and Percussion —Experiments of Rumford—Water boiled by Friction—Consumption of Heat in Work . . . . PAGE 1

APPENDIX TO CHAPTER I.

Notes on the Thermo-electric Pile and Galvanometer . . . . PAGE 17

CHAPTER II.

The Nature of Heat—The Material Theory—The Dynamical Theory—Thermal Effects of Air in Motion—Generation of Heat by Rotation between the Poles of a Magnet—Experiments of Rumford, Davy, and Joule —The Mechanical Equivalent of Heat—Heat generated by Projectiles— —Heat which would be generated by stopping the Earth's Motion—Meteoric Theory of the Sun's Heat—Flame in its Relation to the Dynamical Theory . . . . PAGE 23

APPENDIX TO CHAPTER II.

Extracts from Bacon and Rumford . . . . . . PAGE 53

CHAPTER III.

Expansion: the Solid, Liquid, and Gaseous Forms of Matter—New Hypotheses regarding the Constitution of Gases—Coefficient of Expansion—Heat imparted to a Gas under Constant Pressure—Heat imparted to a Gas at Constant Volume—Mayer's Calculation of the Mechanical Equivalent of Heat—Dilatation of Gases without Refrigeration—Absolute Zero of Temperature—Expansion of Liquids and Solids: Anomalous Deportment of Water and Bismuth—Energy of the Force of Crystallisation—Thermal Effect of stretching Wires—Anomalous Deportment of India-rubber . . . . PAGE 59

APPENDIX TO CHAPTER III.

Additional Data concerning Expansion—Extracts from Sir H. Davy's first Scientific Memoir: Fusion of Ice by Friction, etc. . . . . PAGE 91

CHAPTER IV.

The Trevelyan Instrument—Gore's Revolving Balls—Influence of Pressure on Fusing Point—Liquefaction and Lamination of Ice by Pressure—Dissection of Ice by a Calorific Beam—Liquid Flowers and their Central Spot—Mechanical Properties of Water purged of Air—The Boiling Point of Liquids: Influencing Circumstances—Conversion of Heat into Work in the Steam-engine: the Geysers of Iceland . . . PAGE 99

CHAPTER V.

Application of the Dynamical Theory to the Phenomena of Specific and Latent Heat—Definition of Energy: Potential and Dynamic Energy—Energy of Molecular Forces—Experimental Illustrations of Specific and Latent Heat—Mechanical Values of the Acts of Combination, Condensation, and Congelation in the case of Water—Solid Carbonic Acid—The Spheroidal State of Liquids—Floating of a Spheroid on its own Vapour—Freezing of Water and Mercury in a Red-hot Crucible . . . PAGE 130

CHAPTER VI.

Convection of Heated Air—Winds—The Upper and Lower 'Trades'—Effect of the Earth's Rotation on the Direction of Wind—Influence of Aqueous Vapour upon Climate—Europe the Condenser of the Western Atlantic—Rainfall in Ireland—The Gulf-Stream—Formation of Snow—Formation of Ice from Snow—Glaciers—Phenomena of Glacier Motion—Regelation—Moulding of Ice by Pressure—Ancient Glaciers . PAGE 163

CHAPTER VII.

Conduction a Transmission of Motion—Good Conductors and Bad Conductors—Conductivity of the Metals for Heat: Relation between the Conductivity of Heat and that of Electricity—Influence of Temperature on the Conduction of Electricity—Influence of Molecular Constitution on the Conduction of Heat—Relation of Specific Heat to Conduction—Philosophy of Clothes: Rumford's Experiments—Influence of Mechanical Texture on Conduction—Incrustations of Boilers—The Safety Lamp—Conductivity of Liquids and Gases: Experiments of Rumford and Despretz—Cooling Effect of Hydrogen Gas—Experiments of Magnus on the Conductivity of Gases . . . . PAGE 190

CHAPTER VIII.

Cooling a Loss of Motion: to what is this Motion imparted?—Experiments on Sound bearing on this question—Experiments on Light bearing on this question—The Theories of Emission and Undulation—Length of Waves and Number of Impulses of Light—Physical Cause of Colour—Invisible Rays of the Spectrum—The Calorific Rays beyond the Red—The Chemical Rays beyond the Blue—Definition of Radiant Heat—Reflection of Radiant Heat from Plane and Curved Surfaces: Laws the same as those of Light—Conjugate Mirrors . . . . PAGE 228

APPENDIX TO CHAPTER VIII.

On Singing Flames . . . . PAGE 254

CHAPTER IX.

Law of Diminution with the Distance—The Waves of Sound Longitudinal; those of Light Transversal—When they oscillate, the Molecules of different Bodies communicate different Amounts of Motion to the Ether—Radiation the Communication of Motion to the Ether; Absorption the Acceptance of Motion from the Ether—Those Surfaces which Radiate well Absorb well—A close Woollen Covering facilitates Cooling—Preservative Influence of Gold-leaf—The Atoms of Bodies intercept certain Waves, and allow others to pass—Transparency and Diathermancy—Diathermic Bodies Bad Radiators—Definition of the term 'Quality' as applied to Radiant Heat—The Rays which pass without Absorption do not Heat the Medium—Proportion of Luminous and Obscure Rays in various Flames . . . . PAGE 267

CHAPTER X.

Absorption of Heat by Gaseous Matter—Apparatus employed—Early difficulties—Diathermancy of Air and of the Transparent Elementary Gases—

Athermancy of Olefiant Gas and of other Compound Gases—Absorption of Radiant Heat by Vapours—Radiation of Heat by Gases—Reciprocity of Radiation and Absorption—Influence of Molecular Constitution on the Passage of Radiant Heat—Transmission of Heat through Opaque Bodies—Heat-Spectrum detached from Light-Spectrum by an Opaque Prism . . . . PAGE 294

APPENDIX TO CHAPTER X.

Calibration of the Galvanometer . . . . PAGE 330

CHAPTER XI.

Action of Odorous Substances upon Radiant Heat—Action of Ozone upon Radiant Heat—Determination of the Radiation and Absorption of Gases and Vapours without any Source of Heat external to the Gaseous Body—Dynamic Radiation and Absorption—Radiation through the Earth's Atmosphere—Influence of the Aqueous Vapour of the Atmosphere on Radiant Heat—Connection of the Radiant and Absorbent Power of Aqueous Vapour with Meteorological Phenomena . . . . . PAGE 333

APPENDIX TO CHAPTER XI.

Further Details of the Action of Humid Air . . . . . PAGE 370

CHAPTER XII.

Absorption of Heat by Volatile Liquids—Absorption of Heat by the Vapours of those Liquids at a Common Pressure—Absorption of Heat by the same Vapours when the Quantities of Vapour are Proportional to the Quantities of Liquid—Comparative View of the Action of Liquids and their Vapours upon Radiant Heat—Physical Cause of Opacity and Transparency—Influence of Temperature on the Transmission of Radiant Heat—Changes of Position through Changes of Temperature—Radiation from Flames—Influence of Oscillating Period on the Transmission of Radiant Heat—Explanation of Certain Results of Melloni and Knoblauch . PAGE 378

CHAPTER XIII.

Discovery of Dark Solar Rays—Herschel and Müller's Experiments—Rise of Intensity with Temperature—Heat of Electric Spectrum—Ray-Filters; Sifting the Electric Light—Transmutation of Rays—Thermal Image rendered Luminous—Combustion and Incandescence by Dark Rays—

Fluorescence and Caloresceuce—Dark Solar Rays—Dark Lime-light Rays—Franklin's Experiment on Colours—Its Analysis and Explanation . . . . PAGE 413

CHAPTER XIV.

Dew:—A Clear Sky and Calm but Damp Atmosphere necessary for its copious Formation—Dewed Substances Colder than Undewed Ones—Dewed Substances better Radiators than Undewed Ones—Dew is the Condensation of the Atmospheric Vapour on Substances which have been Chilled by Radiation—Lunar Radiation—Constitution of the Sun—The Bright Lines in the Spectra of the Metals—An Incandescent Vapour absorbs the Rays which it can itself emit—Kirchhoff's Generalisation—Fraunhofer's Lines—Solar Chemistry—Emission of the Sun—Herschel and Pouillet's Experiments—Mayer's Meteoric Theory—Theories of Helmholtz and Thomson—Effect of the Tides on the Earth's Rotation—Energies of the Solar System—Helmholtz, Thomson, Waterston—Relation of the Sun to Animal and Vegetable Life . . . . PAGE 457

CHAPTER XV.

Action of Ether Waves of Short Period upon Gaseous Matter—Clouds formed by Actinic Decomposition—Colour produced by Small Particles—Polarisation of Light by Nebulous Matter—The Blue of the Sky and the Polarisation of its Light . . . . PAGE 504

Remarks on the Convertibility of Natural Forces . . . . PAGE 543
On Cometary Theory . . . . PAGE 547
Polarisation of Heat . . . . PAGE 553
Concluding Additions; anticipations of the Mechanical Theory of Heat . . . . PAGE 555
Production of fire by savages . . . . PAGE 557
Morning chill producing snow in a room . . . . . . PAGE 557

INDEX . . . . PAGE 559

HEAT

AS

A MODE OF MOTION.

CHAPTER I.

INSTRUMENTS—GENERATION OF HEAT BY FRICTION, COMPRESSION, AND PERCUSSION—EXPERIMENTS OF RUMFORD—WATER BOILED BY FRICTION—CONSUMPTION OF HEAT IN WORK.

APPENDIX:—NOTES ON THE THERMO-ELECTRIC PILE AND GALVANOMETER.

(1) THE aspects of Nature provoke in man the spirit of enquiry. As the eye is made for seeing, and the ear for hearing, so the human mind is formed for exploring and understanding the relationship of natural phenomena, the Science of our day being the direct issue of an intellect thus endowed. One great characteristic of Natural Knowledge is its growth; all its results are fruitful, every new discovery becoming instantly the germ of fresh investigation. But no nobler example of this growth can be adduced than the expansion and development, during the last five-and-twenty years, of the great subject which is now to occupy our attention.

In scientific manuals, only scanty reference was, until lately, made to the modern philosophy of Heat, and thus the public knowledge regarding it remained below the attainable level. The reserve, however, was natural, for the subject is an entangled one, and, in the pursuit of it, we must be prepared to encounter difficulties. In the whole range of Natural Science, however, there are none more worthy of being overcome—none the subjugation of which ensures a greater reward to the worker. For the various agencies of Nature are so connected, that in mastering the laws and relations of Heat, we make clear to our minds the interdependence of natural powers generally. Let us then commence our labours with heart and hope; let us familiarise ourselves with the latest facts and conceptions regarding this all-pervading agent, and seek diligently the links of law which connect the facts and give unity to their most diverse appearances. If we succeed here, we shall satisfy, to an extent unknown before, that love of order and of beauty which, no doubt, is implanted in the mind of every person here present. From the heights at which we aim we shall have nobler glimpses of the system of Nature than could possibly be obtained, if I, while acting as your guide in the region which we this day enter, were to confine myself to its lower levels and already trodden roads.

(2) It is my first duty to make you acquainted with some of the instruments intended to be employed in the examination of this question. Some means must be devised of making the indications of heat and cold visible to you, and for this purpose an ordinary thermometer would be useless. You could not see its action; and I am anxious that you should see, with your own eyes, the facts on which our subsequent philosophy is to be based. I wish to give you the material on which an independent judgment may be founded; to enable you to reason as I reason if you deem me right, to correct me if I go astray, and to censure me if you find me dealing unfairly with my subject. To secure these ends I have been obliged to abandon the use of a common thermometer, and to resort to the little instrument which you see before me on the table.

(3) This instrument A B (fig. 1) is called a *thermo-electric pile.** It acts thus:—The heat which the pile receives generates an electric current; and an electric current has the power of deflecting a freely suspended magnetic needle, to which it flows parallel. Before you is placed such a needle *m n* (fig. 1), surrounded by a covered copper wire, the free ends of which, *w w*, are connected with the thermo-electric pile. The needle is suspended by a fibre, *s s*, of unspun silk, and protected by a glass

FIG. 1.

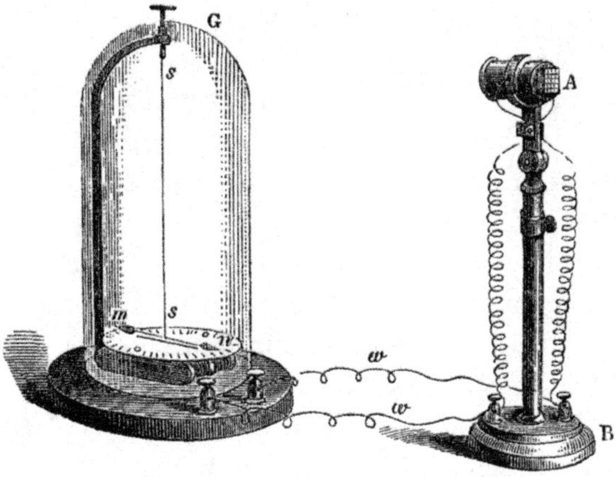

shade, G, from all disturbance by currents of air. To one end of the needle is fixed a piece of red, and to the other end a piece of blue paper. All of you see these pieces of paper, and when the needle moves, its motion will be clearly visible to the most distant person in this room. This instrument is called a *galvanometer.*†

* A brief description of the thermo-electric pile is given in the Appendix to this Chapter.

† In the actual arrangement the galvanometer here described stood on a stool in front of the lecture table, the wires *w w* being sufficiently long to reach from the table to the stool. For a further description of the instrument see the Appendix to this Chapter.

(4) At present the needle is quite at rest, and points to the zero-mark on the graduated disc underneath it. This shows that there is no current passing. I breathe for an instant against the naked face A of the pile—a single puff of breath is sufficient for my purpose—the needle starts off and passes through an arc of 90°. It would go farther did we not limit its swing by fixing, edgeways, a thin plate of mica at this point. This action of the needle is produced by the small amount of warmth communicated by my breath to the face of the pile, and no ordinary thermometer could give so large and prompt an indication. Take notice of the direction of the deflection; the red end of the needle moved from me towards you. We will let the heat waste itself; it will do so in a very short time, and you notice, as the pile cools, that the needle returns to its first position. Observe now the effect of *cold* on the same face of the pile. After chilling this plate of metal by placing it on ice, I wipe the metal, and touch with it the face of the pile. A moment's contact suffices to produce a prompt and energetic deflection of the needle. But mark the direction of the deflection. When the pile was warmed, the red end of the needle moved from me towards you; now the same end moves from you towards me. The important point here established is, that from the direction in which the needle moves we can, with certainty, infer whether cold or heat has been communicated to the pile; and the energy with which the needle moves—the promptness with which it is driven aside from its position of rest—gives us some idea of the quantity of heat or cold imparted in different cases. On a future occasion we shall learn how to express the relative quantities of heat communicated to the pile with numerical accuracy; for the present a general knowledge of the action of our instruments is sufficient.

(5) My desire now is to connect heat with the more

familiar forms of force, furnishing you, in the first place, with a store of facts illustrative of the generation of heat by mechanical processes. In the next room are some pieces of wood, which my assistant will hand to me. The temperature of that room is slightly lower than the temperature of this one, and hence the wood which is now before me must be slightly colder than the face of the pile. Let us prove this. The face of the pile being placed against the piece of wood, the red end of the needle moves from you towards me, thus showing that the contact has chilled the instrument. I now carefully rub the face of the pile along the surface of the wood,—'carefully,' because the pile is brittle, and rough usage would destroy it;—mark what occurs. The prompt and energetic motion of the needle towards you declares that the face of the pile has been heated by this small amount of friction. The needle, you observe, goes quite up to 90° on the side opposite to that towards which it moved before the friction was applied.

(6) These experiments, which illustrate the development of heat by mechanical means, must be to us what a boy's school exercises are to him. In order to fix them in our minds, and obtain due mastery over them, we must repeat them and vary them in many ways. In this task you have now to accompany me. This flat piece of brass is attached to a cork, which, when taken hold of, preserves the brass from all contact with my warm hand. When the brass is placed against the face of the pile, the needle moves, showing that the metal is cold. I now rub the brass on the surface of this cold piece of wood, and lay it once more against the pile. It is so hot, that if allowed to remain in contact with the instrument, the current generated would dash the needle violently against its stops, and probably derange its magnetism. You see the strong deflection which even an instant's contact can produce. Indeed

when a boy at school, I have often blistered my hand by a brass button which had been rubbed energetically against a form. This razor has been cooled by contact with ice; and along this hone, without oil, I rub the cool razor as if to sharpen it. On placing the razor against the face of the pile, the steel, which a moment ago was cold, is declared hot. Similarly, I take this knife and knife-board, which are both cold, and rub the knife along the board. The knife, placed against the pile, declares itself to be hot. I pass this cold saw through this cold piece of wood, and place, in the first instance, the surface of the wood against which the saw has rubbed, in contact with the pile. The needle instantly moves in a direction which shows the wood to be heated. Allowing the needle to return to zero, I apply the saw itself to the pile. It also is hot. These are the simplest and most commonplace examples of the generation of heat by friction, and they are chosen for this reason. Mean as they appear, they are illustrations of a principle which determines the polity of the whole material universe.

(7) We have now to consider the development of heat by compression. This piece of deal is cooled below the temperature of the room, and gives, when placed in contact with our pile, the deflection which indicates cold. I introduce the wood between the plates of a small hydraulic press, and squeeze it forcibly. When, after compression, the wood is brought into contact with the pile, the galvanometer declares that heat has been developed by the act of compression. Precisely the same thing occurs when this block of lead is fixed between the plates of the press and squeezed thus to flatness.

(8) And now for the effect of percussion. I place a cold lead bullet upon this cold anvil, and strike it with a cold sledge-hammer. The sledge descends with a certain mechanical force, and its motion is suddenly arrested by

the bullet and anvil; apparently the force of the sledge is destroyed. But when we examine the lead we find it is heated, and we shall by-and-by learn that if we could gather up all the heat generated by the shock of the sledge, and apply it without loss mechanically, we should be able, by means of it, to lift the hammer to the height from which it fell.

Another experiment is here arranged, which is almost too delicate to be performed with the large apparatus necessary to render lecture experiments visible, but which, nevertheless, is easily executed with proper instruments. This small basin contains a quantity of mercury which has been cooled in the next room. One of the faces of the thermo-electric pile is coated with varnish, to defend it from the mercury, which would otherwise destroy the pile. Thus protected it may, as you observe, be plunged into the liquid metal. The deflection of the needle proves that the mercury is cold. These two glasses, A and B (fig. 2), are swathed thickly round by listing, to prevent the warmth of my hands from reaching the mercury. I pour the cold mercury from the one glass into the other, and back. It falls with a certain mechanical force, its motion is destroyed, but heat is developed. The amount of heat generated by a single pouring out is extremely small; the exact amount might be easily determined, but we shall defer quantitative considerations for the present; so we

FIG. 2.

will pour the mercury from glass to glass ten or fifteen times. Now mark the result when the pile is plunged into the liquid. The needle moves, and its motion declares that the mercury, which at the beginning of the experiment was cooler, is now warmer than the pile. We here introduce into the lecture-room an effect which occurs at the base of every waterfall. There are friends before me who have stood amid the foam of Niagara. Had they, when there, dipped sufficiently sensitive thermometers into the water at the top and bottom of the cataract, they would have found the latter warmer than the former. The sailor's tradition, also, is theoretically correct; the sea is rendered warmer by a storm, the mechanical dash of its billows being ultimately converted into heat.

(9) Whenever friction is overcome, heat is produced, and the heat produced is the exact measure of the force expended in overcoming the friction. The heat is simply the primitive force in another form, and if we wish to avoid this conversion, we must abolish the friction. We put oil upon the surface of a hone, we grease a saw, and are careful to lubricate the axles of our railway carriages. What is the real meaning of these acts? Let us obtain general notions first, and aim at strict accuracy afterwards. It is the object of a railway engineer to urge his train from one place to another; he wishes to apply the force of his steam, or of the furnace which gives tension to his steam, to this particular purpose. It is not his interest to allow any portion of that force to be converted into another form of force which would not promote the attainment of his object. He does not want his axles heated, and hence he avoids as much as possible expending his power in heating them. In fact, he has obtained his force from heat, and it is not his object to reconvert by friction the force thus obtained into its primitive form. For every degree of temperature generated in his axles, a definite amount would be with-

drawn from the urging force of his engine. There is no absolute loss. Could we gather up all the heat generated by the friction, and could we apply it all mechanically, we should, by it, be able to impart to the train the precise amount of speed which it had lost by the friction. Thus every one of those railway porters whom you see moving about with his can of yellow grease, and opening the little boxes which surround the carriage axles, is, without knowing it, illustrating a principle which forms the very solder of Nature. In so doing, he is unconsciously affirming both the convertibility and the indestructibility of force. He is practically asserting that mechanical energy may be converted into heat, and that when so converted it cannot still exist as mechanical energy; but that for every degree of heat developed in the axles, a strict and proportional equivalent of the *locomotive force* of the engine disappears. All the force of our locomotives is derived from heat, and all of it eventually becomes heat. To maintain the proper speed, the friction of the train must be continually overcome, and the force spent in overcoming it is entirely converted into heat. An eminent writer* has compared the process to one of distillation: the energy of heat in the furnace passes into the mechanical motion of the train, and this motion reappears as heat in the wheels, axles, and rails. When a station is approached, say at the rate of thirty miles an hour, a brake is applied, and smoke and sparks issue from the wheel on which it presses. The train is brought to rest—How? Simply by converting the entire moving force which it possessed at the moment the brake was applied, into heat.

(10) So also with regard to the greasing of a saw by a carpenter. He applies the muscular force of his arm with the express object of cutting through the wood. He wishes

* Robert Julius Mayer of Heilbronn, in the Kingdom of Würtemburg.

to tear the wood asunder, to overcome its mechanical cohesion by the teeth of his saw. When the saw moves stiffly, on account of the friction against its flat surface, the same amount of effort may produce a much smaller effect than when the implement moves without friction. But in what sense smaller? Not absolutely so, but smaller as regards the act of sawing. The force not expended in sawing is not lost, it is converted into heat; and I gave you an example of this a few minutes ago. Here again, if we could collect the heat engendered by the friction, and apply it to the urging of the saw, we should make good the precise amount of work which the carpenter, by neglecting the lubrication of his implement, had simply converted into another form of power.

(11) We warm our hands by rubbing, and in the case of frostbite we thus restore the necessary heat to the injured parts. Savages have the art of producing fire by the skilful friction of well-chosen pieces of wood. It is easy to char wood in a lathe by friction. By friction a lucifer-match is raised to the temperature of ignition. From the feet of the labourers on the flinty roads of Hampshire sparks issue copiously on a dark night, the collision of their iron-shod shoes against the flints producing fire. The same effect is often produced by the omnibus horses in the streets of London. In the common flint and steel the particles of the metal struck off are so much heated by the collision that they take fire and burn in the air. But the heat precedes the combustion. Davy found that when a gunlock with a flint was discharged in vacuo, no sparks were produced, but the particles of steel struck off, when examined under the microscope, showed signs of fusion.* Here is a large rock-crystal: I have only to draw this small one briskly over it, to produce light and

* Works of Sir H. Davy, vol. ii. p. 8.

heat. Here are two quartz-pebbles: I have only to rub them together to make them luminous.

(12) Aristotle refers to the heating of arrows by the friction of the air; a rifle-bullet, in passing through air, is also warmed by friction. The most probable theory of shooting stars is that they are small planetary bodies revolving round the sun, which are caused to swerve from their orbits by the attraction of the earth, and are raised to incandescence by friction against our atmosphere. Chladni propounded this view, and Dr. Joule has shown that the atmospheric friction is competent to produce the effect. He may, moreover, be correct in believing that the greater portion of our aërolites are scattered into fragments by heat, and the earth thus spared a terrible bombardment.* These bodies move at planetary rates; the orbital velocities of the four Interior planets are as follows:—

	Miles per Second
Mercury	30·40
Venus	22·24
Earth	18·91
Mars	15·32

while the velocity of the aërolites varies from 18 to 36 miles a second. The friction engendered by this enormous speed is certainly competent to produce the effects ascribed to it.

(13) More than sixty-four years ago Count Rumford, who was one of the founders of the Royal Institution, executed a series of experiments on the generation of heat by friction, which, viewed by the light of to-day, are of the highest interest and importance. Indeed the services which the Founders and Professors of this Institution have rendered to the philosophy of natural forces can never be forgotten. Thomas Young laid the foundations of the Undulatory Theory of light, which, in its fullest applica-

* Philosophical Magazine, 4th Series, vol. xxxii. p. 349.

tion, embraces our present theory of heat. Davy entertained substantially the same views regarding heat as those which I am now endeavouring to approach and elucidate. Faraday established the laws of equivalence between chemistry and electricity, and his magneto-electric discoveries were the very first seized upon by Joule in illustration of the mutual convertibility of heat and mechanical action.* Rumford, in a paper of great power both as regards reasoning and experiment, advocated in 1798† the doctrine regarding the nature of heat which the recent experiments of eminent men have placed upon a secure basis. While engaged in the boring of cannon at Munich, he was so forcibly struck by the large amount of heat developed in the process, that he was induced to devise a special apparatus for the examination of the generation of heat by friction. He had constructed a hollow cylinder of iron, into which fitted a solid plunger, which was caused to press against the bottom of the cylinder. A box which surrounded the cylinder contained 18¾ lbs. of water, in which a thermometer was placed. The original temperature of the water was 60° F. The cylinder was turned by horse-labour, and an hour after the friction had commenced the temperature of the water was 107°, having risen 47°. Half-an-hour afterwards he found the temperature to be 142°. The action was continued, and at the end of two hours the temperature was 178°. At the end of two hours and twenty minutes it was 200°, and at two hours and thirty minutes from the commencement *the water actually boiled!*

Rumford's description of the effect of this experiment on those who witnessed it, is quite delightful. 'It would be difficult,' he says, 'to describe the surprise and astonishment expressed in the countenances of the by-

* Philosophical Magazine, 4th Series, vol. xxiii. pp. 265, 347, 435.

† An abstract of this paper is given in the Appendix to Chapter II.

standers on seeing so large a quantity of water heated, and actually made to boil, without any fire. Though there was nothing that could be considered very surprising in this matter, yet I acknowledge fairly that it afforded me a degree of childish pleasure which, were I ambitious of the reputation of a grave philosopher, I ought most certainly rather to hide than to discover.'* I am sure we can dispense with the application of any philosophy which would stifle such emotion as Rumford here avowed. In connection with this striking experiment, Dr. Joule† has estimated the amount of mechanical force expended in producing the heat, and obtained a result which 'is not very widely different' from that which greater knowledge and more refined experiments enabled Joule himself to obtain, as regards the numerical equivalence of heat and work.

(14) It would be absurd on my part to attempt here a repetition of the experiment of Count Rumford with all

FIG. 3.

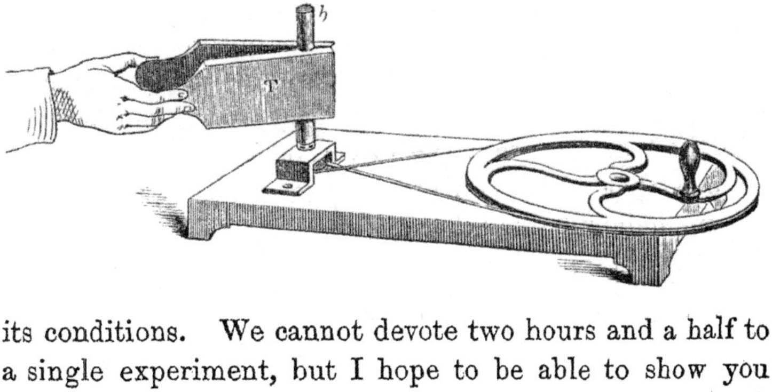

its conditions. We cannot devote two hours and a half to a single experiment, but I hope to be able to show you substantially the same effect in two minutes and a half. Here is a brass tube (*b*, fig. 3), four inches long, and of three-quarters of an inch interior diameter. It is

* Rumford's Essays, vol. ii. p. 484.

† Philosophical Transactions, vol. cxl. p. 62.

stopped at the bottom, and screwed on to a whirling table, by means of which the upright tube can be caused to rotate very rapidly. These two pieces of oak are united by a hinge, in which are two semicircular grooves, intended to embrace the brass tube. Thus the pieces of wood form a kind of tongs, T, the gentle squeezing of which produces friction when the tube rotates. I partially fill the tube with cold water, stop it with a cork to prevent the splashing out of the liquid, and now put the machine in motion. As the action continues, the temperature of the water rises, and now the tube is too hot to be held in the fingers. Continuing the action a little longer, the cork is driven out with explosive violence, the steam which follows it producing by its precipitation a small cloud in the atmosphere.

(15) In all the cases hitherto introduced to your notice, heat has been *generated* by the expenditure of mechanical force. Our experiments have shown that where mechanical force is expended heat is produced; and I wish now to bring before you the converse experiment, and show you the *consumption* of heat in mechanical work. This strong vessel (V, fig. 4) is filled at the present moment with compressed air. It has lain here for some hours, so that the temperature of the air within the vessel is now the same as that of the air of the room without it. At the present moment this inner air is pressing against the sides of the vessel, and if this cock be opened a portion of the air will rush violently out. The word 'rush,' however, but vaguely expresses the true state of things; the air which issues is driven out by the air behind it; this latter accomplishes the work of urging forward the stream of air. And what will be the condition of the *working air* during this process? It will be chilled. The air executes work, and the only agent it can call upon to perform the work is the heat to which the elastic force with which it

presses against the sides of the vessel is entirely due. A portion of this heat will be consumed, and a lowering of temperature will be the consequence. Observe the experiment. I will turn the cock *c* (fig. 4), and allow the current

FIG. 4.

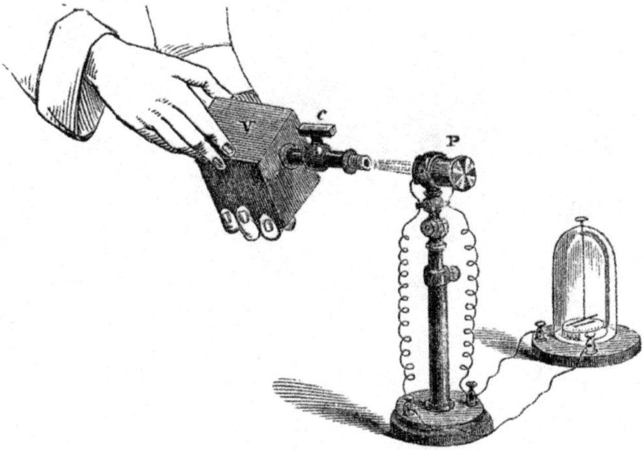

of air from the vessel V to strike against the face of the pile P. The magnetic needle instantly responds; its red end is driven towards me, thus declaring that the pile has been *chilled* by the current of air.

(16) The effect is different when air is urged from the nozzle of a common bellows against the thermo-electric pile. In the last experiment the mechanical work of urging the air forward was performed by the air itself, and a portion of its heat was consumed in the effort. In the case of the bellows, it is my muscles which perform the work. The upper board of the bellows is raised, and the air rushes in. The boards are then pressed with a certain force, and the air rushes out. The expelled air, slightly warmed by compression, strikes the face of the pile; its motion also is in part stopped, and an amount of heat equivalent to the destruction of this motion is instantly

generated. Thus you observe, that when I direct, with the bellows (fig. 5), a current of air against the pile, the red end of the needle moves towards you, thereby showing that the face of the pile is, in this instance, *warmed* by the air.* Here moreover is a bottle of soda-water, slightly warmer

FIG. 5.

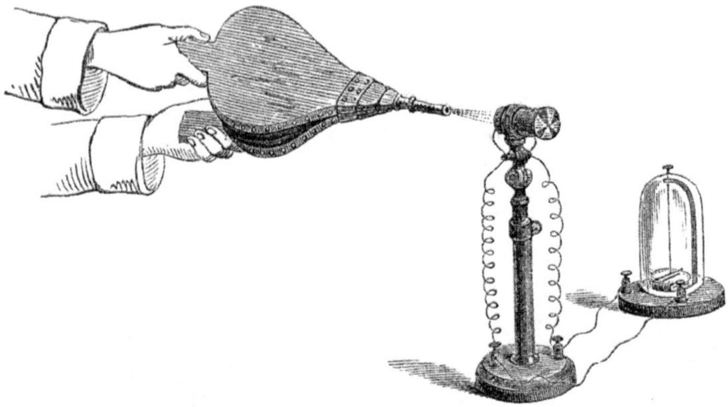

than the pile, as you see by the deflection it produces. Cut the string which holds it, the cork is driven out by the elastic force of the carbonic acid gas: the gas performs work, in so doing it consumes heat, and now the deflection produced by the bottle is that of cold. The truest romance is to be found in the details of daily life; and here, in operations with which every child is familiar, we shall gradually discern the illustration of principles from which all material phenomena flow.

* In this experiment it is necessary to bring the nozzle of the bellows near the pile and to blow strongly. When the nozzle is distant the air, which issues warm under the pressure exerted on the bellows, is chilled by its own expansion. It may be even caused to precipitate its aqueous vapour.

APPENDIX TO CHAPTER I.

NOTE
ON THE CONSTRUCTION OF THE THERMO-ELECTRIC PILE.

LET AB (fig. 6) be a bar of antimony, and BC a bar of bismuth, and let both bars be soldered together at B. Let the free ends A and C be united by a piece of wire, ADC. On warming the place of junction, B, an electric current is generated, the direction of which is from bismuth to antimony (or against the alphabet), across the junction, and from antimony to bismuth (or with the alphabet), through the connecting wire, ADC. The arrows indicate the direction of the current.

FIG. 6.

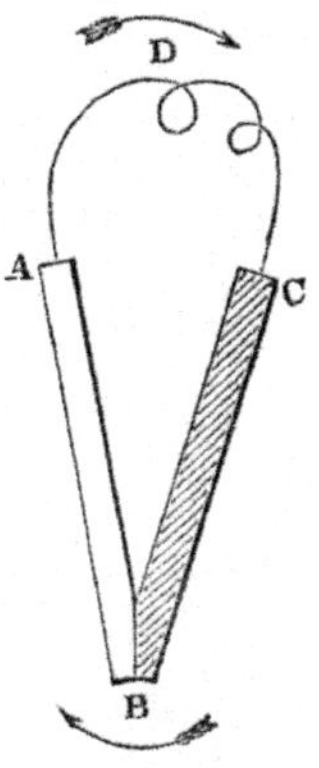

If the junction B be *chilled*, a current is generated opposed in direction to the former. The figure represents what is called a thermo-electric pair or couple.

By the union of several thermo-electric pairs a more powerful current can be generated than would be obtained from a single pair. Fig. 7 (next page), for example, represents such an arrangement, in which the shaded bars are supposed to be all of bismuth, and the unshaded ones of antimony; on warming all the junctions, B, B, &c., a current is generated in each, and the sum of these currents, all of which flow in the same direction, will produce a stronger resultant current than that obtained from a single pair.

The V formed by each pair need not be so wide as it is shown in fig. 7; it may be contracted without prejudice to the couple. And if it is desired to pack several pairs into a small compass, each separate couple may be arranged as in fig. 8, where the black lines represent small bismuth bars, and the white ones small bars of antimony. They are soldered together at the ends,

FIG 7.

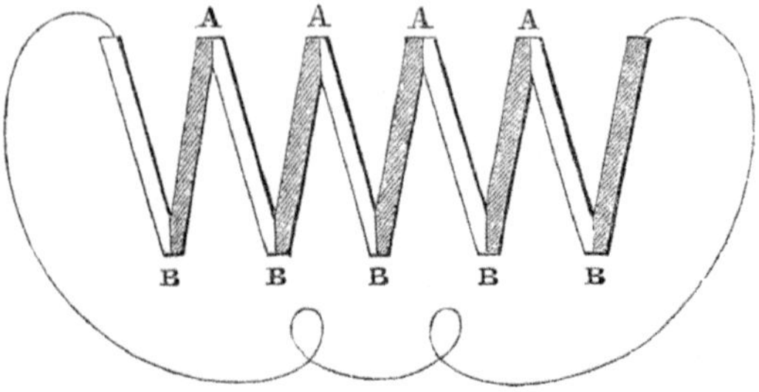

and throughout their length are usually separated by strips of paper merely. A collection of pairs thus compactly set together constitutes a thermo-electric pile, a drawing of which is given in fig. 9.

FIG. 8.

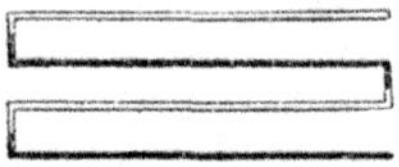

The current produced by heat being always from bismuth to antimony across the heated junction, a moment's inspection of fig. 7 will show that when any one of the junctions, A, A, is heated, a current is generated opposed in direction to that generated when the heat is applied to the junctions B, B. Hence, in the case of the thermo-electric pile, the effect of heat falling upon its two opposite faces is to produce currents in opposite directions. If the temperature of the two faces be alike, they neutralise each other, no matter how highly they may be heated absolutely; but if one of them be warmer than the other, a current is produced. The current is thus due to a *difference* of temperature between the two faces of the pile, and within certain limits the strength of the current is exactly proportional to this difference.

FIG. 9.

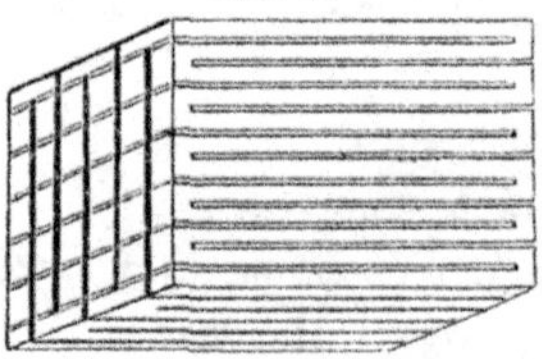

From the junction of almost any other two metals, thermo-electric currents may be obtained, but they are most readily generated by the union of bismuth and antimony.*

* The discovery of thermo-electricity is due to Thomas Seebeck, Professor

NOTE ON THE CONSTRUCTION OF THE GALVANOMETER.

The existence and direction of an electric current are shown by its action upon a freely suspended magnetic needle.

But such a needle is held in the magnetic meridian by the magnetic force of the earth. Hence, to move a single needle, the current must overcome the magnetic force of the earth.

Very feeble currents are incompetent to do this in a sufficiently sensible degree. The following two expedients are, therefore, combined to render sensible the action of such feeble currents:—

The wire through which the current flows is coiled so as to surround the needle several times; the needle must swing freely within the coil. The action of the single current is thus multiplied.

The second device is to neutralise the directive force of the earth, without prejudice to the magnetism of the needle. This is accomplished by using two needles instead of one, attaching them to a common vertical stem, and bringing their opposite poles over each other, the north end of the one needle and the south end of the other being thus turned in the same direction. The double needle is represented in fig. 10.

Fig. 10.

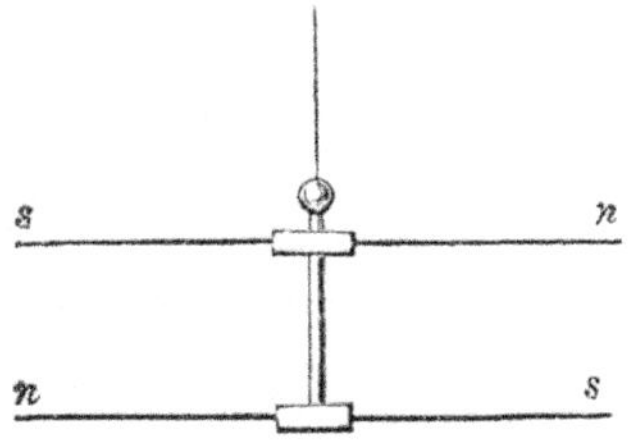

It must be so arranged that one of the needles shall be within the coil through which the current flows, while the other needle swings freely above the coil, the vertical connecting piece passing through an appropriate slit in the coil. Were both the needles within the coil, the same current would urge them in opposite directions, and thus one needle would neutralise the other. But when one is within and the other without, the current urges both needles in the same direction.

in the University of Berlin. Nobili constructed the first thermo-electric pile; but in Melloni's hands it became an instrument so important as to supersede all others in researches on radiant heat. To this purpose it will be applied on future occasions.

The way to prepare such a pair of needles is this. Magnetise both of them to saturation; then suspend them in a vessel, or under a shade, to protect them from air-currents. The system will probably set in the magnetic meridian, one needle being in almost all cases stronger than the other; weaken the stronger needle carefully by the touch of a second smaller magnet. When the needles are precisely equal in strength, they will set at *right angles to the magnetic meridian.*

It might be supposed that when the needles are equal in strength, the directive force of the earth would be completely annulled, that the double needle would be perfectly *astatic*, and perfectly neutral as regards direction; obeying simply the torsion of its suspending fibre. This would be the case if the

FIG. 11.

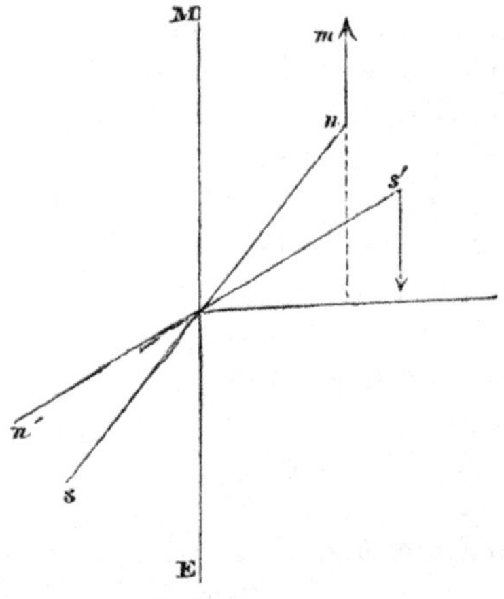

magnetic axes of both needles could be caused to lie with mathematical accuracy in the same vertical plane. In practice this is almost impossible; the axes always cross each other. Let *n s*, *n' s'* (fig. 11) represent the axes of two needles thus crossing, the magnetic meridian being parallel to M E; let the pole *n* be drawn by the earth's attractive force in the direction *n m*; the pole *s'* being urged by the repulsion of the earth in a precisely opposite direction. When the poles *n* and *s'* are of exactly equal strength it is manifest that the force acting on the pole *s'*, in the case here supposed, would have the advantage as regards leverage, and would therefore overcome the force acting on *n*. The crossed needles would therefore turn away still further from the magnetic meridian, and a little reflection will show that they cannot come to rest until the line which bisects the angle enclosed by the needles is at right angles to the magnetic meridian.

This is the test of perfect equality as regards the magnetism of the needles; but in bringing the needles to this state of perfection we have often to pass through various stages of obliquity to the magnetic meridian. In these cases the superior strength of one needle is compensated by an advantage, as regards leverage, possessed by the other. By a happy accident a touch is sometimes sufficient to make the needles perfectly equal; but many hours are often expended in securing this result. It is only of course in very delicate experiments that this perfect equality is needed; but in such experiments it is essential.

Another grave difficulty has beset experimenters, even after the perfect magnetisation of their needles has been accomplished. Such needles are sensitive to the slightest magnetic action, and the covered copper wire, of which the galvanometer coils are formed, usually contains a trace of iron sufficient to deflect the prepared needle from its true position. I have had coils in which this deflection amounted to thirty degrees; and in the splendid instruments used by Professor Du Bois Raymond, in his researches on animal electricity, the deflection by the coil is sometimes even greater than this. Melloni encountered this difficulty, and proposed that the wires should be drawn through agate holes, thus avoiding all contact with iron or steel. The disturbance has always been ascribed to a trace of iron contained in the copper wire. Pure silver has also been proposed instead of copper.

To pursue his beautiful thermo-electric researches in a satisfactory manner, Professor Magnus, of Berlin, obtained pure copper by a most laborious electrolytic process, and after the metal had been obtained it required to be melted eight times in succession before it could be drawn into wire. In fact the impurity of the coil entirely vitiated the accuracy of the instruments, and almost any amount of labour would be well expended in removing this great defect.

My own experience of this subject is instructive. I had a beautiful instrument constructed a few years ago by Sauerwald, of Berlin, the coil of which, when no current flowed through it, deflected my double needle fully thirty degrees from the zero line. It was impossible to attain quantitative accuracy with this instrument.

I had the wire removed by Mr. Becker, and English wire used in its stead; the deflection fell to three degrees.

This was a great improvement, but not sufficient for my purpose. I commenced making enquiries about the possibility of obtaining pure copper, but the result was very discouraging. When almost despairing, the following thought occurred to me:— The action of the coil must be due to the admixture of iron with the copper, for pure copper is diamagnetic, it is feebly *repelled* by a strong magnet. The magnet therefore occurred to me as a means of instant analysis; I could tell by it, in a moment, whether my wire was free from magnetic metal or not.

The wire of M. Sauerwald's coil was strongly attracted by the magnet. The wire of Mr. Becker's coil was also attracted, though in a much feebler degree.

Both wires were covered with green silk: I removed this, but the Berlin wire was still attracted; the English wire, on the contrary, when presented *naked* to the magnet was feebly *repelled*; it was truly diamagnetic, and contained no sensible trace of iron. Thus the whole annoyance was fixed upon the green silk; some iron compound had been used in the dyeing of it, and to this the deviation of the needle from zero was manifestly due.

I had the green coating removed and the wire overspun with white silk, clean hands being used in the process. A perfect galvanometer is the result; the needle, when released from the action of the current, returns accurately to zero, and is perfectly free from all magnetic action on the part of the coil. In fact, while we have been devising agate plates and other learned methods to get rid of the nuisance of a magnetic coil, the means of doing so are at hand. Let the copper wire be selected by the magnet, and no difficulty will be experienced in obtaining specimens magnetically pure.

CHAPTER II.

THE NATURE OF HEAT—THE MATERIAL THEORY—THE DYNAMICAL THEORY—THERMAL EFFECTS OF AIR IN MOTION—GENERATION OF HEAT BY ROTATION BETWEEN THE POLES OF A MAGNET—EXPERIMENTS OF RUMFORD, DAVY, AND JOULE—THE MECHANICAL EQUIVALENT OF HEAT—HEAT GENERATED BY PROJECTILES—HEAT WHICH WOULD BE GENERATED BY STOPPING THE EARTH'S MOTION—METEORIC THEORY OF THE SUN'S HEAT—FLAME IN ITS RELATION TO THE DYNAMICAL THEORY.

APPENDIX:—EXTRACTS FROM BACON AND RUMFORD.

(17) THE development of heat by mechanical action was illustrated by suitable experiments when we last assembled here. But experimental facts alone cannot satisfy the human mind; we desire to know the cause of the fact; we search after the principle by the operation of which the phenomena are produced. Why should heat be generated by mechanical action, and what is the real nature of the agent thus generated? Two rival theories have been offered in answer to these questions, which are named respectively the *material theory*, and the *dynamical*, or *mechanical*, *theory* of heat. For a long time, however, the former of these—the material theory—had the greater number of adherents. Within certain limits it involved conceptions of a very simple kind, and this simplicity secured its general acceptance. The material theory supposes heat to be a kind of matter—a subtle fluid stored up in the inter-atomic spaces of bodies. The laborious Gmelin, for example, in his Handbook of Chemistry, defines heat to be 'that substance whose en-

trance into our bodies causes the sensation of warmth, and its egress the sensation of cold.' * He also speaks of heat combining with bodies as one ponderable substance does with another; and many other eminent chemists treat the subject from the same point of view.

(18) The development of heat by mechanical means, inasmuch as its generation seemed unlimited, was a great difficulty with the materialists; but they were acquainted with the fact (which shall be amply elucidated on a future occasion), that different bodies possess different powers of holding heat, if such a term may be employed. Take, for example, the two liquids, water and mercury, and warm a pound of each of them, say from fifty degrees to sixty. The absolute quantity of heat required by the water to raise its temperature ten degrees is fully thirty times the quantity required by the mercury. Technically speaking the water is said to have a greater *capacity* for heat than the mercury has, and this term 'capacity' suggests the views of those who invented it. The water was supposed to possess the power of storing up the caloric or matter of heat;—of hiding it, in fact, to such an extent that it required thirty measures of this caloric to produce the same sensible effect on water that one measure would produce upon mercury.

(19) All substances possess, in a greater or less degree, this apparent power of storing up heat. Lead, for example, possesses it; and our experiment with the lead bullet, in which heat was generated by compression, was explained by those who held the material theory in the following way. The uncompressed lead, they said, has a higher capacity for heat than the compressed substance; the size of its atomic storehouse is diminished by compression, and hence, when the lead is squeezed, a portion of that heat which, previous to compression, was hidden,

* English translation, vol. i. p. 22.

must make its appearance, for the compressed substance can no longer hold it all. In some similar way the experiments on friction and percussion were accounted for. The idea of calling *new heat* into existence was rejected by the believers in the material theory. According to their views, the quantity of heat in the universe is as constant as the quantity of ordinary matter, and the utmost we can do by mechanical and chemical means, is to store up this heat, or to drive it from its lurking-places into the open day.

(20) The dynamical theory, or, as it is sometimes called, the mechanical theory of heat, discards the idea of materiality as applied to heat. The supporters of this theory do not believe heat to be matter, but an accident or condition of matter; namely, *a motion of its ultimate particles*. From the direct contemplation of some of the phenomena of heat, a profound mind is led almost instinctively to conclude that heat is a kind of motion. Bacon held a view of this kind,* and Locke stated a similar view with singular felicity. 'Heat,' he says, 'is a very brisk agitation of the insensible parts of the object, which produces in us that sensation from whence we denominate the object hot: so what in our sensation is *heat*, in the object is nothing but *motion*.' The experiments of Count Rumford † on the boring of cannon have been already referred to. Now he showed that the hot chips cut from his cannon did not change their capacity for heat: he, moreover, collected the scales and powder produced by the abrasion of his metal, weighed them, and demanded whether it could be believed that the vast amount of heat which he had

* See Appendix to this Chapter.

† I have particular pleasure in directing the reader's attention to an abstract of Count Rumford's memoir on the Generation of Heat by Friction, contained in the Appendix to this Chapter. Rumford, in this memoir, annihilates the material theory of heat. Nothing more powerful has since been written.

generated had been all squeezed out of that modicum of crushed metal. 'You have not,' he might have urged on those who maintained this view, 'given yourselves the trouble to enquire whether any change whatever has been produced by friction in the capacity of the metal for heat. You are quick in inventing reasons to save your theory from destruction, but slow to enquire whether these reasons are not merely the finespun fancies of your own brains.' Theories are indispensable, but they sometimes act like drugs upon the mind. Men grow fond of them as they do of dram-drinking, and feel discontented and irascible when the stimulant to the imagination is taken away.

(21) At this point an experiment of Davy comes forth in its true significance.* Ice is solid water, and the solid has only one-half the capacity for heat that liquid water possesses. A quantity of heat which would raise a pound of ice ten degrees in temperature, would raise a pound of water only five degrees. Further, simply to liquefy a mass of ice, an enormous amount of heat is necessary, this heat being so utterly absorbed or rendered 'latent' as to make no impression upon the thermometer. The question of 'latent heat' shall be fully discussed in its proper place: what I am desirous of impressing on you at present is, that, taking the materialists on their own ground, *liquid water*, at its freezing temperature, possesses a vastly greater amount of heat than *ice* at the same temperature.

(22) Davy reasoned thus: 'If I, by friction, liquefy ice, a substance will be produced which contains a far greater absolute amount of heat than the ice; and in this case, it cannot with any show of reason be affirmed that I merely render sensible heat which had been previously insensible in the frozen mass. Liquefaction in this case will conclusively demonstrate a *generation* of heat.' He made the experiment, and liquefied the ice by pure friction; and the

* Works of Sir H. Davy, vol. ii. p. 11.

result has been regarded as the first which really proved the immateriality of heat.

(23) When a hammer strikes a bell the motion of the hammer is arrested, but not destroyed; it has been shivered into vibrations, which impart motion to the air and affect the auditory nerve as sound. So also when our sledge-hammer descended upon our lead bullet, the descending motion of the sledge was arrested: but it was not destroyed. *The motion was transferred to the atoms of the lead*, and announced itself to the proper nerves as heat. The theory, then, which Rumford so powerfully advocated, and which Davy so ably supported,* is that heat is a kind of molecular motion; and that by friction, percussion, and compression, this motion may be generated, as well as by combustion. This is the theory which it shall be my aim to develope until your minds attain to perfect clearness regarding it. At the outset you must exercise patience. We are entering a jungle and striking into the brambles in rather a random fashion at first. But we shall thus make ourselves acquainted with the general character of our work, and with due persistence cut through all entanglements at last.

(24) You have already witnessed the effect of projecting a current of compressed air against the face of the thermo-electric pile (page 15). The instrument was chilled by the current of air. Now *heat* is known to be developed when air is compressed; and I have been asked repeatedly how this heat was disposed of in the case of

* In Davy's first scientific memoir he calls heat a repulsive motion, which he says may be augmented in various ways. 'First, by the transmutation of mechanical into repulsive motion; that is, by friction or percussion. In this case the mechanical motion lost by the masses of matter in friction is the repulsive motion gained by their corpuscles;' an extremely remarkable passage. I have given further extracts from this paper in the Appendix to Chapter III.

the condensed air employed in the experiment referred to. Pray listen to my reply. Supposing the vessel which contained the air to be formed of a substance perfectly impervious to heat, and supposing all the heat developed by my arm, in compressing the air, to be retained within the vessel, *that* quantity of heat would be exactly competent to undo what had been done, and to restore the compressed air to its original volume and temperature. But this vessel V (fig. 12) is not impervious to heat, and it

FIG. 12.

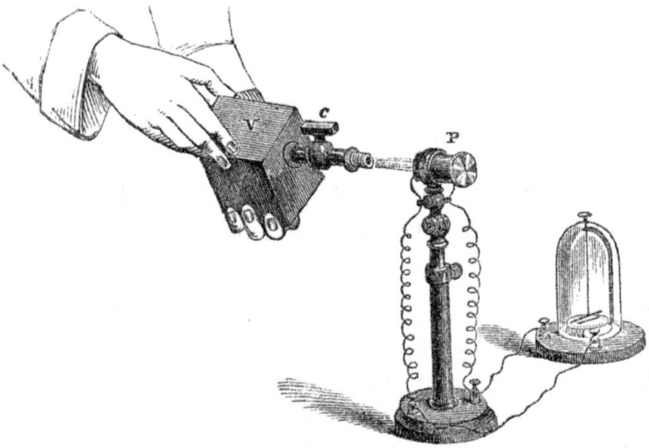

was not my object to draw upon the heat developed by my arm; after condensation, therefore, the vessel was allowed to rest till all the heat generated by the condensation was dissipated, and the temperature of the air within and without the vessel was the same. When, therefore, the air rushed out, it had not the heat to draw upon which had been developed during compression. The heat from which it derived its elastic force was only sufficient to keep it at the temperature of the surrounding air. In doing its work a portion of this heat, equivalent to the work done, was consumed, and the issuing air was consequently chilled. Do not be disheartened if this reasoning should not appear quite clear to you. We are now in comparative darkness, but as we proceed light will gradually appear, and irradiate retrospectively our present gloom.

(25) Let me now make evident to you that heat is developed by the compression of air. Here is a strong cylinder of glass T U (fig. 13), accurately bored, and quite smooth within. Into it this piston fits air-tight, so that, by driving the piston down, the air underneath it is forcibly compressed; and when the air is thus compressed, heat is suddenly generated. Tinder may be ignited by this heat. Here, moreover, is a morsel of cotton wool, wetted with an inflammable liquid, the bisulphide of carbon. I throw the bit of wet cotton into the glass syringe, and instantly eject it. It has left behind it a residue of vapour. On compressing the air suddenly, the heat developed is sufficient to ignite the vapour, and you see a flash of light within the syringe. Nor is it necessary to eject the cotton; I replace it in the tube, and urge the piston downwards; you see the flash as before. If with a narrow glass tube the fumés generated by the combustion of the vapour be in every instance blown away, without once removing the cotton from the syringe, the experiment may be repeated and the flash of light obtained twenty times in succession.

FIG. 13.

(26) The chilling effect of a current of compressed air on the thermo-electric pile has been already illustrated. Here is another illustration of the thermal effect produced in air by its own mechanical action. This tin tube is stopped at both ends, and connected with an air-pump. The tube is at present full of air, and the

face of the thermo-electric pile rests against its curved surface. The galvanometer declares that the face of the pile in contact with the tube has been warmed by the latter. I was prepared for this result, having reason to know that the air within the tube is slightly warmer than that without. We will now cause this air to perform work, and then examine the thermal condition of the vessel in which the work has been performed. We will work the pump; the cylinders of the machine will be emptied, and the air within this tube will be driven into the exhausted cylinders by its own elastic force. This is the work that it performs, and as the tube is exhausted you will see the needle, which is now deflected so considerably in the direction of heat, descend to zero, and pass quite up to 90° in the direction of cold. The pump is now in action, and you observe the result. The needle falls as predicted, and its advance in the direction of cold is only arrested by its concussion against the stops.

(27) Three strokes of the pump suffice to chill the tube so as to send the needle up to 90°;* let it now come to rest. It would require more time than we can afford to allow the tube to assume the temperature of the air around it; but the needle is now sensibly at rest at a good distance on the cold side of zero. Turning on this cock, the air will rush in, and each of its atoms will hit the inner surface of the tube like a projectile. The mechanical motion of the atoms will be thereby annihilated, but an amount of heat equivalent to this motion will be generated. This heat will be sufficient to re-warm the tube, to undo the present deflection, and to send the needle up on the heat side of zero. The air is now entering, and you see the

* The galvanometer used in this experiment was that which I employ in my original researches: it is an exceedingly delicate one. When here introduced, its dial was illuminated by the electric light; and an image of it, two feet in diameter, was projected on a screen.

effect: the needle moves, and goes quite up to 90° on that side which indicates the heating of the pile.*

(28) I have now to direct your attention to an important effect connected with this chilling of the air by rarefaction. On the plate of the air-pump is placed a large glass receiver, which is now filled with the air of this room. This air, and, indeed, all air, unless it be dried artificially, contains a quantity of aqueous vapour which, as vapour, is perfectly invisible. A certain temperature is requisite to maintain the vapour in this invisible state, and if the air be chilled so as to bring it below this temperature, the vapour will instantly condense, and form a visible cloud. Such a cloud, which you will remember is not *vapour*, but *liquid water* in a state of fine division, will form within this glass vessel R (fig. 14), when the air is pumped out of it; and to make this effect visible to everybody present, to those right and left of me, as well as to those in front, these eight little gas jets are arranged in a semicircle which half surrounds the receiver. Each person present sees one or more of these jets on looking through the receiver; and when the cloud forms, the dimness which it produces will at once declare its presence. The pump is now quickly worked, and a very few strokes suffice to precipitate the vapour. It spreads throughout the entire receiver, and many of you see a

* In this experiment a mere line along the surface of the tube was in contact with the face of the pile, and the heat had to propagate itself through the tin envelope to reach the instrument. Previously to adopting this arrangement I had the tube pierced, and a separate pile, with its naked face turned inwards, cemented air-tight into the orifice. The pile came thus into direct contact with the air, and its entire face was exposed to the action. The effects thus obtained were very large; sufficient, indeed, to swing the needle quite round. My desire to complicate the subject as little as possible induced me to abandon the cemented pile, and to make use of the instrument with which my audience had already become familiar. With the arrangement actually adopted the effects were, moreover, so large, that I drew only on a portion of my power.

colouring of the cloud, as the light shines through it, similar to that observed sometimes, on a large scale, around the moon. When the air is allowed to re-enter the vessel, it is heated, exactly as in the experiment with our tin tube;

FIG. 14.

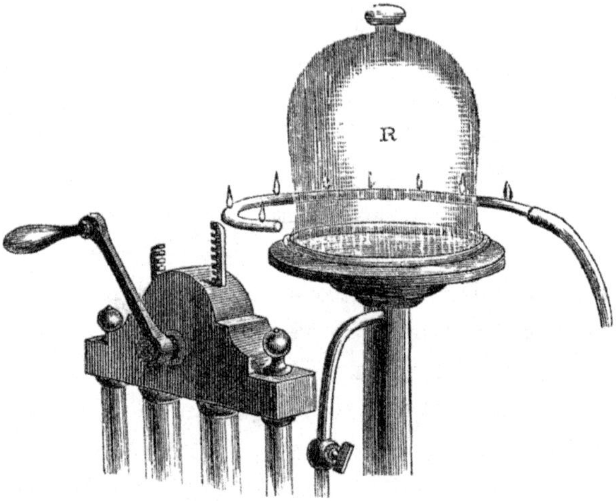

the cloud melts away, and the perfect transparency of the air within the receiver is restored.*

(29) Sir Humphry Davy refers, in his Chemical Philosophy, to a machine at Schemnitz, in Hungary, in which

* A far more beautiful mode of demonstration was subsequently resorted to. Removing the lens from the camera of an electric lamp, the rays from the coal-points issued divergent. A large plano-convex lens was placed in front, so as to convert the divergent cone into a convergent one, and caused the cone to pass through the receiver. Its track was at first invisible, but two or three strokes of the pump precipitated the vapour, and then the track of the beam through the receiver resembled a white solid bar. After crossing the receiver, the light fell upon a white screen, and exhibited splendid diffraction colours when the cloud formed. In my recent experiments, clouds of far greater density, permanence, and splendour of colour than those obtainable from aqueous vapour have been produced. Aqueous hydrochloric acid yields such clouds. See Proceedings of Royal Society, vol. xvii. p. 317.

air was compressed by a column of water 260 feet in height. When a stopcock was opened so as to allow the air to escape, a degree of cold was produced which not only precipitated the aqueous vapour diffused in the air, but caused it to congeal in a shower of snow, while the pipe from which the air issued became bearded with icicles. 'Dr. Darwin,' writes Davy, 'has ingeniously explained the production of snow on the tops of the highest mountains, by the precipitation of vapour from the rarefied air which ascends from plains and valleys. The Andes, placed almost under the line, rise in the midst of burning sands; about the middle height is a pleasant and mild climate; the summits are covered with unchanging snows.'

(30) And now I would request your attention to an experiment, in which heat will be developed by what must appear to many of you a very mysterious agency, and indeed the most instructed amongst us know, in reality, very little about the subject. I wish to develope heat by what might be regarded as friction *against pure space.* And indeed it may be, and probably is, due to a kind of friction against the inter-stellar medium, to which we shall have occasion to refer more fully by-and-by.

(31) Here is a mass of iron—part of a link of a huge chain cable—which is surrounded by these multiple coils of copper wire c c (fig. 15), and which can instantly be converted into a powerful magnet by sending an electric current through the wire. You see, when thus excited, how powerful it is. This poker clings to it, and these chisels, screws, and nails cling to the poker. Turned upside down, this magnet will hold a half-hundredweight attached to each of its poles, and probably a score of the heaviest people in this room if suspended from the weights. At the proper signal my assistant will interrupt the electric current. The iron falls and all the magic disappears: the magnet now is mere common

Fig. 15.

iron. On the ends of the magnet are placed two pieces of iron P P—moveable poles, as they are called—which, when the magnet is unexcited, can be brought within any required distance of each other. When the exciting current passes, these pieces of iron virtually form parts of the magnet. Between them I will place a substance which the magnet, even when exerting its utmost power, is incompetent to attract. This substance is simply a piece of silver—in fact, a silver medal. When it is brought close to the excited magnet, no attraction ensues. Indeed what little force—and it is so little as to be utterly insensible in these experiments—the magnet really exerts upon the silver, is repulsive instead of attractive.

(32) Suspending this medal between the poles P P of the magnet, I send the current through the coil. The medal hangs between the poles; it is neither attracted nor repelled, but if we seek to move it we encounter resistance. To turn the medal round this resistance must be overcome, the silver moving as if it were surrounded by a viscous fluid. This extraordinary effect may also be rendered manifest in another way. Causing this rectangular plate of copper to pass quickly to and fro like a saw between the poles P P, with their points turned towards it; you seem, though you can see nothing, to be sawing through a mass of cheese or butter.* No effect of this kind is noticed when the magnet is not active: the copper saw then encounters nothing but the infinitesimal resistance of the air.

(33) Thus far you have been compelled to take my statements for granted, but an experiment is here arranged which will make this strange action of the magnet on the silver medal strikingly manifest to you all. Above the suspended medal, and attached to it by a bit of wire, is

* An experiment of Faraday's. He also was the first to arrest by a magnet the motion of a spinning cube of copper.

a little reflecting pyramid M, formed of four triangular pieces of looking-glass; both the medal and the reflector are suspended by a thread which was twisted in its manufacture, and which will untwist itself when the weight which it sustains is set free. When from this electric lamp a strong beam of light is caused to fall upon the little pyramid, the light is reflected, and you see these long luminous spokes moving through the dusty air of the room as the mirror turns.

(34) Let us start it from a state of rest. The beam now passes through the room and strikes against the white wall. As the mirror commences rotating, the patch of light moves, at first slowly, over the wall and ceiling. The motion quickens, and now you can no longer see the distinct patches of light, but instead of them you have this splendid luminous band fully twenty feet in diameter drawn upon the wall by the quick rotation of the reflected beams. At the word of command the magnet will be excited. See the effect: the medal seems struck dead by the magnet, the band suddenly disappears, and there you have the single patch of light upon the wall. This strange result is produced without any visible change in the space between the two poles. Observe the slight motion of the image on the wall: the tension of the string is struggling with an unseen antagonist and producing that motion. It is such as would be produced if the medal, instead of being surrounded by air, were immersed in a pot of treacle. On destroying the magnetic power, the viscous character of the space between the poles instantly disappears; the medal begins to twirl as before; there are the revolving beams, and there is now the luminous band. I again excite the magnet: the beams are struck motionless, and the band disappears.

(35) By the force of the hand this resistance can be overcome and the medal turned round; but to turn it

force must be expended. What becomes of that force? It is converted into heat. The medal, if forcibly compelled to turn, will become heated. Many of you are acquainted with the grand discovery of Faraday, that electric currents are developed when a conductor of electricity is set in motion between the poles of a magnet. We have these currents here, and they are competent to heat the medal. But what *are* these currents? How are they related to the space between the magnetic poles—how to the muscular force which is expended in their generation? We do not yet know, but we shall know by-and-by. It does not in the least lessen the interest of the experiment if the force of my arm, previous to appearing as heat, appears in another form—in the form of electricity. The result is the same: the heat developed ultimately is the exact equivalent of the power employed to move the medal in the excited magnetic field.

(36) I wish now to make evident to all here present this development of heat. Here is a solid metal cylinder, the

FIG. 16.

core of which is composed of a metal more easily melted than its outer case. The outer case is copper, and this is filled by a hard but fusible alloy. The cylinder is set upright between the conical poles P P (fig. 16) of the magnet. A string S S passes from the cylinder to a whirling table, by which the cylinder may be caused to spin round. It might turn till doomsday with the magnet unexcited, and not produce the effect sought; but when the magnet is in action an amount of heat will be developed sufficient to melt the core of that cylinder, and, if successful, I will pour the liquid metal out before you. Two minutes will suffice for the experiment. The cylinder is now rotating, its upper end being open. We will permit it to remain open until the liquid metal is seen spattering over the poles of the magnet. The metallic spray is already there, though a minute has scarcely elapsed since the commencement of the experiment. I now stop the motion for a moment, cork up the end of the cylinder, so as to prevent the loss of the metal, and let the action continue for half a minute longer. The entire mass of the core is now melted. I withdraw the cylinder, remove the cork, and thus pour out before you the liquefied alloy.*

(37) It is now time to consider more closely than we have hitherto done the relation of the heat developed by mechanical action to the force which produces it. Doubtless this relation floated in many minds before it received either correct enunciation or experimental proof. The celebrated Montgolfier entertained the idea of the equivalence of heat and mechanical work; and the idea has been developed by his nephew M. Séguin, in his work 'On the Influence of Railways,' printed in 1839. Those who reflect on vital processes—on the changes which occur in the animal body—and the relation of the forces involved in food to muscular force, are led naturally to entertain the idea of interdependence between these forces. It is therefore not a matter of surprise that the man who was one of the first, if not the first, to raise the idea of the equivalence between heat and mechanical energy, and of the mutual convertibility of natural powers generally, to

* The development of heat by causing a conductor to revolve between the poles of a magnet was first effected by Mr. Joule (Phil. Mag. vol. xxiii. 3rd Series, year 1843, pp. 355 and 439), and his experiment was afterwards revived in a striking form by M. Foucault. The artifice above described, of fusing the core out of the cylinder, renders the experiment very effective in a lecture-room.

true philosophic clearness in his own mind, was a physician. Dr. Mayer, of Heilbronn in Germany, enunciated in 1842 * the relation which subsists between the forces of inorganic nature. He first calculated the 'mechanical equivalent of heat,' and followed up, as will be shown in due time, the statement of the principle by its fearless application. But the intuitions of Mayer required experimental proof; and to Dr. Joule, of Manchester, belongs the honour of being the first to give a decisive demonstration of the correctness of the dynamical theory.† Entirely independent of Mayer, and undismayed by the coolness with which his first labours appear to have been received, he persisted for years in his attempts to prove the invariability of the relation of heat to ordinary mechanical power. He placed water in a suitable vessel, agitated the water by paddles, and determined both the amount of heat developed by the stirring of the liquid, and the amount of labour expended in its production. He did the same with mercury and with sperm oil. He also caused disks of cast iron to rub against each other, and measured the heat produced by their friction, and the force expended in overcoming it. He urged water through capillary tubes, and determined the amount of heat generated by the friction of the liquid against the sides of the tubes. And his experiments leave

* Liebig's Annalen, vol. xlii. p. 233; Phil. Mag. 4th Series, vol. xxiv. p. 371; and in *résumé*, Phil. Mag. vol. xxv. p. 378. I am indebted to Mr. Wheatstone for the perusal of a rare and curious pamphlet by G. Rebenstein, with the following (translated) title: 'Progress of our Time. Generation of Heat without Fuel; or, Description of a Mechanical Process, based on physical and mathematical proofs, by which Caloric may be extracted from Atmospheric Air, and in a high degree concentrated. The cheapest Substitute for Fuel in most cases where combustion is necessary.' Rebenstein deduces from the experiments of Dulong the quantity of heat evolved in the compression of a gas. No glimpse of the dynamical theory is, however, to be found in his paper; his heat is *matter* (*Wärmestoff*) which is squeezed out of the air as water is out of a sponge.

† Phil. Mag., Aug. 1863. Mr. Joule's experiments on the mechanical equivalent of heat extend from 1843 to 1849.

no shadow of doubt upon the mind that, under all circumstances, the quantity of heat generated by the same amount of force is fixed and invariable. A given expenditure of force, in causing the iron disks to rotate against each other, produced precisely the same amount of heat as when it was applied to agitate water, mercury, or sperm oil. At the end of the experiments, the *temperatures* in the respective cases would of course be very different; the temperature of water, for example, would be only $\frac{1}{30}$th of that of mercury, because, as we already know, the capacity of water for heat is thirty times that of mercury. Dr. Joule took this into account in reducing his experiments, and found, as already stated, that, however the temperatures might differ, in consequence of the different capacities for heat of the substances employed, *the absolute amount of heat* generated by the same expenditure of power was in all cases the same.

(38) In this way it was proved that the quantity of heat necessary to raise one pound of water one degree Fahrenheit in temperature, is equal to that generated by a pound weight, falling from a height of 772 feet against the earth. Conversely, the amount of heat necessary to raise a pound of water one degree in temperature, would, if all applied mechanically, be competent to raise a pound weight 772 feet high, or it would raise 772 lbs. one foot high. The term 'foot-pound' has been introduced to express in a convenient way the lifting of one pound to the height of a foot. Thus, the quantity of heat necessary to raise the temperature of a pound of water one degree Fahrenheit being taken as a standard, 772 foot-pounds constitute what is called *the mechanical equivalent of heat.* If the degrees be centigrade, 1,390 foot-pounds constitute the equivalent.*

* In 1843 an essay entitled 'Theses concerning Force' was presented to the Royal Society of Copenhagen by a Danish philosopher named

(39) In order to imprint upon your minds the thermal effect produced by a body falling from a height, I will go through the operation of allowing a lead ball to fall from our ceiling upon this floor. That the ball is at the present moment slightly colder than the air of this room is proved by bringing the lead into contact with the thermo-electric pile; the deflection of the needle indicates cold. On the floor is placed a slab of iron, intended to receive the lead, and also cooler than the air of the room. At the top of the house is an assistant, who will pull up the ball by means of a string. He will not touch the ball, nor will he allow it to touch anything else. The lead now falls, and is received upon the plate of iron. The amount of heat generated by a single descent is very small, because the height is inconsiderable; we will, therefore, allow the ball to be drawn up and to descend three or four times in succession. We have now arrived at the fourth collision. I

Colding. At this early date M. Colding sought to ascertain the quantity of heat generated by the friction of various metals against each other and against other substances, and to determine the amount of mechanical work consumed in its generation. In an account of his researches given by himself in the Philosophical Magazine (vol. xxvii. p. 56), he states that the result of his experiments, nearly 200 in number, was that the heat disengaged was always in proportion to the mechanical energy lost. Independently of the materials by which the heat was generated, M. Colding found that an amount of heat competent to raise a pound of water 1° C. would raise a weight of a pound 1,148 feet high; a most remarkable result. M. Colding starts from the principle that 'as the forces of nature are something spiritual and immaterial—entities whereof we are cognisant only by their mastery over nature—those entities must of course be very superior to everything material in the world; and as it is obvious that it is through them only that the wisdom we perceive and admire in nature expresses itself, these powers must evidently be in relation to the spiritual, immaterial, and intellectual power itself that guides nature in its progress; but if such is the case it is consequently quite impossible to conceive of these forces as anything naturally mortal or perishable. Surely, therefore, the forces ought to be regarded as absolutely imperishable.' The case of M. Colding shows how a speculation, though utterly unphysical, may, by stimulating experiment, be the means of developing important physical results.

place the ball, which at the commencement produced cold, again upon the pile, and the immediate deflection of the needle in the opposite direction declares the lead to be heated; this heat is due entirely to the destruction of the moving energy which the ball possessed when it struck the plate of iron. According to our theory, the common mechanical motion of the lead, as a mass, has been transferred to the atoms of the mass, producing among them the agitation we call heat.

(40) We can readily calculate the amount of heat generated in this experiment. The space fallen through by the ball in each instance is twenty-six feet. The heat generated is proportional to the height through which the body falls. Now a ball of lead in falling through 772 feet would generate heat sufficient to raise its own temperature 30° F., its 'capacity' being $\frac{1}{30}$th of that of water: hence, in falling through 26 feet, which is in round numbers $\frac{1}{30}$th of 772, the heat generated would, if all concentrated in the lead, raise its temperature one degree. This is the amount of heat generated by a single descent of the ball, and four times this amount would, of course, be generated by four descents. The heat generated is not, however, all concentrated in the ball; a small portion of it belongs to the iron on which the ball falls.

(41) It is needless to say, that if motion be imparted to a body by other means than gravity, the destruction of this motion also produces heat. A rifle bullet when it strikes a target is intensely heated. The mechanical equivalent of heat enables us to calculate with accuracy the amount of heat generated by the bullet, when its velocity is known. This is a point worthy of our attention, and in dealing with it permit me to address myself to those of my audience who are unacquainted even with the elements of mechanics. Everybody knows that the greater the height is from which a body falls, the greater is the force with which it strikes

the earth, and that this is entirely due to the greater velocity imparted to the body in falling from the greater height. The velocity imparted to the body is not, however, proportional to the height from which it falls. If the height be augmented four-fold, the velocity is augmented only two-fold; if the height be augmented nine-fold, the velocity is augmented only three-fold; if the height be augmented sixteen-fold, the velocity is augmented only four-fold; or, expressed generally, the height is proportional to the square of the velocity.

(42) But the heat generated by the collision of the falling body increases simply as the height; consequently, the heat generated *increases as the square of the velocity.*

(43) If therefore we double the velocity of a projectile, we augment the heat generated, when its moving force is destroyed, four-fold; if we treble its velocity, we augment the heat nine-fold; if we quadruple the velocity, we augment the heat sixteen-fold, and so on.

(44) The velocity imparted to a body by gravity in falling through 772 feet is, in round numbers, 223 feet a second; that is to say, immediately before the body strikes the earth, this is its velocity. Six times this quantity, or 1,338 feet a second, would not be an inordinate velocity for a rifle bullet.

(45) But a rifle bullet, if formed of lead, moving at a velocity of 223 feet a second, would generate on striking a target an amount of heat which, if concentrated in the bullet, would, as already shown, raise its temperature 30° F.; with 6 times this velocity it would generate 36 times the amount of heat; hence 36 times 30, or 1,080°, would represent the augmentation of the temperature of the bullet on striking a target with a velocity of 1,338 feet a second, if all the heat generated were confined to the bullet itself. This amount of heat would be sufficient to fuse a portion of the lead. Were the ball iron instead of lead, the

heat generated, under the conditions supposed, would be competent to raise the temperature of the ball only about $\frac{1}{3}$rd of 1,080°, because the capacity of iron for heat is about three times that of lead.

(46) From these considerations it is manifest that if we know the velocity and weight of any projectile, we can calculate with ease the amount of heat developed by the destruction of its moving force. For example, knowing as we do the weight of the earth and the velocity with which it moves through space, a simple calculation enables us to state the exact amount of heat which would be developed, supposing the earth to strike against a target strong enough to stop its motion. We could tell, for example, the number of degrees which this amount of heat would impart to a globe of water equal to the earth in size. Mayer and Helmholtz have made this calculation, and found that the quantity of heat which would be generated by this colossal shock would be quite sufficient, not only to fuse the entire earth, but to reduce it, in great part, to vapour. Thus, by the simple stoppage of the earth in its orbit, 'the elements' might be caused 'to melt with fervent heat.' The amount of heat thus developed would be equal to that derived from the combustion of fourteen globes of coal, each equal to the earth in magnitude. And if, after the stoppage of its motion, the earth should fall into the sun, as it assuredly would, the amount of heat generated by the blow would be equal to that developed by the combustion of 5,600 worlds of solid carbon.

(47) Knowledge such as that which you now possess has caused philosophers, in speculating on the mode in which the sun's power is maintained, to suppose the solar heat and light to be caused by the showering down of meteoric matter upon the sun's surface.* The Zodiacal

* Mayer propounded this hypothesis in 1848, and worked it out to a great extent. It was afterwards enunciated independently by Mr. Water-

Light is supposed to be a cloud of meteorites, and from it, it has been imagined, the rain of meteoric matter was derived. Now, whatever be the value of this speculation, it is to be borne in mind that the pouring down of meteors in the way indicated would be competent to produce the light and heat of the sun. I shall develope the theory on a future occasion. With regard to its probable truth or fallacy, it is not necessary that I should offer an opinion; the theory deals with a cause which, if in sufficient operation, would certainly be competent to produce the effects ascribed to it.

(48) Let me now pass from the sun to something less,—to the opposite pole of nature, if the expression be permitted. And here that divine power of the human intellect which annihilates mere magnitude in its dealings with *law*, comes conspicuously into play. Our theory is applicable not only to suns and planets, but equally so to atoms. Most of you know the scientific history of the diamond—that Newton, antedating intellectually the discoveries of modern chemistry, pronounced it to be an unctuous or combustible substance. Everybody now knows that this brilliant gem is composed of the same substance as common charcoal, graphite, or plumbago. A diamond is pure carbon, and carbon burns in oxygen. Here is a diamond, held fast in a loop of platinum wire; heating the gem to redness in this flame, I plunge it into this jar, which contains oxygen gas. See how it brightens on entering the jar of oxygen, and now it glows, like a little star, with a pure white light. How are we to figure the action here going on? Exactly as you would present to your minds the idea of meteorites showering down upon the sun. The conceptions are, in quality, the same, and to the intellect the one is not more difficult than the other. You are to

ston, and admirably developed by Professor William Thomson (Transactions of the Royal Soc. of Edinb., 1853). See also Chapter XIV.

figure the atoms of oxygen showering against this diamond on all sides. They are urged towards it by what is called chemical affinity; but this force, made clear, presents itself to the mind as pure attraction, of the same mechanical quality, if I may use the term, as gravity. Every oxygen atom as it strikes the surface, and has its motion of translation destroyed by its collision with the carbon, assumes the motion which we call heat: and this heat is so intense, the attractions exerted at these molecular distances are so mighty, that the crystal is kept white-hot, and the compound, formed by the union of its atoms with those of the oxygen, flies away as carbonic acid gas.

(49) Let us now pass from the diamond to ordinary flame. Before you is an ignited jet of gas. What is the constitution of the jet? Within the flame we have a core of gas as yet unburnt, and outside the flame we have the oxygen of the air. The surface of the core of gas is in contact with the air, and here it is that the atoms clash together and produce light and heat by their collision. But the exact constitution of the flame is worthy of our special attention, and for our knowledge of this we are indebted to one of Davy's most beautiful investigations. Coal-gas is what we call a hydro-carbon; it consists of carbon and hydrogen in a state of chemical union. From this transparent gas escape the soot and lampblack which we notice when the combustion is incomplete. Soot and lampblack are there now, but they are compounded with other substances to a transparent form. Here, then, we have a surface of this compound gas, in presence of the oxygen of our air: we apply heat, and the attractions are instantly so intensified that the gas bursts into flame. The oxygen has a choice of two partners, and it closes with that for which it has the strongest attraction. It first unites with the hydrogen, and sets the carbon free. Innumerable solid particles of carbon thus scattered in the

midst of the burning hydrogen are raised to a state of intense incandescence; they become white-hot, and mainly to them the *light* of our lamps is due. The carbon, however, in due time, closes with the oxygen, and becomes, or ought to become, carbonic acid; but in passing from the hydrogen with which it was first combined, to the oxygen with which it enters into final union, it exist for a time in the solid state, and then gives us the splendour of its light.

FIG. 17.

(50) The combustion of a candle is in principle the same as that of a jet of gas. Here we have a rod of wax or tallow (fig. 17), through which passes a cotton wick. You ignite the wick; it burns, melts the tallow at its base, the liquid ascends through the wick by capillary attraction, it is converted by the heat into vapour, and this vapour is a hydro-carbon, which burns exactly like the gas. In this case also you have unburnt vapour within, common air without, while between both is a shell, which forms the battle-ground of the clashing atoms. where they develope their light and heat. There is hardly anything more beautiful than a burning candle: the hollow basin partially filled with melted matter at the base of the wick; the creeping up of the liquid; its vaporisation; the structure of the flame; its shape tapering to a point; the converging air-currents which rush in to supply its needs. Its beauty, its brightness, its mobility, have made it a favourite type of spiritual essences; and its dissection by Davy, far from diminishing the pleasure with which we look upon a flame, has rendered it more than ever an object of wonder to the enlightened mind.

(51) You ought now to be able to picture clearly before your minds the structure of a candle-flame. You ought to see the unburnt core within and the burning shell which envelopes this core. From the core, through this shell, the substance of the candle is incessantly passing and escaping to the surrounding air. In the case of a candle also we have a hollow cone of burning matter. Imagine this cone cut across horizontally; a burning ring would be exposed. We can practically cut the flame of a candle thus across. I bring this piece of white paper down upon the candle, until the paper almost touches the wick. Observe the upper surface of the paper; it becomes charred, but how? Corresponding to the burning ring of the candle,

FIG. 18.

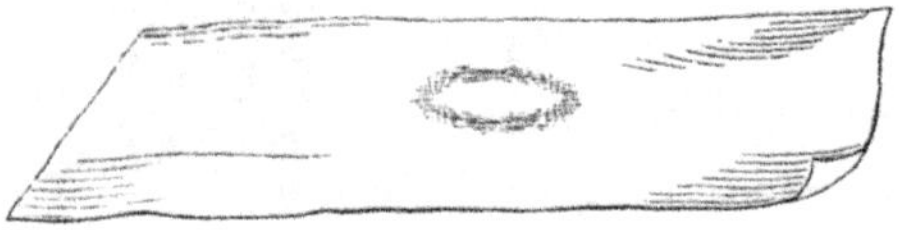

we have a charred ring upon the paper (fig. 18). Operating in the same manner with a jet of gas, we obtain the ring which it produces. Within the ring the paper is intact, for at this place the unburnt vapour of the candle, or the unburnt gas of the jet, impinges against the surface, and no charring can occur.

(52) To the existence, then, of solid carbon particles the light of our lamps is mainly due. But the existence of these particles, in the single state, implies the absence of oxygen to seize hold of them. If, at the moment of their liberation from the hydrogen, oxygen were present to seize upon them, their state of singleness would be abolished, and we should no longer have their light. For this reason, when we mix a sufficient quantity of air with the gas issuing from a jet, and thus cause the oxygen to penetrate to the very heart of the jet, the light disappears.

(53) Professor Bunsen has invented a burner for the express purpose of destroying, by quick combustion, the solid carbon particles. The burner from which the ga escapes is introduced into a tube; this tube is perforated nearly on a level with the burner, and through the orifices the air enters, mingles with the gas, and the mixture issues from the top of the tube. Fig. 19 represents a form of this burner; the gas is discharged into the perforated chamber *a*, where air mingles with it, and both ascend the tube *a b* together; *d* is a rose-burner, which may be used to vary the shape of the flame. I ignite the mixture, but the flame produces hardly any light. . Heat is the thing here aimed at, and this lightless flame is much hotter than an ordinary flame, because the combustion is much quicker, and therefore more intense.* If the orifices in *a* be stopped, the supply of air is cut off, and the flame at once becomes luminous: we have now the ordinary case of a core of unburnt gas surrounded by a burning shell. The illuminating power of a gas may, in fact, be estimated by the quantity of air necessary to prevent the precipitation of the solid carbon particles; the richer the gas, the more air will be required to produce this effect.

FIG. 19.

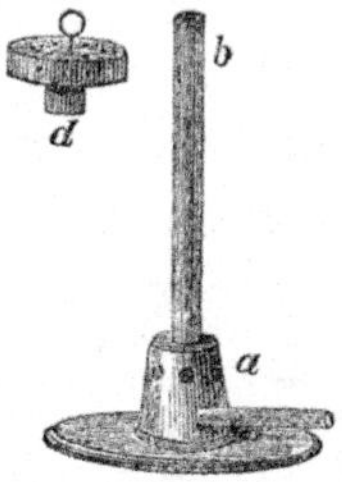

(54) An interesting observation may be made almost any windy Saturday evening in the streets of London, on the sudden and almost total extinction of the light of the gas jets, exposed chiefly in butchers' shops. When the wind blows, the oxygen is carried mechanically to the very heart of the flame, and the white light instantly vanishes to a pale and ghastly blue. During festive illu-

* Not hotter, nor nearly so hot, to a body exposed to its radiation; but very much hotter to a body *plunged in the flame.*

minations the same effect may be observed; the absence of the light being due, as in the case of Bunsen's burner, to the presence of a sufficient amount of oxygen to consume, instantly, the carbon of the flame.

(55) To determine the influence of height upon the rate of combustion, was one of the problems set before me in my journey to the Alps in 1859. Fortunately for science, I invited Dr. Frankland to accompany me on the occasion, and to undertake the experiments on combustion, while I proposed devoting myself to observations on solar radiation. The plan pursued was this; six candles were purchased at Chamouni and carefully weighed; they were then allowed to burn for an hour in the Hôtel de l'Union, and the loss of weight was determined. The same candles were taken to the summit of Mont Blanc, and, on the morning of Aug. 21, 1859, were allowed to burn for an hour in a tent, which perfectly sheltered them from the action of the wind. The aspect of the six flames at the summit surprised us both. They seemed the mere ghosts of the flames which the same candles were competent to produce in the valley of Chamouni—pale, feeble, and suggesting to us a greatly diminished energy of combustion. The candles being carefully weighed on our return, the unexpected fact was revealed, that the quantity of stearine consumed above was almost precisely the same as that consumed below. Thus, though the light-giving power of the flame was diminished in an extraordinary degree by the elevation, the energy of the combustion was the same above as it was below. This curious result is to be ascribed mainly to the mobility of the air at this great height, The particles of oxygen could penetrate the flame with comparative freedom, thus destroying its light, and making atonement for the smallness of their number by the promptness of their action. I find, indeed, that by reducing the density of ordinary

atmospheric air to one-half, we nearly double the mobility of its atoms.

(56) Dr. Frankland has made these experiments the basis of a very interesting memoir.* He shows that the quantity of a candle consumed in a given time is, within wide limits, independent of the density of the air; and the reason is, that although by compressing the air we augment the number of active particles in contact with the flame, we, almost in the same degree, diminish their mobility and retard the combustion. When an excess of air, moreover, surrounds the flame, its chilling effect will tend to prolong the existence of the carbon particles in a solid form, and even to prevent their final combustion. One of the most interesting facts established by Dr. Frankland is, that by condensing the air around it, the pale and smokeless flame of a spirit-lamp may be rendered as bright as that of coal-gas, and, by pushing the condensation sufficiently far, the flame may actually be rendered smoky, the oxygen present being too sluggish to effect the complete combustion of the carbon.

(57) But to return to our theory of combustion: it is to the clashing together of the oxygen of the air and the constituents of our gas and candles, that the light and heat of our flames are due. When steel filings are scattered in this Bunsen's flame, you see the starlike scintillations produced by the combustion of the steel. Here the steel is first heated, till the attraction between it and the oxygen of the air becomes sufficiently strong to cause them to combine, and these rocket-like flashes are the result of their collision. It is the impact of atoms of oxygen against atoms of sulphur which produces the heat and flame observed when sulphur is burned in oxygen or in air: to the collision of the same atoms against phosphorus are due

* Philosophical Transactions for 1861.

the intense heat and dazzling light which result from the combustion of phosphorus in oxygen gas. It is the collision of chlorine and antimony which produces the light and heat observed when these bodies are mixed together; and it is the clashing of sulphur and copper which produces incandescence when these substances are heated together in a Florence flask. In short, all cases of combustion are to be ascribed to the collision of atoms which have been urged together by their mutual attractions.

APPENDIX TO CHAPTER II.

EXTRACTS FROM THE TWENTIETH APHORISM OF THE SECOND BOOK OF THE 'NOVUM ORGANUM.'

WHEN I say of motion that it is the genus of which heat is a species, I would be understood to mean, not that heat generates motion, or that motion generates heat (though both are true in certain cases), but that heat itself, its essence and quiddity, is motion and nothing else; limited, however, by the specific differences which I will presently subjoin, as soon as I have added a few cautions for the sake of avoiding ambiguity. . . . .

Nor, again, must the communication of heat, or its transitive nature, by means of which a body becomes hot when a hot body is applied to it, be confounded with the form of heat. For heat is one thing, and heating is another. Heat is produced by the motion of attrition without any preceding heat. . . . .

Heat is an expansive motion, whereby a body strives to dilate and stretch itself to a larger sphere or dimension than it had previously occupied. This difference is most observable in flame, where the smoke or thick vapour manifestly dilates and expands into flame.

It is shown also in all boiling liquid, which manifestly swells, rises, and bubbles, and carries on the process of self-expansion, till it turns into a body far more extended and dilated than the liquid itself, namely, into vapour, smoke, or air.

* * * * * * *

The third specific difference is this, that heat is a motion of expansion, not uniformly of the whole body together, but in the smaller parts of it; and at the same time checked, repelled and beaten back, so that the body acquires a motion alternative, perpetually quivering, striving and struggling, and irritated by repercussion, whence springs the fury of fire and heat.

Again, it is shown in this, that when the air is expanded in a calender glass, without impediment or repulsion, that is to say uniformly and equably, there is no perceptible heat. Also when wind escapes from confinement, although it burst forth with the greatest violence, there is no very great heat perceptible; because the motion is of the whole, without a motion alternating in the particles.

And this specific difference is common also to the nature of cold; for in cold contractive motion is checked by a resisting tendency to expand, just as in heat the expansive action is checked by a resisting tendency to contract. Thus, whether the particles of a body work inward or outward, the mode of action is the same.

* * * * * * *

Now from this our first vintage it follows, that the form or true definition of heat (heat, that is, in relation to the universe, not simply in relation to man) is, in a few words, as follows: *Heat is a motion, expansive, restrained, and acting in its strife upon the smaller particles of bodies.* But the expansion is thus modified: *while it expands all ways, it has at the same time an inclination upwards.* And the struggle in the particles is modified also; *it is not sluggish, but hurried and with violence.**

ABSTRACT OF COUNT RUMFORD'S ESSAY, ENTITLED, 'AN ENQUIRY CONCERNING THE SOURCE OF THE HEAT WHICH IS EXCITED BY FRICTION.'

[*Read before the Royal Society, January* 25, 1798.]

Being engaged in superintending the boring of cannon in the workshops of the military arsenal at Munich, Count Rumford was struck with the very considerable degree of heat which a brass gun acquires, in a short time, in being bored, and with the still more intense heat (much greater than that of boiling water) of

* Bacon's Works, vol. iv.: Spedding's Translation.

the metallic chips separated from it by the borer;—he proposed to himself the following questions:—

'Whence comes the heat actually produced in the mechanical operation above mentioned?

'Is it furnished by the metallic chips which are separated from the metal?'

If this were the case, then the *capacity for heat* of the parts of the metal so reduced to chips ought not only to be changed, but the change undergone by them should be sufficiently great to account for *all* the heat produced. No such change, however, had taken place; for the chips were found to have the same capacity as slices of the same metal cut by a fine saw, where heating was avoided. Hence, it is evident, that the heat produced could not possibly have been furnished at the expense of the latent heat of the metallic chips. Rumford describes these experiments at length, and they are conclusive.

He then designed a cylinder for the express purpose of generating heat by friction, by having a blunt borer forced against its solid bottom, while the cylinder was turned round its axis by the force of horses. To measure the heat developed, a small round hole was bored in the cylinder for the purpose of introducing a small mercurial thermometer. The weight of the cylinder was 113·13 lbs. avoirdupois.

The borer was a flat piece of hardened steel, 0·63 of an inch thick, 4 inches long, and nearly as wide as the cavity of the bore of the cylinder, namely, 3½ inches. The area of the surface by which its end was in contact with the bottom of the bore was nearly 2½ inches. At the beginning of the experiment the temperature of the air in the shade, and also that of the cylinder, was 60 degrees Fahr. At the end of 30 minutes, and after the cylinder had made 960 revolutions round its axis, the temperature was found to be 130 degrees.

Having taken away the borer, he now removed the metallic dust, or rather scaly matter, which had been detached from the bottom of the cylinder by the blunt steel borer, and found its weight to be 837 grains troy. 'Is it possible,' he exclaims, 'that the very considerable quantity of heat produced in this experiment—a quantity which actually raised the temperature of above 113 pounds of gun-metal at least 70 degrees of Fahrenheit's thermometer—could have been furnished by so inconsiderable a quantity of metallic dust, and this merely in consequence of *a change* in its capacity for heat?'

'But, without insisting on the improbability of this supposition, we have only to recollect that from the results of actual and decisive experiments, made for the express purpose of ascertaining that fact, the capacity for heat of the metal of which great guns are cast is *not sensibly changed* by being reduced to the form of metallic chips, and there does not seem to be any reason to think that it can be much changed, if it be changed at all, in being reduced to much smaller pieces by a borer which is less sharp.'

He next surrounded his cylinder by an oblong deal box, in such a manner that the cylinder could turn water-tight in the centre of the box, while the borer was pressed against the bottom of the cylinder. The box was filled with water until the entire cylinder was covered, and then the apparatus was set in action. The temperature of the water on commencing was 60 degrees.

'The result of this beautiful experiment,' writes Rumford, 'was very striking, and the pleasure it afforded me amply repaid me for all the trouble I had had in contriving and arranging the complicated machinery used in making it. The cylinder had been in motion but a short time, when I perceived, by putting my hand into the water, and touching the outside of the cylinder, that heat was generated.

'At the end of one hour the fluid, which weighed 18·77 lbs., or 2½ gallons, had its temperature raised 47 degrees, being now 107 degrees.

'In thirty minutes more, or one hour and thirty minutes after the machinery had been set in motion, the heat of the water was 142 degrees.

'At the end of two hours from the beginning, the temperature was 178 degrees.

'At two hours and twenty minutes it was 200 degrees, and at two hours and thirty minutes it ACTUALLY BOILED!'

It is in reference to this experiment that Rumford made the remarks regarding the surprise of the bystanders, which I have quoted in Chapter I.

He then carefully estimates the quantity of heat possessed by each portion of his apparatus at the conclusion of the experiment, and, adding all together, finds a total sufficient to raise 26·58 lbs.

of ice-cold water to its boiling-point, or through 180 degrees Fahrenheit. By careful calculation, he finds this heat equal to that given out by the combustion of 2,303·8 grains (= $4\frac{8}{10}$ oz. troy) of wax.

He then determines the '*celerity*' with which the heat was generated, summing up thus: 'From the results of these computations, it appears that the quantity of heat produced equably, or in a continuous stream, if I may use the expression, by the friction of the blunt steel borer against the bottom of the hollow metallic cylinder, was *greater* than that produced in the combustion of nine *wax candles*, each three quarters of an inch in diameter, all burning together with clear bright flames.'

'One horse would have been equal to the work performed, though two were actually employed. Heat may thus be produced merely by the strength of a horse, and, in a case of necessity, this heat might be used in cooking victuals. But no circumstances could be imagined in which this method of procuring heat would be advantageous; for more heat might be obtained by using the fodder necessary for the support of a horse as fuel.'

[This is an extremely significant passage, intimating as it does, that Rumford saw clearly that the force of animals was derived from the food; *no creation of force* taking place in the animal body.]

'By meditating on the results of all these experiments, we are naturally brought to that great question which has so often been the subject of speculation among philosophers, namely, What is heat—is there any such thing as an *igneous fluid*? Is there anything that, with propriety, can be called caloric?'

'We have seen that a very considerable quantity of heat may be excited by the friction of two metallic surfaces, and given off in a constant stream or flux *in all directions*, without interruption or intermission, and without any signs of *diminution* or *exhaustion*. In reasoning on this subject we must not forget *that most remarkable circumstance,* that the source of the heat generated by friction in these experiments appeared evidently to be *inexhaustible*. [The italics are Rumford's.] It is hardly necessary to add, that anything which any *insulated* body or system of bodies can continue to furnish *without limitation* cannot possibly be a *material substance*; and it appears to me to be extremely difficult, if not quite impossible, to form any distinct idea of anything capable of being excited and communicated in those experiments, except it be MOTION.'

When the history of the dynamical theory of heat is written, the man who, in opposition to the scientific belief of his time, could experiment and reason upon experiment, as Rumford did in the investigation here referred to, cannot be lightly passed over. Hardly anything more powerful against the materiality of heat has been since adduced, hardly anything more conclusive in the way of establishing that heat is, what Rumford considered it to be, *Motion*.

CHAPTER III.

EXPANSION: THE SOLID, LIQUID, AND GASEOUS FORMS OF MATTER—NEW HYPOTHESES REGARDING THE CONSTITUTION OF GASES—COEFFICIENT OF EXPANSION—HEAT IMPARTED TO A GAS UNDER CONSTANT PRESSURE—HEAT IMPARTED TO A GAS AT CONSTANT VOLUME—MAYER'S CALCULATION OF THE MECHANICAL EQUIVALENT OF HEAT—DILATATION OF GASES WITHOUT REFRIGERATION—ABSOLUTE ZERO OF TEMPERATURE—EXPANSION OF LIQUIDS AND SOLIDS: ANOMALOUS DEPORTMENT OF WATER AND BISMUTH—ENERGY OF THE FORCE OF CRYSTALLISATION—THERMAL EFFECT OF STRETCHING WIRES—ANOMALOUS DEPORTMENT OF INDIA-RUBBER.

APPENDIX: ADDITIONAL DATA CONCERNING EXPANSION—EXTRACTS FROM SIR H. DAVY'S FIRST SCIENTIFIC MEMOIR: FUSION OF ICE BY FRICTION, ETC.

(58) ON the occasion of our first meeting here a sledge-hammer was permitted to descend upon a lump of lead, and the lead was found to have been heated by the blow. Formerly it was assumed that the force of the hammer was simply lost by the concussion. In elastic bodies it was supposed that a portion of the force was restored by the elasticity of the body, which caused the descending mass to rebound; but in the collision of inelastic bodies it was taken for granted that the force of impact was lost. This, according to our present notions, was a fundamental mistake; we now admit no loss, but assume, that when the motion of the descending hammer ceases, it is simply a case of *transference*, instead of annihilation. The motion of the mass, as a whole, has been transformed into a motion of the molecules of the mass. This motion of heat, however, though intense, is executed within limits too minute, and the moving particles are too small, to be visible. Here the imagination must help us.

In the case of solid bodies, while the force of cohesion still holds the molecules together, you must conceive a power of vibration, within certain limits, to be possessed by the molecules. You must suppose them oscillating to and fro, and the greater the amount of heat we impart to the body, or the greater the amount of mechanical action which we invest in it by percussion, compression, or friction, the greater will be the rapidity, and the wider the amplitude, of the atomic oscillations.

(59) Now, nothing is more natural than that particles thus vibrating, and ever as it were seeking wider room, should urge each other apart, and thus cause the body of which they are the constituents, to expand in volume. This, in general, is the consequence of imparting heat to bodies—expansion of volume. We shall closely consider the few apparent exceptions by-and-by. By the force of cohesion, then, the particles are held together; by the force of heat they are pushed asunder: here are the two antagonistic powers on which the molecular aggregation of the body depends. Let us suppose the communication of heat to continue; every increment of heat pushes the particles more widely apart; but the force of cohesion, like all other known forces, acts more and more feebly as the distance between the particles which are the seat of the force is augmented. As, therefore, the heat strengthens, its opponent grows weak, until, finally, the particles are so far loosened from the rigid thrall of cohesion, that they are at liberty, not only to vibrate to and fro across a fixed position, but also to roll or glide around each other. Cohesion is not yet destroyed, but it is so far modified, that the particles, while still offering resistance to being torn directly asunder, have their lateral mobility over each other's surfaces secured. *This is the liquid condition of matter.*

(60) In the interior of a mass of liquid the motion of every atom is controlled by the atoms which surround it.

But when we develope heat of sufficient power even within the body of a liquid, the molecules break the last fetters of cohesion, and fly asunder to form bubbles of vapour. If, moreover, one of the surfaces of the liquid be quite free, that is to say, uncontrolled either by a liquid or solid, it is quite easy to conceive that some of the vibrating superficial molecules will be jerked quite away from the liquid, and will fly with a certain velocity through space. *Thus freed from the influence of cohesion, we have matter in the vaporous or gaseous form.*

(61) My object here is to familiarise your minds with the general conception of atomic motion. I have spoken of the vibration of the molecules of a solid as causing its expansion; the molecules have been thought by some to revolve round each other, and the communication of heat, by augmenting their centrifugal force, was supposed to push them more widely asunder.* To this spiral spring is attached a weight; if the weight be twirled round in the air, it tends to fly away from me, the spring stretches to a certain extent, and, as the speed of revolution is augmented, the spring stretches still more, the distance between my hand and the weight being thus increased. And imagine the motion to continue till the spring snapped; the ball attached to it would fly off along a tangent to its former orbit, and thus represent an atom freed, by heat, from the force of cohesion, which is rudely represented by that of our spring. The ideas of the most well-informed philosophers are as yet uncertain regarding the exact nature of the motion of heat; but the great point, at present, is to regard it as motion of some kind, leaving its more precise character to be dealt with in future investigations.

* This was the hypothesis of Sir Humphry Davy. (See Appendix to this Chapter.) We are indebted to Mr. Rankine for the complete mathematical development of a 'Theory of Molecular Vortices.' (Phil. Mag., 1851, vol. ii. p. 509.)

(62) We might extend the notion of revolving atoms to gases also, and deduce their phenomena from a motion of this kind. But I have just thrown out an idea regarding gaseous molecules, which is at present very ably maintained:* the idea, namely, that such molecules fly in straight lines through space. This may be called the hypothesis of *translation*, in contradistinction to Davy's hypothesis of *revolution*. Everybody must have remarked how quickly the perfume of an odorous body fills a room, and this fact harmonises with the idea of the direct projection of the molecules. It may, however, be proved, that if the theory of rectilinear motion be true, the molecules must move at the rate of several hundred feet a second. Hence it might be objected that, according to the above hypothesis, odours ought to spread much more quickly than they are observed to do.

(63) The answer to this objection is, that the odorous particles have to make their way through a crowd of air atoms, with which they come into incessant collision. On an average, the distance through which a molecule can travel in common air, without striking against an atom of air, is infinitesimal, and hence the propagation of a perfume through air is enormously retarded by the air itself. It is well known that when a free communication is opened between the surface of a liquid and a vacuum, the vacuous space is much more speedily filled with the vapour of the liquid than when air is present.

(64) It is not difficult to determine the average velocities with which the particles of various gases move according to the hypothesis of translation. Taking, for example, a gas at the pressure of an atmosphere, or of 15 lbs. per square inch, and placing it in a vessel a cubic inch in size and shape; from the weight of the gas we can

* By Joule, Krönig, Maxwell; and, in a series of masterly papers, by Clausius.

calculate the velocity with which its particles must strike each side of the vessel in order to counteract a pressure of 15 lbs. It is manifest at the outset, that the lighter the gas is, the greater must be its velocity to produce the required effect. According to Clausius (Phil. Mag., 1857, vol. xiv. p. 124), the following are the average velocities of the atoms of oxygen, nitrogen, and hydrogen, at the temperature of melting ice—

Oxygen	1,514 feet per second.
Nitrogen	1,616 " "
Hydrogen	6,050 " "

As far back as 1848, Mr. Joule deduced from this hypothesis the velocity of hydrogen atoms, and found it to be 6,055 feet per second.

(65) According to this hypothesis, then, we are to figure a gaseous body as one whose molecules are flying in straight lines through space, impinging like little projectiles upon each other, and striking against the boundaries of the space which they occupy. I place this bladder, half filled with air, under the receiver of the air-pump, and remove the air from the receiver. The bladder swells; the air within it appears quite to fill it, so as to remove all its folds and creases. How is this expansion of the bladder produced? According to our present theory, it is produced by the shooting of atomic projectiles against its interior surface, driving the envelope outwards, until its tension is able to cope with their force. When air is admitted into the receiver, the bladder shrivels to its former size; and here we must figure the discharge of the atoms against the outer surface of the bladder, driving the envelope inwards, causing, at the same time, the atoms within to concentrate their fire, until finally the force from within equals that from without, and the envelope remains quiescent. All the impressions, then, which we derive from heated air or vapour are, according to his hypothesis

due to the impact of gaseous molecules. They stir the nerves in their own peculiar way, the nerves transmit the motion to the brain, and the brain declares it to be heat. Thus the impression one receives on entering the hot room of a Turkish bath, is caused by the atomic cannonade which is there maintained against the surface of the body. I would state this as a hypothesis advocated by eminent men, without expressing either assent or dissent myself.

(66) If, instead of placing this bladder under the receiver of an air-pump, and withdrawing the external air, I augment, by heat, the projectile force of the particles within it, these particles, though comparatively few in number, will strike with such impetuous energy against the inner surface as to cause the envelope to retreat: the bladder swells and becomes apparently filled with air; holding the bladder close to the fire, all its creases are removed. The bladder here intercepts the radiant heat of the fire, becomes warm, and then communicates its heat, by contact, to the air within.

(67) This, then, is a simple illustration of the expansive force of heat, and I have made an arrangement intended to show you the same fact in another manner. This flask, F (fig. 20), is empty, except as regards air, intended to be heated by placing a spirit-lamp underneath it. From the flask a bent tube passes to this dish, containing a coloured liquid. In the dish, a glass tube, *t t*, two feet long, is inverted, closed at the top, but with its open end downwards. You know that the pressure of the atmosphere is competent to keep a column of liquid in this tube, and here you have it quite filled to the top with the liquid. The tube passing from the flask F is caused to turn up exactly underneath the open end of this upright tube, so that if a bubble of air should issue from the former, it will ascend the latter. I now heat the flask, and the air

expands, for the reasons already given; bubbles are driven from the end of the bent tube, and they ascend in the

FIG. 20.

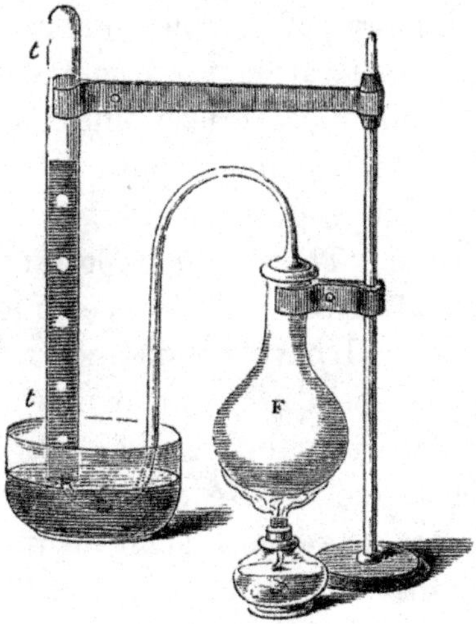

tube *t t*. The air speedily depresses the coloured liquid, until now, in the course of a very few seconds, the whole column of liquid has been superseded by air.

(68) It is perfectly manifest that the air, thus expanded by heat, is lighter than the unexpanded air. Our flask, at the conclusion of this experiment, is lighter than it was at the commencement, by the weight of the air transferred from it into the upright tube. Supposing, therefore, a light bag to be filled with such air, it is plain that the bag would, with reference to the heavy air outside it, be like a drop of oil in water. The oil, being lighter than the water, will ascend through the latter: so also our bag, filled with heated air, will ascend in the atmosphere; and this is the principle of the so-called fire-balloon. My assistant will ignite some tow in this vessel, over it he will place a funnel, and over the funnel I will hold the mouth of this paper balloon. The heated air ascending from the burning tow enters the balloon, and causes it to

swell; its tendency to rise is already manifest. On letting it go, it sails aloft till it strikes the ceiling of the room.

(69) But we must not be content with regarding these phenomena in a general way; without exact quantitative determinations, our discoveries would soon confound and bewilder us. We must now enquire what is the amount of expansion which a given quantity of heat is able to produce in a gas? This is an important point, and demands our special attention. In speaking of the volume of a gas, we should have no distinct notion of its real quantity if its temperature were omitted, so largely does the volume vary with the temperature. Take, then, a measure of gas at the precise temperature of water when it begins to freeze, or of ice when it begins to melt, that is to say, at a temperature of 32° Fahr. or 0° Cent., and raise that volume of gas one degree in temperature, *the pressure on every square inch of the envelope which holds the gas being preserved constant.* The volume of the gas will become expanded by a quantity which we may call a; raise it another degree in temperature, its volume will be expanded by $2a$, a third degree will cause an expansion of $3a$, and so on. Thus we see, that for every degree which we add to the temperature of the gas, it is expanded by the same amount. What is this amount? No matter what volume the gas may possess at the freezing temperature, by raising it one degree *Fahrenheit* above the freezing point we augment its volume by $\frac{1}{490}$th of its own amount; while by raising it one degree *Centigrade* we augment the volume by $\frac{1}{273}$rd of its own amount. A cubic foot of gas, for example, at 0° C. becomes, on being heated to 1°, $1\frac{1}{273}$ cubic foot, or, expressed in decimals—

1 vol. at 0° C. becomes $1 + \cdot 00366$ at 1° C.;
at 2° C. becomes $1 + \cdot 00366 \times 2$;
at 3° C. becomes $1 + \cdot 00366 \times 3$, and so on.

The constant number ·00366, which expresses the fraction of its own volume, which a gas, at the freezing temperature, expands on being heated one degree, is called the *coefficient of expansion* of the gas. Of course if we use the degrees of Fahrenheit, the coefficient will be smaller in the proportion of 9 to 5.

(70) It is a very remarkable and significant fact that all permanent gases expand by almost precisely the same amount for every degree added to their temperature. We can deduce from this with extreme probability the important conclusion, that where heat causes a gas to expand, the work it performs consists solely in overcoming the constant pressure from without—that, in other words, the heat is not interfered with by the mutual attraction of the gaseous molecules. For if this were the case, we should have every reason to expect, in the case of different gases, the same irregularities of expansion which we observe in liquids and solids. I said intentionally 'by *almost* precisely the same amount,' for many gases which are permanent at all ordinary temperatures deviate slightly from the rule. This will be seen from the following table:—

Name of Gas	Coefficient of Expansion
Hydrogen	0·00366
Air	0·00367
Carbonic oxide	0·00367
Carbonic acid	0·00371
Protoxide of nitrogen	0·00372
Sulphurous acid	0·00390

Here hydrogen, air, and carbonic oxide agree very closely; still there *is* a slight difference, the coefficient for hydrogen being the least. We remark in the other cases a greater deviation from the rule; and it is particularly to be noticed that the gases which deviate most are those which are nearest their point of liquefaction. The first three gases in the table never have been liquefied, all the others have. These are, in fact, *imperfect* gases, occupying a kind of intermediate place between the liquid and the perfect gaseous condition.

(71) This much made clear, we shall now approach, by slow degrees, an interesting but difficult subject. Suppose a quantity of air to be contained in a very tall cylinder, A B (fig. 21), the transverse section of which is one square inch in area. Let the top A of the cylinder be open to the air, and let P be a piston, which, for reasons to be explained immediately, I will suppose to weigh two pounds one ounce, and which can move air-tight and *without friction* up or down in the cylinder. At the commencement of the experiment, let the piston be at the point P of the cylinder, and let the height of the cylinder from its bottom B to the point P be 273 inches, the air underneath the piston being at a temperature of 0° C. Then, on heating the air from 0° to 1° C., the piston will rise one inch; it will now stand at 274 inches above the bottom. If the temperature be raised two degrees, the piston will stand at 275; if raised three degrees, it will stand at 276; if raised ten degrees, it will stand at 283; if 100 degrees, it will stand at 373 inches above the bottom; finally, if the temperature were raised to 273° C., it is quite manifest that 273 inches would be added to the height of the column, or, in other words, that by heating the air to 273 C., *its volume would be doubled.*

FIG. 21.

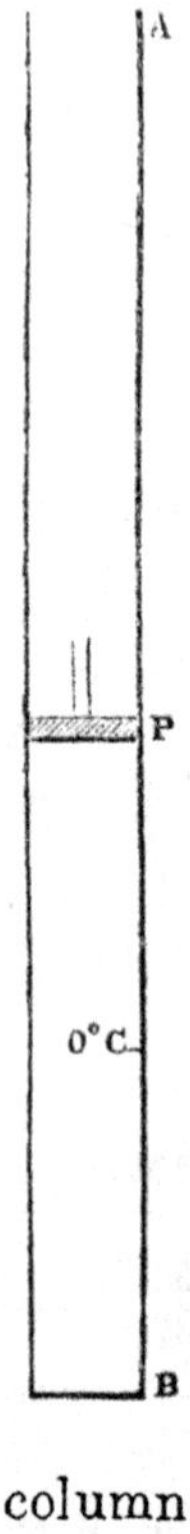

(72) The gas, in this experiment, executes work. In expanding from P upwards it has to overcome the downward pressure of the atmosphere, which amounts to 15 lbs. on every square inch, and also the weight of the piston itself, which is 2 lbs. 1 oz. Hence, the section of the cylinder being one square inch in area, in expanding from P to P′ the work done by the gas is equivalent to the

raising a weight of 17 lbs. 1 oz., or 273 ounces, to a height of 273 inches. It is just the same as what it would accomplish, if the air above P were entirely abolished, and a piston weighing 17 lbs. 1 oz. were placed at P.

(73) Let us now alter our mode of experiment, and instead of allowing our gas to expand when heated, let us oppose its expansion by augmenting the pressure upon it. In other words, let us keep *its volume constant* while it is being heated. Suppose, as before, the initial temperature of the gas to be 0° C., the pressure upon it, including the weight of the piston P, being, as formerly, 273 ounces. Let us warm the gas from 0° C. to 1° C.; what weight must we add at P in order to keep its volume constant? Exactly one ounce. But we have supposed the gas, at the commencement, to be under a pressure of 273 ounces, and the pressure it sustains is the measure of its elastic force; hence, by being heated one degree, the elastic force of the gas has augmented by $\frac{1}{273}$rd of what it possessed at 0°. If we warm it 2°, 2 ozs. must be added to keep its volume constant; if 3°, 3 ozs. must be added. And if we raise its temperature 273°, we should have to add 273 ozs.; that is, we should have to *double the original pressure* to keep the volume constant.

(74) It is simply for the sake of clearness, and to avoid fractions, that I have supposed the gas to be under the original pressure of 273 ozs. No matter what the pressure may be, the addition of 1° C. to its temperature produces an augmentation of $\frac{1}{273}$rd of the elastic force which the gas possesses at the freezing temperature; and by raising its temperature 273°, while its volume is kept constant, its elastic force is doubled. Let us now compare this experiment with the last one. *There* we heated a certain amount of gas from 0° to 273° C., and doubled its volume by so doing, the double volume being attained by lifting a weight of 273 ozs. through a height

of 273 inches. *Here* we heat the same amount of gas from 0° to 273°, but we do not permit it to lift any weight. We keep its volume constant. The quantity of matter heated in both cases is the same; the *temperature* to which it is heated is the same; but are the *absolute quantities* of heat imparted in both cases the same? By no means. Supposing that to raise the temperature of the gas, whose *volume* is kept constant, 273°, 10 grains of combustible matter are necessary; then to raise the temperature of the gas, whose *pressure* is kept constant, an equal number of degrees, would require the consumption of 14¼ grains of the same combustible matter. *The heat produced by the combustion of the additional 4¼ grains, in the latter case, is entirely consumed in lifting the weight.* Using the accurate numbers, the quantity of heat applied when the volume is constant, is to the quantity applied when the pressure is constant, in the proportion of

1 to 1·421.

(75) This extremely important fact constitutes the basis from which the mechanical equivalent of heat was first calculated. And here we have reached a point which is worthy of, and which will demand, your entire attention. I will endeavour to make this calculation before you.

FIG. 22.

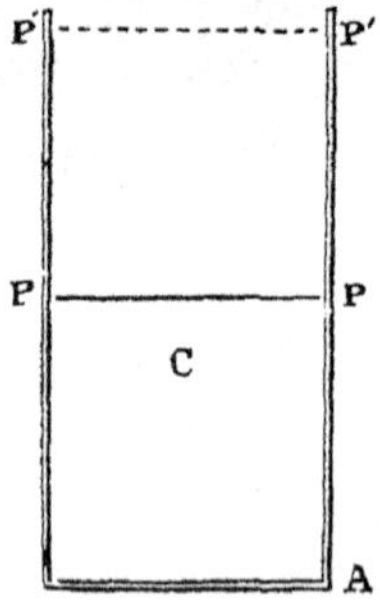

(76) Let C (fig. 22) be a cylindrical vessel with a base one square foot in area. Let PP mark the upper surface of a cubic foot of air at a temperature of 0° C., or 32° Fahr. The height A P will be then one foot. Let the air be heated till its volume is doubled; to effect this it must, as before explained, be raised 273° C., or 490° F. in temperature; and, when expanded, its upper surface will stand at P′ P′, one foot above its initial position. But in rising from P P to P′ P′ it has forced

back the atmosphere, which exerts a pressure of 15 lbs. on every square inch of its upper surface; in other words it has lifted a weight of $144 \times 15 = 2{,}160$ lbs. to a height of one foot.

(77) The 'capacity' for heat of the air thus expanding is 0·24; water being unity. The weight of our cubic foot of air, is 1·29 oz.; hence the quantity of heat required to raise 1·29 oz. of air 490° Fahr. would raise a little less than one-fourth of that weight of water 490°. The exact quantity of water equivalent to our 1·29 oz. of air is $1{\cdot}29 \times 0{\cdot}24 = 0{\cdot}31$ oz.

(78) But 0·31 oz. of water, heated to 490°, is equal to 152 ozs. or $9\frac{1}{2}$ lbs. heated 1°. Thus the heat imparted to our cubic foot of air, in order to double its volume, and enable it to lift a weight of 2,160 lbs. one foot high, would be competent to raise $9\frac{1}{2}$ lbs. of water one degree in temperature.

(79) The air has here been heated *under a constant pressure*, and we have learned that the quantity of heat necessary to raise the temperature of a gas under constant pressure a certain number of degrees, is to that required to raise the temperature of the gas the same number of degrees, *when its volume is kept constant*, in the propor tion of 1·42 : 1; hence we have the statement—

$$1{\cdot}42 : 1 = \overset{\text{lbs.}}{9{\cdot}5} : \overset{\text{lbs.}}{6{\cdot}7},$$

which shows that the quantity of heat necessary to augment the temperature of our cubic foot of air, at constant volume, 490°, would heat 6·7 lbs. of water 1°.

(80) Deducting 6·7 lbs. from 9·5 lbs., we find that the excess of heat imparted to the air, in the case where it is permitted to expand, is competent to raise 2·8 lbs. of water 1° in temperature.

(81) As explained already, this excess is employed to lift a weight of 2,160 lbs. one foot high. Dividing 2,160

by 2·8, we find that a quantity of heat sufficient to raise 1 lb. of water 1° Fahr· in temperature, is competent to raise a weight of 771·4 lbs. a foot high.

(82) This method of calculating the mechanical equivalent of heat was followed by Dr. Mayer, a physician in Heilbronn, Germany, in the spring of 1842.

(83) Mayer's first paper contains merely an indication of the way in which he had found the equivalent. In it were enunciated the convertibility and indestructibility of force, and its author referred to the mechanical equivalent of heat, merely in illustration of his principles. The essay was evidently a kind of preliminary note, from which date might be taken. Mayer's subsequent labours conferred dignity on the theory which they illustrated. In 1845 he published an Essay on Organic Motion and Nutrition, of extraordinary merit and importance. This was followed in 1848 by an Essay on Celestial Dynamics, in which, with remarkable boldness, sagacity, and completeness, he developed the meteoric theory of the sun. And this was followed by a fourth memoir in 1851, which also bears the stamp of intellectual power. Taking him all in all, the right of Dr. Mayer to stand, as a man of true genius, in the front rank of the founders of the dynamical theory of heat, cannot be disputed.

(84) On the 21st of August, 1843, Dr. Joule* communicated to the British Association, then meeting at Cork, a paper which was devoted, in part, to the determination of the 'mechanical value of heat.' Joule's publication had been preceded by a long course of experiments, so that his first work and Mayer's were really contemporaneous. This elaborate investigation gave the following weights raised one foot high, as equivalent to the warming of 1 lb. of water 1° Fahrenheit, the thermometric scale employed by Dr. Joule.

* Phil. Mag. 1843, vol. xxiii. p. 435.

1.	896 lbs.	5.	1,026 lbs.
2.	1,001 „	6.	587 „
3.	1,040 „	7.	742 „
4.	910 „	8.	860 „

(85) From the passage of water through narrow tubes, Joule deduced an equivalent of

770 foot pounds.

(86) In 1844 he deduced from experiments on the condensation of air, the following equivalents to 1 lb. of water heated 1° Fahr.

823 foot pounds.
795 „
820 „
814 „
760 „

(87) As the experience of the experimenter increased, we find that the coincidence of his results became closer. In 1845 Dr. Joule deduced from experiments with water, agitated by a paddle-wheel, an equivalent of

890 foot pounds.

(88) Summing up his results in 1845, and taking the mean, he found the equivalent to be

817 foot pounds.

(89) In 1847 he found the mean of two experiments to give as equivalent,

781·8 foot pounds.

(90) Finally, in 1849, applying all the precautions suggested by seven years' experience, he obtained the following numbers for the mechanical equivalent of heat:—

772·692,	from friction of water,	mean of	40	experiments.
774·083	„ „ mercury,	„	50	„
774·987	„ „ cast-iron,	„	20	„

These experiments rank among the most memorable that have ever been executed in physical science. They form of themselves a strict demonstration of the dynamical theory of heat.

(91) For reasons assigned in his paper, Joule fixes the exact equivalent at

772 foot pounds.

(92) According to the method pursued by Mayer, in 1842, the equivalent is

771·4 foot pounds.

Such a coincidence relieves the mind of every shade of uncertainty regarding the correctness of our present mechanical equivalent of heat.

(93) The immortal investigations here briefly referred to place Dr. Joule in the foremost rank of physical philosophers. Mayer's labours have, in some measure, the stamp of a profound intuition, which rose, however, to the energy of undoubting conviction in the author's mind. Joule's labours, on the contrary, are an experimental demonstration. Mayer *thought* his theory out, and rose to its grandest applications; Joule *worked* his theory out, and gave it for ever the solidity of demonstrated truth. True to the speculative instinct of his country, Mayer drew large and weighty conclusions from slender premises; while the Englishman aimed, above all things, at the firm establishment of facts. The future historian of science will not, I think, place these men in antagonism. To each belongs a reputation which will not quickly fade, for the share he has had, not only in establishing the dynamical theory of heat, but also in leading the way towards a right appreciation of the general energies of the universe.

(94) Let us now check our conclusion regarding the influence which the performance of work has on the temperature of a gas. Is it not possible to allow a gas to expand, without performing work? This question is answered by the following important experiment, which was first made by Gay Lussac. These two copper vessels, A, B (fig. 23),

FIG. 23.

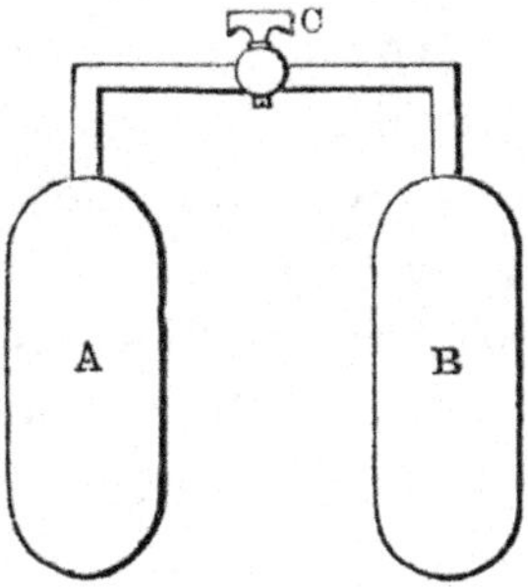

are of the same size; one of which, A, is exhausted, and the other, B, filled with air. I turn the cock C; the air rushes out of B into A, until the same pressure exists in both vessels. Now the air, in driving its own particles out of B, performs work, and experiments which we have already made inform us that the air which remains in B must be chilled. The particles of air enter A with a certain velocity, to generate which the heat of the air in B has been sacrificed; but they immediately strike against the interior surface of A, their motion of translation is annihilated, and the exact quantity of heat lost by B appears in A. The contents of A and B mixed together, give air of the original temperature. There is no work performed, and there is no heat lost. With the dynamical theory of heat in view, Dr. Joule made this experiment by compressing twenty-two atmospheres of air into one of his vessels, while the other was exhausted. On surrounding both vessels by water, kept properly agitated, no augmentation of its temperature was observed, when the gas was allowed to stream from one vessel into the other.* In like manner, suppose the top of the cylinder (fig. 21) to be closed, and the half above the piston P a perfect vacuum; and suppose the air in the lower half to be heated up to 273° C., its volume being kept constant. If the pressure were removed, the air would expand and fill the cylinder; the lower portion of the column would thereby be chilled, but the upper portion would be heated, and mixing both portions together, we should have the whole column at a temperature of 273°.† In this case, we raise the temperature of the gas from 0° to 273°, and afterwards allow it to double its volume; the temperatures of the gas at the commencement, and at the end, are the same as when the gas expands against a constant pressure, or lifts a constant weight; but the absolute quantity of heat in the latter case is 1·421 times that employed in the former, because, in the one case, the gas performs mechanical work, and in the other not.

* Phil. Mag. 1845, vol. xxvi. p. 378.

† I have recently found a case mentioned by Faraday (Researches in Chemistry and Physics, p. 221), where the effect referred to in the text was, in substance, observed. Faraday's explanation of the effect is a most instructive instance of the application of the material theory of heat. The observation was made at the Portable Gas Works, in 1827. 'It frequently happens,' writes Faraday, 'that gas previously at the pressure of thirty atmospheres is suddenly allowed to enter these long gas vessels (cylinders), at which time a curious effect is observed. That end of the cylinder at which the gas enters becomes very much cooled, whilst, on the contrary, the other end acquires a considerable rise of temperature. *The effect is produced by change of capacity in the gas*; for as it enters the vessel from the parts in which it was previously confined, at a pressure of thirty atmospheres, it suddenly expands, has its capacity for heat increased, falls in temperature, and consequently cools that part of the vessel with which it first comes in contact. But the part which has thus taken heat from the vessel being thrust forward to the further extremity of the cylinder by the successive portions which enter, is there compressed by them, *has its capacity diminished*, and now gives out that heat, or a part of it, which it had a moment before absorbed.' I have italicised the phrases which express the old notion. The difference in capacity here assumed is now known to have no existence.

(95) We are taught by this experiment that mere rarefaction is not of itself sufficient to produce a lowering of the mean temperature of a mass of air. It was, and is still, a current notion, that the mere expansion of a gas produces refrigeration, no matter how that expansion may be effected. The coldness of the higher atmospheric regions was accounted for by reference to the expansion of the air. It was thought that what we have called the 'capacity for heat' was greater in the case of the rarefied than of the unrarefied gas, and that chilling must therefore be the consequence of rarefaction. But the refrigeration which accompanies expansion is, in reality, due to the consumption of heat in the performance of work. Where no work is performed, there is no absolute refrigeration. All this needs reflection to arrive at clearness, but every effort of this kind which you make will render your subsequent efforts easier; and should you fail, at present, to gain clearness of comprehension, I repeat my recommendation of patience. Do not quit this portion of the subject without an effort to comprehend it—wrestle with it for a time, but do not despair if you fail to arrive at clearness.

(96) I have now to direct your attention to one other interesting question. We have seen the elastic force of our gas augmented by an increase of temperature. In an inflexible envelope we have, for every degree of temperature, a certain definite increment of elastic force, due to the augmented energy of the gaseous projectiles. Reckoning from 0° C. upwards, we find that every degree added to the temperature produces an augmentation of elastic force, equal to $\frac{1}{273}$rd of that which the gas possesses at 0°, and, hence, that by adding 273° we double the elastic force. Supposing the same law to hold good when we reckon from 0° *downwards*—that for every degree of temperature *withdrawn* from the gas we diminish its elastic force, or the motion which produces it, by $\frac{1}{273}$rd of what

it possesses at 0°, it is manifest that at a temperature of 273° Centigrade below 0° we should cease to have any elastic force whatever. The motion to which the elastic force is due must here vanish, and we reach what is called the *absolute zero of temperature.*

No doubt, practically, every gas deviates from the above law of contraction before it sinks so low, and it would become solid before reaching—273° C., or the absolute zero. This is considerably below any temperature which we have as yet been able to obtain.

I will not subject your minds to any further strain in connection with this subject to-day, but will now pass on to illustrate experimentally the expansion of liquids by heat.

(97) Here is a Florence flask filled with alcohol, and tightly corked; through the cork a tube, *t′* (fig. 24), passes water-tight, and the liquid rises a foot or so in this tube. When this flask is heated, the alcohol will expand, and it will rise in the tube. But you must see it rising, and to enable you to do so, the tube *t t′* is placed in front of the electric lamp E, a strong beam of light being sent across it, at the place *t′*, where the liquid column ends; the tube and column are thus illuminated. In front of the tube is placed this lens L, its distance being arranged so that it shall cast an enlarged image, *i i*, of the column upon the screen. You see clearly where the column ends, and if it moves you will be able to see its motion. It is needless to say that the image upon the screen is inverted, and that when the liquid expands, the top of the column will *descend* along the screen. I fill this beaker, B, with hot water, and raise the beaker so that the hot water shall surround the Florence flask. Observe the experiment from the commencement: the flask is now in the hot water, and the head of our column on the screen *ascends*, as if the liquid contracted. Now it stops and commences descending, and it

will continue to do so permanently. But why the first ascent? It is not due to the contraction of the liquid, but *to the momentary expansion of the flask*, to which the heat is first communicated. The glass expands before the heat can fairly reach the liquid, and hence the column falls; but

FIG. 24.

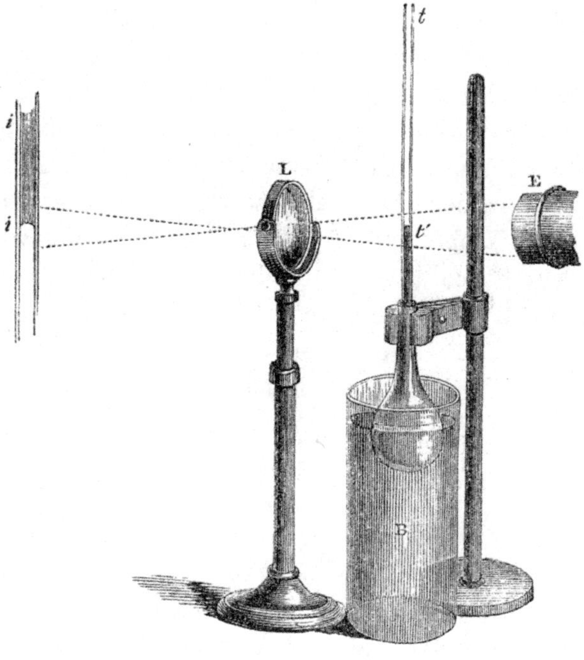

the expansion of the liquid soon exceeds that of the glass, and the column rises. Two things are here illustrated; the expansion of the solid glass by heat, and the fact that the observed dilatation of the liquid does not give us its true augmentation of volume, but only the difference of dilatation between itself and the glass.

(98) Here is another flask filled with water, exactly equal in size to the former, and furnished with a similar tube. I place it in the same position, and repeat with it

the experiment made with the alcohol. You see, first of all, the transitory effect due to the expansion of the glass, and afterwards, the permanent expansion of the liquid; but you can observe that the latter proceeds much more slowly than in the case of alcohol; the alcohol expands more rapidly than the water. Now we might examine a hundred liquids in this way, and find them all expanding by heat, and we might thus be led to conclude that expansion by heat is a law without exception; but we should err in this conclusion. It is really to illustrate an exception of this kind that this flask of water has been introduced. Let us cool the flask by plunging it into a substance somewhat colder than water, when it first freezes. This substance is obtained by mixing pounded ice with salt. You see the column gradually sinking, the heat is being given up to the freezing mixture, and the water contracts. The contraction is now very slow, and now it stops altogether. A slight motion commences in the opposite direction, and now the liquid *is visibly expanding*. By stirring the freezing mixture, we bring colder portions of it into contact with the flask; the colder the mixture, the quicker the expansion. Here then we have Nature pausing in her ordinary course, and reversing her ordinary habits. The fact is, that the water goes on contracting till it reaches a temperature of 39° Fahr., or 4° Cent., at which point the contraction ceases. This is the *point of maximum density* of water; from this downwards, to its freezing point, the liquid expands; and when it is converted into ice, the expansion is sudden and considerable. Ice, we know, swims upon water, being lightened by this expansion. If heat be now applied, the series of changes are reversed; the column descends, showing the *contraction of the liquid by heat*. After a time the contraction ceases, and permanent expansion sets in.

(99) The force with which these molecular changes are

effected is all but irresistible. The changes usually occur under conditions which allow us no opportunity of observing the energy involved in their accomplishment. But, to give you an example of this energy, a quantity of water is confined in this iron bottle. The iron is fully half an inch thick, and the quantity of water is small, though sufficient to fill the bottle. The bottle is closed by a screw firmly fixed in its neck. Here is a second bottle of the same kind, prepared in a similar manner. I place both of them in this copper vessel, and surround them with a freezing mixture. They cool gradually, the water within approaches its point of maximum density; no doubt, at this moment, the water does not quite fill the bottle, a small vacuous space exists within. But soon the contraction ceases, and expansion sets in; the vacuous place is slowly filled, the water gradually changes from liquid to solid; in doing so it requires more room, which the rigid iron refuses to grant. But its rigidity is powerless in the presence of the atomic forces. These atoms are giants in disguise, and the sound you now hear indicates that the bottle is shivered by the crystallising molecules—the other bottle follows; and here are the fragments of the vessels, showing their thickness, and impressing you with the might of that energy by which they have been thus riven.*

(100) You have now no difficulty in understanding the effect of frosty weather upon the water pipes of your houses. Before you is a number of pieces of such pipes, all rent. You become first sensible of the damage when the thaw sets in, but the mischief is really done at the time of freezing; the pipes then burst, and through the rents the water escapes, when the solid within is liquefied.

(101) It is hardly necessary for me to say a word on

* Metal cylinders, an inch in thickness, are unable to resist the decomposing force of a small galvanic battery. M. Gassiot has burst many such cylinders by electrolytic gas.

the importance of this property of water in the economy of nature. Suppose a lake exposed to a clear wintry sky; the superficial water is chilled, contracts, becomes thus heavier, and sinks by its superior weight, its place being supplied by the lighter water from below. In time this is chilled, and sinks in its turn. Thus a circulation is established, the cold dense water descending, and the lighter and warmer water rising to the top. Supposing this to continue, even after the first pellicles of ice were formed at the surface; the ice would sink,* and the process would not cease until the entire water of the lake would be solidified. Death to every living thing in the water would be the consequence. But just when matters become critical, Nature steps aside from her ordinary proceeding, causes the water to expand by cooling, and the cold water to swim like a scum on the surface of the warmer water underneath. Solidification ensues, but the solid is much lighter than the subjacent liquid, and the ice forms a protecting roof over the living things below.

(102) Such facts naturally and rightly excite the emotions; indeed, the relations of life to the conditions of life —the general adaptations of means to ends in Nature, excite, in the profoundest degree, the interest of the philosopher. But in dealing with natural phenomena, the feelings must be carefully watched. They often lead us unconsciously to overstep the bounds of fact. Thus, I have heard this wonderful property of water referred to as an irresistible proof of design, unique of its kind, and

* Sir William Thomson has raised a point which deserves the grave consideration of theoretic geologists: Supposing the constituents of the earth's crust to *contract* on solidifying, as the experiments of Bischof indicate, a breaking in and sinking of the crust would assuredly follow its formation. Under these circumstances, it is extremely difficult to conceive that a solid shell should be formed, as is generally assumed, round a liquid nucleus. Mr. Nasmyth, however, informs me that molten rocks *expand* on solidifying.

suggestive of pure benevolence. 'Why,' it is urged, 'should this case of water stand out isolated, if not for the purpose of protecting Nature against herself?' The fact, however, is, that the case is not an isolated one. You see this iron bottle, rent from neck to bottom; when broken with this hammer, you see a core of metal within. This is the metal bismuth, which, when in a molten condition, was poured into this bottle, and the bottle being closed by a screw, exactly as in the case of the water. The metal cooled, solidified, expanded, and the force of its expansion was sufficient to burst the bottle. There are no fish here to be saved, still the molten bismuth acts exactly as the water acts. Once for all, I would say that the natural philosopher, as such, has nothing to do with purposes and designs. His vocation is to enquire *what* Nature is, not *why* she is; though he, like others, and he, more than others, must stand at times rapt in wonder at the mystery in which he dwells, and towards the final solution of which his studies furnish him with no clue.

(103) We must now pass on to the expansion of solid bodies by heat, which may be illustrated in this way: Here are two wooden stands, A and B (fig. 25), with plates of brass, *p p'*, riveted against them. These two bars are of equal length, one of them is brass, the other iron, and, as you observe, they are not sufficiently long to stretch from stand to stand. They are therefore supported on two little projections of wood attached to the stands at *p* and *p'*. One of the plates of brass, *p*, is connected with one pole of a voltaic battery, D, and from the other, *p'*, a wire proceeds to the little instrument C, in front of the table; and again, from that instrument, a wire returns direct to the other pole of the battery. The instrument in front consists merely of an arrangement to support a spiral C of platinum wire, which will glow with a pure white light when the current from D passes through it.

At the present moment the only break in the circuit is due to the insufficient length of the bars of brass and iron to bridge the space from stand to stand. Underneath the

FIG. 25.

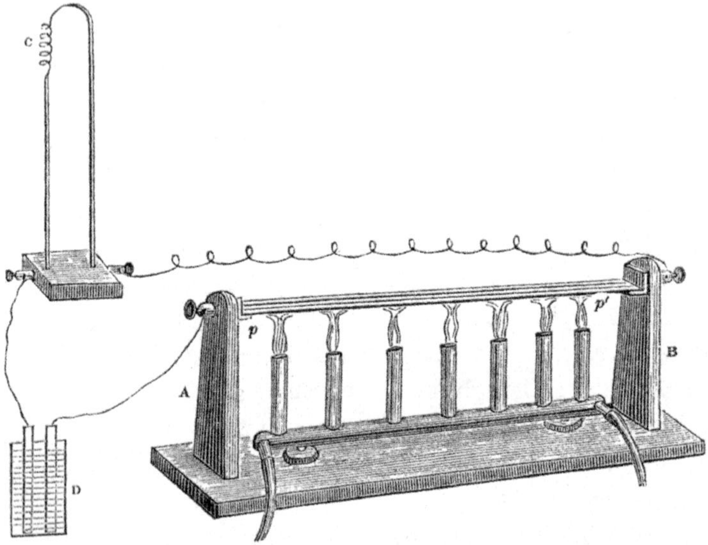

bars is a row of gas jets, which I will now ignite; the bars are heated, the metals expand, and in a few moments they will stretch quite across from plate to plate. When this occurs, the current will pass, and the fact of the gap being bridged will be declared by the sudden glowing of the platinum spiral. It is still non-luminous, the bridge not being yet complete; but now the spiral brightens up, showing that one, or both, of these bars have expanded so as to stretch quite across from stand to stand. Which of the bars is it? On removing the iron, the platinum still glows: I restore the iron, and remove the brass; the light disappears. It was the brass then that bridged the gap. So that we have here an illustration, not only of the general fact of expansion, but also of the fact that different bodies expand in different degrees.

(104) The expansion of both brass and iron is very small; and various instruments have been devised to measure the expansion. Such instruments bear the general name of pyrometers. But before you is a means of multiplying the effect, far more powerful than the ordinary pyrometer. On a mirror connected with the top of this solid upright bar of iron two feet long, is thrown a beam of light from the electric lamp, which beam is reflected to the upper part of the wall. If the bar shorten, the mirror will turn in one direction: if it lengthen, the mirror will turn in the opposite direction. Every movement of the mirror, however slight, is multiplied by this long index of light; which, besides its length, has the advantage of moving with twice the angular velocity of the mirror. Even the human breath, projected against this massive bar of iron, produces a sensible motion of the beam; and if it be warmed for a moment with the flame of a spirit-lamp, the luminous index will travel downwards, the patch of light upon the wall moving through a space of full thirty feet. I withdraw the lamp, and allow the bar to cool; it contracts, and the patch of light reascends the wall: the contraction is hastened by throwing a little alcohol on the bar of iron, the light moves more speedily upwards, and now it occupies a place near the ceiling, as at the commencement of the experiment.*

(105) It has been stated that different bodies possess different powers of expansion; † that brass, for example, expands more, on being heated, than iron. Of these two rulers, one is of brass and the other of iron, and they are riveted together so as to form, at this temperature, a straight

* The piece of apparatus with which this experiment was made, is intended for a totally different purpose. I therefore indicate its principle merely.

† The coefficients of expansion of a few well-known substances are given in the Appendix to this Chapter.

compound ruler. But when the temperature is changed, the ruler is no longer straight. If heated, it bends in one direction: if cooled, it bends in the opposite direction. When heated, the brass expands most, and forms the convex side of the curved ruler. When cooled, the brass contracts most, and forms the concave side of the ruler. Facts like these must, of course, be taken into account in structures where it is necessary to avoid distortion. The force with which bodies expand when heated, is quite irresistible by any mechanical appliances that we can make use of. All these molecular forces, though operating in such minute spaces, are almost infinite in energy. The contractile force of cooling has been applied by engineers to draw leaning walls into an upright position. If a body be brittle, the heating of one portion of it, producing expansion, may so press or strain another portion as to produce fracture. Hot water poured into a glass often cracks it, through the sudden expansion of the interior. It may also be cracked by the contraction produced by intense cold.

(106) Before you are some flasks of very thick glass, which, when blown, were allowed to cool quickly. The external portions became first chilled and rigid. The internal portions cooled more gradually, but they found themselves, on cooling, surrounded, as it were, by a rigid shell, on which they exerted the powerful strain of their contraction. The consequence is, that the superficial portions of these flasks are in such a state of tension that the slightest scratch produces rupture. I throw into this flask a grain of quartz; the mere dropping of the little bit of hard quartz into the flask causes the bottom to fly out of it. Here also are these so-called Rupert drops, or Dutch tears, produced by glass being fused to drops, and suddenly cooled. The external rigid shell has to bear the strain of the inner contraction; but the strain is distributed so equally all over the surface, that no part gives

way. But by simply breaking this filament of glass, which forms the tail of the drop, the solid mass is instantly reduced to powder. I dip the drop into a small flask filled with water, and break the tail outside the flask; the drop is shivered with such force that the shock, transferred through the water, is sufficient to break the bottle in pieces.

(107) A very curious effect of expansion was observed, and explained, some years ago, by the Reverend Canon Moseley. The choir of Bristol Cathedral was covered with sheet lead, the length of the covering being 60 feet, and its depth 19 feet 4 inches. It had been laid on in the year 1851, and two years afterwards it had moved bodily down for a distance of eighteen inches. The descent had been continually going on from the time the lead had been laid down, and an attempt made to stop it by driving nails into the rafters had failed; for the force with which the lead descended was sufficient to draw out the nails. The roof was not a steep one, and the lead would have rested on it for ever, without *sliding* down by gravity. What, then, was the cause of the descent? Simply this. The lead was exposed to the varying temperatures of day and night. During the day the heat imparted to it caused it to expand. Had it lain upon a horizontal surface, it would have expanded equally all round; but as it lay upon an inclined surface, it expanded more freely downwards than upwards. When, on the contrary, the lead contracted at night, its upper edge was drawn more easily downwards than its lower edge upwards. Its motion was therefore exactly that of a common earthworm; it pushed its lower edge forward during the day, and drew its upper edge after it during the night, and thus by degrees it crawled through a space of eighteen inches in two years. Every minor change of temperature during the day and during the night contributed also to the result; indeed

Canon Moseley afterwards found the main effect to be due to these quicker alternations of temperature.

(108) Not only do different bodies expand differently by heat, but the same body may expand differently in different directions. In crystals, the atoms are laid together according to law, and along some lines they are more closely packed than along others. It is also likely that the atoms of many crystalline bodies oscillate more freely and widely in some directions than in others. The consequence of this would be an unequal expansion by heat in different directions. The crystal in my hand (Iceland spar), has been proved by Professor Mitscherlich to expand more along its crystallographic axis than in any other direction. Nay, while the crystal expands as a whole—that is to say, while its volume is augmented by heat—it actually contracts in a direction at right angles to the crystallographic axis. Many other crystals also expand differently in different directions; and, I doubt not, most organic structures would, if examined, exhibit the same fact.

(109) Nature is full of anomalies which no foresight can predict, and which experiment alone can reveal. From the deportment of a vast number of bodies, we should be led to conclude that heat always produces expansion, and that cold always produces contraction. But water steps in, and bismuth steps in, to qualify this conclusion. If a metal be compressed, heat is developed: but if a wire be stretched, cold is developed. Dr. Joule and others have worked experimentally at this subject, and found the above fact all but general.

(110) One striking exception to this rule (there are probably many others) has been known for a great number of years; and I will now illustrate this exception by an experiment. The sheet of india-rubber now handed to me has been placed in the next room to keep it quite cold. Cutting from this sheet a strip three inches long,

and an inch and a half wide, and turning the thermo-electric pile upon its back, I lay upon its exposed face this piece of india-rubber. The deflection of the needle proves that the piece of rubber is cold. I now lay hold of the ends of the strip, suddenly stretch it, and press it, while stretched, on the face of the pile. The needle moves with energy, showing that the stretched rubber has heated the pile.

(111) But one deviation from a rule always carries other deviations in its train. In the physical world, as in the moral, acts are never isolated. Thus with regard to our india-rubber; its deviation from the rule referred to is only part of a series of deviations. In many of his investigations Dr. Joule has been associated with an eminent natural philosopher—Professor William Thomson*—and when Mr. Thomson was made aware of the deviation of india-rubber from an almost general rule, he suggested that the stretched india-rubber might *shorten*, on being heated. The test was applied by Joule, and the shortening was found to take place.† This singular experiment, thrown into a suitable form, will now be performed before you.

(112) To this arm, *a a* (fig. 26), is fastened a length of common vulcanised india-rubber tubing, stretched by a weight, w, of ten pounds, to about three times its former length. The index *i i*, is formed first of a piece of light wood moving freely on a pivot, being prolonged by a stout straight straw. At the end of the straw is placed a spear-shaped piece of paper, which can range over a graduated circle drawn on this black board. The index is now pressed down at *i*, by a projection attached to the weight. If the weight should be lifted by the contraction of the india-rubber, the index will follow, being drawn after it

* Now Sir William Thomson.

† Phil. Mag. 1857, vol. xiv. p. 227.

by a spring, *s s*, which acts upon the short arm of the lever.

FIG. 26.

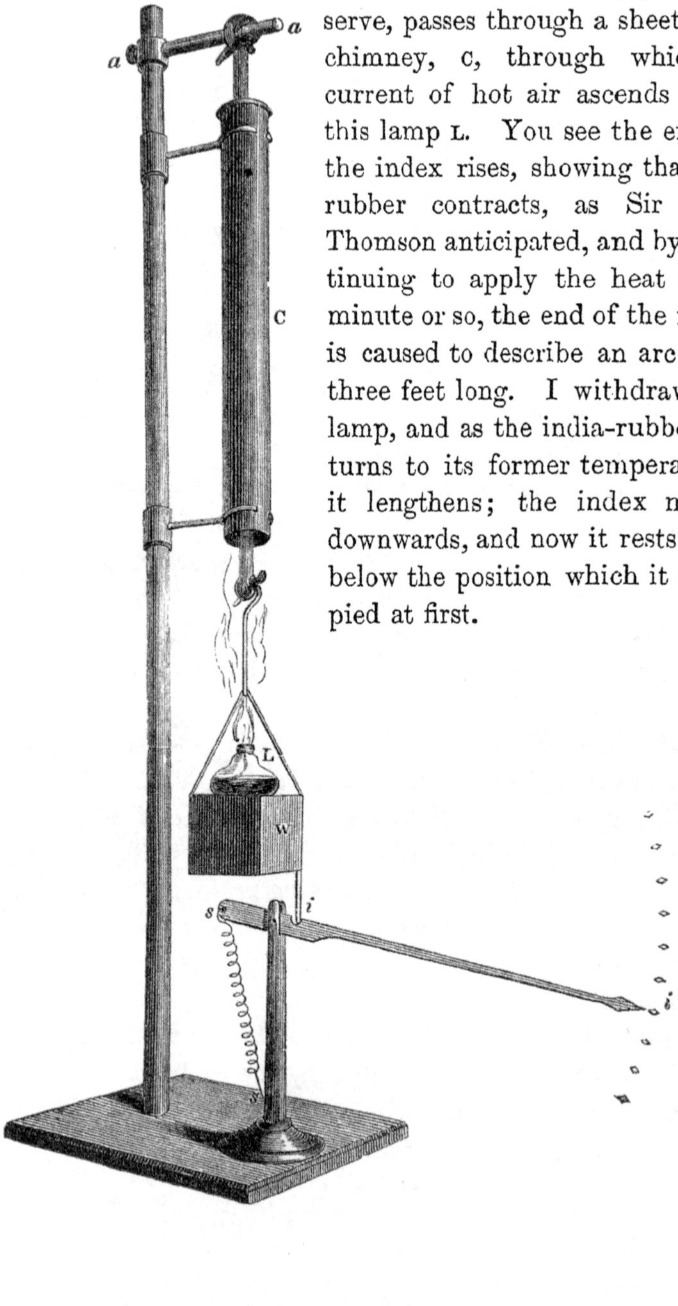

The india-rubber tube, you observe, passes through a sheet-iron chimney, C, through which a current of hot air ascends from this lamp L. You see the effect: the index rises, showing that the rubber contracts, as Sir Wm. Thomson anticipated, and by continuing to apply the heat for a minute or so, the end of the index is caused to describe an arc fully three feet long. I withdraw the lamp, and as the india-rubber returns to its former temperature, it lengthens; the index moves downwards, and now it rests even below the position which it occupied at first.

APPENDIX TO CHAPTER III.

FURTHER REMARKS ON DILATATION.

It is not within the scope of my present intention to dwell in detail on all the phenomena of expansion by heat; but, for the sake of my younger readers, I will supplement this chapter by a few additional remarks.

The linear, superficial, or cubic coefficient of expansion, is that fraction of a body's length, surface, or volume, which it expands on being heated one degree.

Supposing one of the sides of a square plate of metal, whose length is 1, to expand, on being heated one degree, by the quantity a; then the side of the new square is $1+a$, and its area is

$$1+2a+a^2.$$

In the case of expansion by heat, the quantity a is so small, that its square is almost insensible; the square of a small fraction is, of course, greatly less than the fraction itself. Hence, without sensible error, we may throw away the a^2 in the above expression, and then we have the area of the new square

$$1+2a.$$

$2a$, then, is the superficial coefficient of expansion; hence we infer that by multiplying the linear coefficient by 2, we obtain the superficial coefficient.

Suppose, instead of a square, that we had a cube, having a side$=1$; and that on heating the cube one degree, the side expanded to $1+a$; then the volume of the expanded cube would be

$$1+3a+3a^2+a^3.$$

In this, as in the former case, the square of a, and much more the cube of a, may be neglected, on account of their exceeding smallness; we then have the volume of the expanded cube=

$$1+3a;$$

that is to say, the cubic coefficient of expansion is found by trebling the linear coefficient.

The following table contains the coefficients of expansion for a number of well-known substances:—

Copper	0·000017	0·000051	0·000051
Lead	0·000029	0·000087	0·000089
Tin	0·000023	0·000069	0·000069
Iron	0·0000123	0·000037	0·000037
Zinc	0·0000294	0·000088	0·000089
Glass	0·000008	0·000024	0·000024

The first column of figures here gives the linear coefficient of expansion for 1° C.; the second column contains this coefficient trebled, which is the cubic expansion of the substance; and the third column gives the cubic expansion of the same substance, determined directly by Professor Kopp.* It will be seen that Kopp's coefficients agree almost exactly with those obtained by the trebling of the linear coefficients.

The linear coefficient of glass for 1° C. is

0·0000080.

That of platinum is

0·0000088.

Hence glass and platinum expand nearly alike. This is of the greatest importance to chemists, who often find it necessary to *fuse* platinum wires into their glass tubes. Were the coefficients different, the fracture of the glass would be inevitable during the contraction.

The Thermometer.

Water owes its liquidity to the motion of heat; when this motion sinks sufficiently, crystallisation, as we have seen, sets in. The temperature of crystallisation is perfectly constant, if the water be kept under the same pressure. For example, water crystallises in all climates at the sea-level at a temperature of

* Phil. Mag. 1852, vol. iii. p. 268.

32° F., or of 0° C. The temperature of condensation from the state of steam is equally constant, as long as the pressure remains the same. The melting of ice and the freezing of water touch each other, if I may use the expression, at 32° F.; the condensation of steam and the boiling of water under one atmosphere of pressure touch each other at 212°: 32° then is the freezing point of water, and it is the melting point of ice; 212° is the condensing point of steam and the boiling point of water. Both are invariable as long as the pressure remains the same.

Here, then, we have two invaluable standard points of temperature, and they have been used for this throughout the world. The mercurial thermometer consists of a bulb and a stem with capillary bore. The bore ought to be of equal diameter throughout. The bulb and a portion of the stem are filled with mercury. Both are then plunged into melting ice, the mercury shrinks, the column descends, and finally comes to rest. Let the point at which it becomes stationary be marked; it is the *freezing point* of the thermometer. Let the instrument be now removed and thrust into boiling water; the mercury expands, the column rises, and finally attains a stationary height. Let this point be marked; it is the *boiling point* of the thermometer. The space between the freezing point and the boiling point has been divided by Réaumur into 80 equal parts, by Fahrenheit into 180 equal parts, and by Celsius into 100 equal parts, called degrees. The thermometer of Celsius is also called the Centigrade thermometer.

Both Réaumur and Celsius call the freezing point 0°, Fahrenheit calls it 32°, because he started from a zero which he incorrectly imagined was the greatest terrestrial cold. Fahrenheit's boiling point is therefore 212°. Réaumur's boiling point is 80°, while the boiling point of Celsius is 100°.

The length of the degrees being in the proportion of 80 : 100 : 180, or of 4 : 5 : 9, nothing can be easier than to convert one into the other. If you want to convert Fahrenheit into Celsius, multiply by 5 and divide by 9; if Celsius into Fahrenheit, multiply by 9 and divide by 5. Thus 20° of Celsius are equal to 36° Fahrenheit; but if we would know what temperature by Fahrenheit's thermometer corresponds to 20° of Celsius, we must add 32 to the 36, which would make the temperature 20°, as shown by Celsius, equal the temperature 68°, as shown by Fahrenheit.

EXTRACTS FROM SIR H. DAVY'S FIRST SCIENTIFIC MEMOIR, BEARING THE TITLE 'ON HEAT, LIGHT, AND THE COMBINATIONS OF LIGHT.' *

The peculiar modes of existence of bodies—solidity, fluidity, and gazity—depend (according to the calorists) on the quantity of the fluid of heat entering into their composition. This substance insinuating itself between their corpuscles, separating them from each other, and preventing their actual contact, is by them supposed to be the cause of repulsion.

Other philosophers, dissatisfied with the evidences produced in favour of the existence of this fluid, and perceiving the generation of heat by friction and percussion, have supposed it to be motion. Considering the discovery of the true cause of the repulsive power as highly important to philosophy, I have endeavoured to investigate this part of chemical science by experiments; from these experiments (of which I am now about to give a detail) I conclude that heat or the power of repulsion is not *matter*.

The Phenomena of Repulsion are not dependent on a peculiar elastic fluid for their existence, or Caloric does not exist.

Without considering the effects of the repulsive power on bodies, or endeavouring to prove from these effects that it is motion, I shall attempt to demonstrate by experiments, that it is not matter; and in doing this, I shall use the method called by mathematicians, *reductio ad absurdum.*

First, let the increase of temperature produced by friction and percussion be supposed to arise from a diminution of the capacities of the acting bodies. In this case it is evident some change must be induced in the bodies by the action, which lessens their capacities and increases their temperatures.

Experiment.—I procured two parallelopipedons of ice,† of the temperature of 29°, six inches long, two wide, and two-thirds of an inch thick: they were fastened by wires to two bars of iron.

* Sir Humphry Davy's Works, vol. ii.

† The result of this experiment is the same, if wax, tallow, resin, or any substance fusible at a low temperature, be used; even iron may be fused by collision.

By a peculiar mechanism, their surfaces were placed in contact, and kept in a continued and most violent friction for some minutes. They were almost entirely converted into water, which water was collected, and its temperature ascertained to be 35°, after remaining in an atmosphere of a lower temperature for some minutes. The fusion took place only at the plane of contact of the two pieces of ice, and no bodies were in friction but ice.

From this experiment it is evident that ice by friction is converted into water, and according to the supposition, its capacity is diminished; but it is a well-known fact, that the capacity of water for heat is much greater than that of ice; and ice must have an absolute quantity of heat added to it, before it can be converted into water. Friction consequently does not diminish the capacities of bodies for heat.

From this experiment it is likewise evident, that the increase of temperature consequent on friction cannot arise from the decomposition of the oxygen gas in contact, for ice has no attraction for oxygen. Since the increase of temperature consequent on friction cannot arise from the diminution of capacity, or oxydation of the acting bodies, the only remaining supposition is, that it arises from an absolute quantity of heat added to them, which heat must be attracted from the bodies in contact. Then friction must induce some change in bodies, enabling them to attract heat from the bodies in contact.

Experiment.—I procured a piece of clockwork, so constructed as to be set at work in the exhausted receiver; one of the external wheels of this machine came in contact with a thin metallic plate. A considerable degree of sensible heat was produced by friction between the wheel and plate when the machine worked, uninsulated from bodies capable of communicating heat. I next procured a small piece of ice; * round the superior edge of this a

* The temperature of the ice and of the surrounding atmosphere at the commencement of the experiment was 32°, that of the machine was likewise 32°. At the end of the experiment, the temperature of the coldest part of the machine was near 33°, that of the ice and surrounding atmosphere the same as at the commencement of the experiment; so that the heat produced by the friction of the different parts of the machine was sufficient to raise the temperature of near half-a-pound of metal at least one degree; and to convert eighteen grains of wax (the quantity employed) into a fluid.

small canal was made, and filled with water. The machine was placed on the ice, but not in contact with the water. Thus disposed, the whole was placed under the receiver (which had been previously filled with carbonic acid), a quantity of potash (i.e. caustic vegetable alkali) being at the same time introduced.

The receiver was now exhausted. From the exhaustion and from the attraction of the carbonic acid gas by the potash, a vacuum nearly perfect was, I believe, made.

The machine was now set to work; the wax rapidly melted, proving an increase of temperature.

Caloric then was collected by friction; which caloric, on the supposition, was communicated by the bodies in contact with the machine. In this experiment, ice was the only body in contact with the machine. Had this ice given out caloric, the water on the top of it must have been frozen. The water on the top of it was not frozen, consequently the ice did not give out caloric. The caloric could not come from the bodies in contact with the ice, for it must have passed through the ice to penetrate the machine, and an addition of caloric to the ice would have converted it into water.

Heat, when produced by friction, cannot be collected from the bodies in contact, and it was proved, by the first experiment, that the increase of temperature consequent on friction, cannot arise from diminution of capacity or oxydation. But if it be considered as matter, it must be produced in one of these modes. Since (as is demonstrated by these experiments) it is produced in neither of these modes, it cannot be considered as matter. It has therefore been experimentally demonstrated that caloric, or the matter of heat, does not exist.

Solids by long and violent friction become expanded, and if of a higher temperature than our bodies, affect the sensory organ with the peculiar sensation known by the common name of heat.

Since bodies become expanded by friction, it is evident that their corpuscles must move or separate from each other.

Now a motion or vibration of the corpuscles of bodies must be necessarily generated by friction and percussion. Therefore we may reasonably conclude that this motion or vibration is heat, or the repulsive power.

Heat, then, or that power which prevents the actual contact of the corpuscles of bodies, and which is the cause of our peculiar sensations of heat and cold, may be defined as a peculiar motion, probably a vibration, of the corpuscles of bodies, tending to separate them. It may with propriety be called the repulsive motion.

Since there exists a repulsive motion, the particles of bodies may be considered as acted on by two opposing forces; the approximating power (which may, for greater ease of expression, be called attraction) and the repulsive motion. The first of these is the compound effect of the attraction of cohesion, by which the particles tend to come in contact with each other; the attraction of gravitation, by which they tend to approximate to the great contiguous masses of matter, and the pressure under which they exist, dependent on the gravitation of the superincumbent bodies.

The second is the effect of a peculiar motory or vibratory impulse given to them, tending to remove them farther from each other, and which can be generated, or rather increased, by friction or percussion. The effect of the attraction of cohesion, the great approximating cause, on the corpuscles of bodies, is exactly similar to that of the attraction of gravitation on the great masses of matter composing the universe, and the repulsive force is analogous to the planetary projectile force.

In his 'Chemical Philosophy,' pp. 94 and 95, Davy expresses himself thus: 'By a moderate degree of friction, as it would appear from Rumford's experiments, the same piece of metal may be kept hot for any length of time; so that, if the heat be pressed out, the quantity must be inexhaustible. When any body is cooled, it occupies a smaller volume than before; it is evident, therefore, that its parts must have approached each other; when the body has expanded by heat, it is equally evident that its parts must have separated from each other. The immediate cause of the phenomenon of heat, then, is motion; and the laws of its communication are precisely the same as the laws of the communication of motion.

'Since all matter may be made to fill a smaller space by cooling, it is evident that the particles of matter must have space between them; and since every body can communicate the power of expansion to a body of a lower temperature—that is, can give an expansive motion to its particles—it is a probable inference that its own particles are possessed of motion; but as there is no change in the position of its parts, as long as its temperature is uniform, the motion, if it exist, must be a vibratory or undulatory motion, or a motion of the particles round their axes, or a motion of the particles round each other.

'It seems possible to account for all the phenomena of heat if it be supposed that in solids the particles are in a constant state of vibratory motion, the particles of the hottest bodies moving with the greatest velocity, and through the greatest space; that in fluids and elastic fluids, besides the vibratory motion, which must be conceived greatest in the last, the particles have a motion round their own axes with different velocity, the particles of elastic fluids moving with the greatest quickness, and that in ethereal substances the particles move round their own axes, and separate from each other, penetrating in right lines through space. Temperature may be conceived to depend upon the velocity of the vibrations; increase of capacity in the motion being performed in greater space; and the diminution of temperature during the conversion of solids into fluids or gases, may be explained on the idea of the loss of vibratory motion, in consequence of the revolution of particles round their axes, at the moment when the body becomes fluid or aëriform, or from the loss of rapidity of vibration in consequence of the motion of the particles through space.'

CHAPTER IV.

THE TREVELYAN INSTRUMENT—GORE'S REVOLVING BALLS—INFLUENCE OF PRESSURE ON FREEZING POINT—LIQUEFACTION AND LAMINATION OF ICE BY PRESSURE—DISSECTION OF ICE BY A CALORIFIC BEAM—LIQUID FLOWERS AND THEIR CENTRAL SPOT—MECHANICAL PROPERTIES OF WATER PURGED OF AIR—THE BOILING POINT OF LIQUIDS; INFLUENCING CIRCUMSTANCES—CONVERSION OF HEAT INTO WORK IN THE STEAM-ENGINE: THE GEYSERS OF ICELAND.

(113) BEFORE finally quitting the subject of expansion, I wish to show you an experiment which illustrates in a curious and agreeable way the conversion of heat into mechanical energy. The fact to be reproduced was first observed by a gentleman named Schwartz, in one of the smelting works of Saxony. A quantity of silver which had been fused in a ladle was allowed to solidify, and to hasten its cooling it was turned out upon an anvil. Some time afterwards a strange buzzing sound was heard in the locality. The sound was finally traced to the hot silver, which was found quivering upon the anvil. Many years subsequent to this, Mr. Arthur Trevelyan chanced to be using a hot soldering-iron, which he laid by accident against a piece of lead. Soon afterwards, his attention was excited by a most singular sound, which, after some searching, was found to proceed from the soldering-iron. Like the silver of Schwartz, the soldering-iron was in a state of vibration. Mr. Trevelyan made his discovery the subject of a very interesting investigation, He determined the best form to be given to the 'rocker,' as the vibrating mass is now called, and

throughout Europe this instrument is known as 'Trevelyan's Instrument.' Since that time the subject has engaged the attention of Professor J. D. Forbes, Dr. Seebeck, Mr. Faraday, M. Sondhaus, and myself; but to Trevelyan and Seebeck we owe most of our knowledge regarding it.

(114) Here is a rocker made of brass. Its length, A C (fig. 27), is five inches; the width, A B, 1·5 in.; and the

Fig. 27.

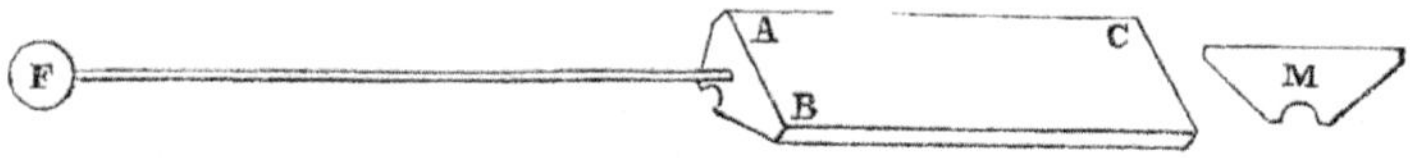

length of the handle, which terminates in the knob F, is ten inches. A groove runs at the back of the rocker, along its centre; the cross-section of the rocker is given at M. We will heat the rocker to a temperature somewhat higher than that of boiling water, and lay it on a block of lead, allowing its knob to rest upon the table. You hear a quick succession of forcible taps. But you cannot see the oscillations of the rocker, to which the taps are due. I therefore place on it this brass rod A B (fig. 28), with a brass ball at each of its ends; the oscillations are thereby rendered much slower, and you can easily follow with the eye the pendulous motion of the rod and balls. This motion will continue as long as the rocker is able to communicate sufficient heat to the carrier on

Fig. 28.

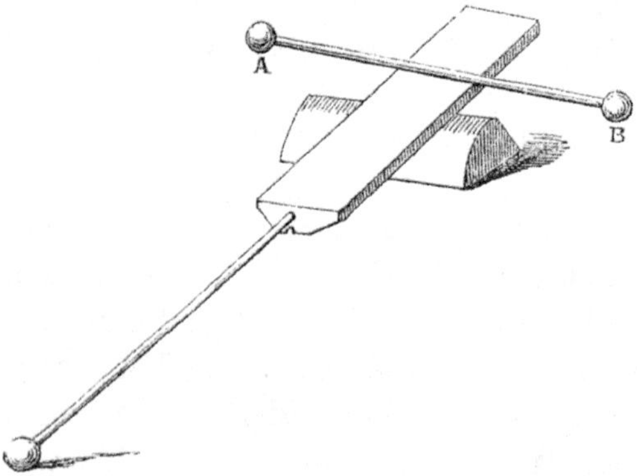

which it rests. Thus we render the vibrations slow, but they can also be rendered quick by using a rocker with a wider groove. The sides of this new rocker do not overhang so much as those of the last; it is virtually a shorter pendulum, and will vibrate more quickly. Placed upon the lead, as before, it fills the room with a clear full note. Its taps are periodic and regular, and they have linked themselves together to produce this music. Here is a third rocker, with a still wider groove, and with it we obtain a shriller tone. You know that the pitch of note augments with the number of the vibrations; this wide-grooved rocker oscillates more quickly than its predecessor, and therefore emits a higher note. By means of a beam of light we obtain an index without weight, which does not retard motion. To the rocker is fastened, by a single screw at its centre, a small disk of polished silver, on which the beam of the electric lamp falls, and from which it is reflected against the screen. When the rocker vibrates, the beam vibrates also, but with twice the angular velocity, and you now see the patch of light drawn out to a band upon the white surface.

(115) What is the cause of these singular vibrations and tones? They are due simply to the sudden expansion by heat of the body on which the rocker rests. Whenever the hot metal comes into contact with its lead carrier, a nipple suddenly juts from the latter, being produced by the heat communicated to the lead at the point of contact. The rocker is thus tilted up, and some other point of it comes into contact with the lead, a fresh nipple is formed and the weight is again tilted. Let A B (fig. 29) be the surface of the lead, and R the cross-section of the hot rocker; tilted to the right, the nipple is formed as at R; tilted to the left, it is formed as at L, the nipple in each case disappearing as soon as the contact with the

rocker ceases. The consequence is, that while its temperature remains sufficiently high, the rocker is tossed to and fro, and the quick succession of its taps against the lead produces a musical sound.

(116) These two pieces of sheet lead are fixed edge-

FIG. 29.

ways in a vice; their edges are about half an inch asunder. A long bar of heated brass is laid across the two edges. It rests first on one edge, which expands at the point of contact and jerks it upwards; it then falls upon the second edge, which also rejects it; and thus it goes on oscillating, and will continue to do so as long as the bar can communicate sufficient heat to the lead. This fire-shovel will answer quite as well as the prepared bar. I balance the heated shovel thus upon the edges of the

FIG. 30.

lead, and it oscillates exactly as the bar did (fig. 30). It may be added, that by properly laying either the poker or fire-shovel upon a block of lead, and supporting the handle so as to avoid friction, you may obtain notes as sweet and musical as any you have heard to-day. A heated hoop placed upon a plate of lead may also be caused to vibrate and sing; and a hot penny-piece or halfcrown may be caused to do the same.

(117) Looked at with reference to the connection of natural forces, this experiment is interesting. The atoms of bodies must be regarded as all but infinitely small, but then they must be regarded as all but infinitely numerous. The augmentation of the amplitude of any oscillating atom by the communication of heat, is insensible; but the summation of an almost infinite number of such augmentations becomes sensible. Such a summation, effected almost in an instant, produces the nipple, and tilts the heavy mass of the rocker. Here we have a direct conversion of heat into common mechanical motion. But the tilted rocker falls again by gravity, and in its collision with the block restores almost the precise amount of heat which was consumed in lifting it. Here we have the direct conversion of common gravitating force into heat. Again, the rocker is surrounded by a medium capable of being set in motion. The air of this room weighs some tons, and every particle of it is shaken by the rocker, and every tympanic membrane, and every auditory nerve present, is similarly shaken. Thus we have *the conversion of a portion of the heat into sound.* And, finally, every sonorous vibration which speeds through the air of this room, and wastes itself upon the walls, seats, and cushions, is converted into the form with which the cycle of actions commenced—namely, into heat.

(118) There is another curious effect, for which we are indebted to Mr. George Gore, which admits of a similar explanation. You see this line of rails. Two strips of brass, S S, S′ S′ (fig. 31), are set edgeways, about an inch asunder. A hollow ball B, of very thin metal, is placed upon the rails. If it be pushed it rolls along them; but

when left alone it is quite still. I connect the two rails, by the wires w w', with the two poles of a voltaic battery. A current now passes down one rail to the metal ball, thence over the ball to the other rail, and finally back to the battery. At the two points of contact of the ball with the rails the current encounters resistance, and wherever a current encounters resistance heat is developed. The heat

Fig. 31.

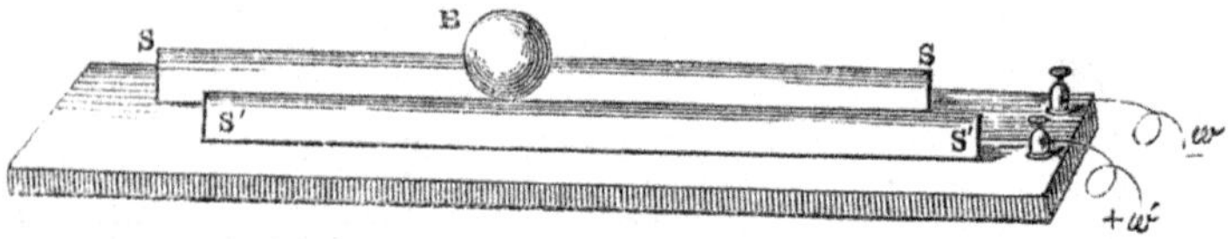

produces an elevation of the rail at these points. Observe the effect: the ball, which a moment ago was tranquil, is now uneasy. It vibrates a little at first without rolling; now it actually rolls a little way, stops, and rolls back again. It gradually augments its excursion, now it has gone farther than was intended: it has rolled quite off the rails, and injured itself by falling on the floor.

(119) In this other apparatus, for which I am indebted to Mr. Gore himself, the rails form a pair of concentric hoops. When the circuit is established, the ball F (fig. 32) rolls round the circle.* Abandoning the electric current, Mr. Gore has obtained the rotation of light balls, by placing them on circular rails of copper heated in a fire, the *rolling* force in this case being the same as the *rocking* force in the Trevelyan instrument.

(120) In the vast majority of cases the passage of bodies from the liquid to the solid state is accompanied by contraction. Here, for example, is a round glass dish containing hot water. Over the water I pour from a ladle a quantity of melted wax. The wax now forms a liquid

* Phil. Mag. vol. xv. p. 521.

layer nearly half an inch thick above the water. We will suffer both water and wax to cool, and when they are cool you will find that the wax, which now overspreads the entire surface, and is attached all round to the glass, will retreat, and we shall finally obtain a cake of wax of considerably smaller area than the dish.

(121) The wax, therefore, in passing from the solid to the liquid state, *expands*. To assume the liquid form, its particles must be pushed more widely apart, a certain play between the particles being necessary to the condition of liquidity. Now, suppose we resist the expansion of the

FIG. 32.

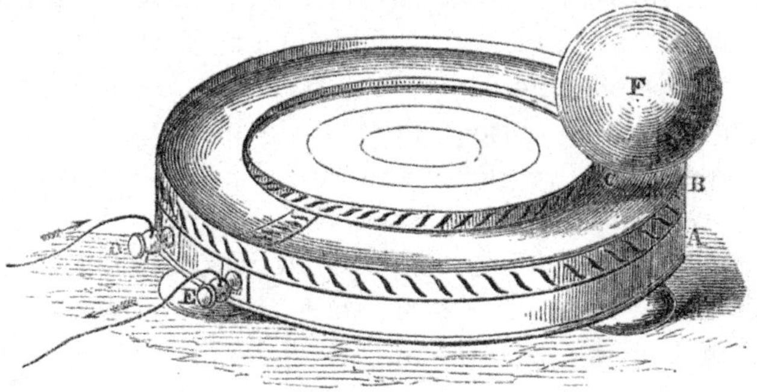

wax by an external mechanical force; suppose we have a very strong vessel completely filled with solid wax, and offering a powerful resistance to the expansion of the mass within it; what would you expect if you sought to liquefy the wax in this vessel? When the wax is free, the heat has only to conquer the attraction of the molecules of the wax, but in the strong vessel it has not only this to conquer, but also the resistance offered by the vessel. By a mere process of reasoning, we should thus be led to infer that a greater amount of heat would be required to melt the wax under pressure, than when it is free; or, in other words, that the point of fusion of the wax will be *elevated* by pressure. This reasoning is completely justified by experiment. Messrs. Hopkins and Fairbairn have, by pressure, raised the melting point of some substances, which, like wax, contract considerably on solidifying, as much as 20° and 30° Fahr.

(122) The experiments here referred to connect themselves with a very remarkable speculation. The earth is known gradually to augment in temperature as we pierce it deeper, and the depth has been calculated at which all known terrestrial bodies would be in a state of fusion. Mr. Hopkins, however, observes that, owing to the enormous pressure of the superincumbent layers, the deeper strata would require a far higher temperature to fuse them than would be necessary to fuse the strata near the earth's surface. Hence he infers that the solid crust must have a considerably greater thickness than that given by a calculation which assumes the fusing points of the superficial and the deeper strata to be the same. Sir William Thomson (Proceedings of the Royal Society, vol. xii. p. 103) expresses the conclusion that 'unless the solid substance of the earth be on the whole of extremely *rigid* material, it must yield (be deformed) by that attraction of the sun and moon which generates the tides so as to very sensibly diminish the actual phenomena of the tides, and of precession and nutation.' Mr. Hopkins had already rejected the conclusion of geologists that the earth could be a molten nucleus covered by a crust only 100 miles in thickness. He concluded that the depth of the crust must be at least 800 miles. Sir William Thomson considers it 'extremely improbable that any crust thinner than 2,000 or 2,500 miles could maintain its figure with sufficient rigidity against the tide-generating forces of sun and moon, to allow the phenomena of the ocean tides and of precession and nutation to be as they now are.'

(123) The deportment of ice is opposed to that of wax.

Ice on liquefying *contracts* ; in the arrangement of its atoms to form a solid, more room is required than they need in the neighbouring liquid state. No doubt this is due to crystalline arrangement; the attracting poles of the molecules are so situated, that when the crystallising force comes into play, the molecules unite so as to leave larger interatomic spaces in the mass. We may suppose them to attach themselves by their corners; and, in turning corner to corner, to cause a recession of the atomic centres. At all events, their centres retreat from each other when solidification sets in. By cooling, then, this power of retreat, and of consequent enlargement of volume, is conferred. It is evident that pressure in this case would resist the expansion which is necessary to solidification, and hence the tendency of pressure, in the case of water, is to keep it liquid. Thus reasoning, we should be led to the conclusion that the fusing points of substances which expand on solidifying are *lowered* by pressure.

(124) Professor James Thomson first investigated this subject from a theoretic point of view, and his conclusions have been completely verified by the experiments of his brother Professor Sir William Thomson.

(125) Let us illustrate these principles by a striking experiment. This square pillar of clear ice is an inch and a half in height, and about a square inch in cross section. At present the temperature of the ice is 0° C. But if the ice be subjected to pressure its point of fusion will be lowered: the compressed ice will melt at a temperature under 0° C., and hence the temperature which it now possesses is in excess of that at which it will melt under pressure. The ice is cut so that its planes of freezing are perpendicular to the height of the pillar. I set the column of ice, L (fig. 33), upright between two slabs of boxwood, B B′, and place the whole between the plates of a small hydraulic press. A beam from an electric lamp now

passes through the ice; the beam having been previously sent through water to deprive it of the power of melting the ice.* The light of the lamp now passes through the substance without causing fusion. In front of the press is placed a lens, and by it a magnified image of the ice is projected upon the screen before you. I now work the arm of the press, and gently squeeze the pillar of ice between the two slabs of boxwood. Dark streaks soon begin to draw themselves across the substance, at right

FIG. 33.

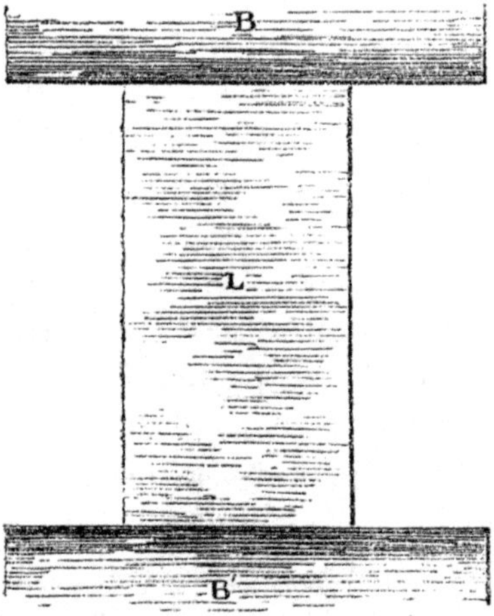

angles to the direction of pressure. Right in the middle of the mass they are appearing; and as the pressure continues, the old streaks expand and new ones are developed. The entire column of ice is now scarred by these transverse striæ. What are they? They are simply *liquid layers* foreshortened, and when you examine this column by looking into it obliquely, you see the surfaces of the layers. We have thus liquefied the ice in planes per-

* The 'sifting' of a calorific beam will be fully explained and illustrated in a subsequent Chapter.

pendicular to the pressure, and these liquid planes interspersed throughout the mass give it this laminated appearance.

(126) Whether as a solid, a liquid, or a gas, water is one of the most wonderful substances in nature. Let us consider it a little farther. At all temperatures above 32° F. or 0° C., the motion of heat is sufficient to keep the molecules of water from rigid union. But at 0° C., the motion is so reduced that the molecules then begin to seize upon each other, aggregating to a solid. This union, however, is a union according to law. To many persons here present a block of ice may seem of no more interest and beauty than a block of glass; but in reality it bears the same relation to glass that an oratorio of Handel does to the cries of a market-place. The ice is music, the glass is noise; the ice is order, the glass is confusion. In the glass, molecular forces constitute an inextricably entangled skein; in the ice they are woven to a symmetric web, the wonderful texture of which I will now try to make evident to you.

(127) How shall I dissect this ice? In the solar beam—or, failing that, in the beam of our electric lamp—we have an anatomist competent to perform this work. I will remove the agent by which this beam was purified in the last experiment, and send the rays direct from the lamp through this slab of pellucid ice. It will take the crystal edifice to pieces by accurately reversing the order of its architecture. Silently and symmetrically the crystallising force built the molecules up, silently and symmetrically the electric beam will take them down. A plate of ice, five inches square and an inch thick, is now in front of the electric lamp, the rays from which pass through the ice. The water employed in our last experiment has been removed, the entire beam being now permitted to impinge

upon the ice. Compare the radiant beam before it enters the ice with the same beam after its passage through the

Fig. 34.

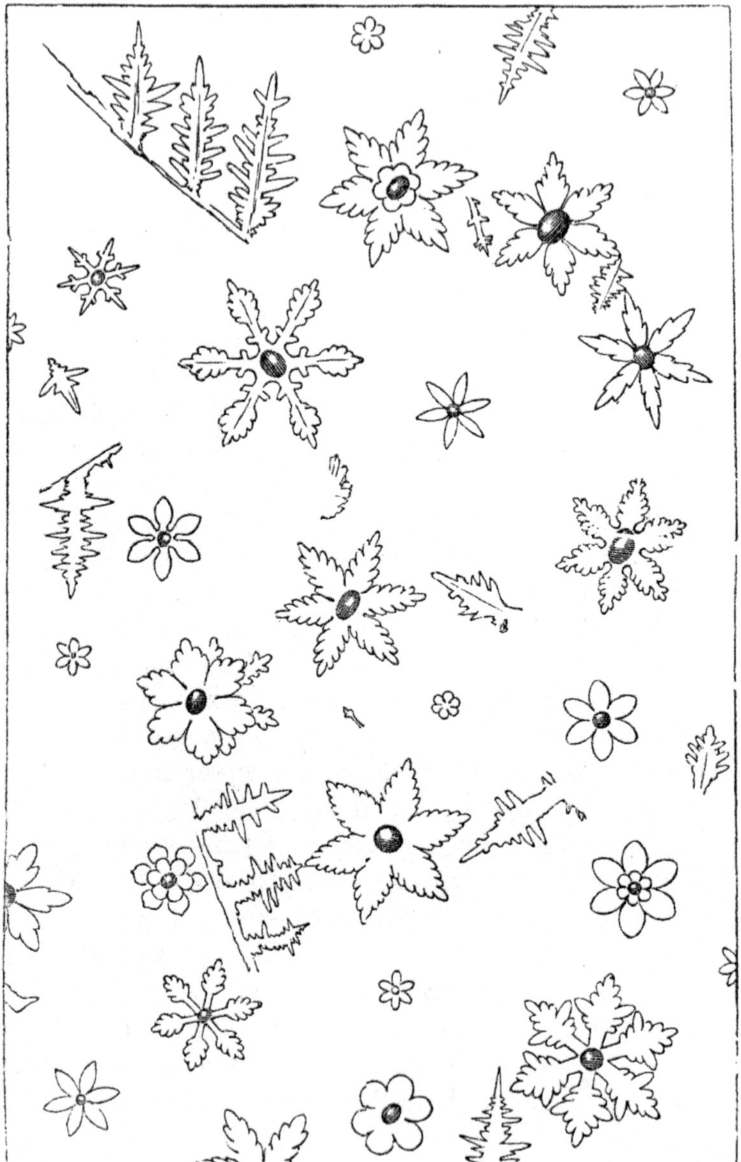

substance: to the eye there is no difference: the light is not sensibly diminished. Not so with the heat. As a thermic agent, the beam, before entering, is far more powerful than after its emergence. A portion of it has been arrested in the ice, and that portion is to be our working anatomist. I place a lens in front of the ice, and cast a magnified image of the slab upon the screen. Observe that image (fig. 34). Here we have a star, and there a star; and as the action continues, the ice appears to resolve itself into stars, each one possessing six rays, each one resembling a beautiful flower of six petals. When the lens is shifted to and fro, new stars are brought into view; and as the action continues, the edges of the petals become serrated, and spread themselves out like fern-leaves upon the screen. Probably few here present were aware of the beauty latent in a block of common ice. And only think of lavish Nature operating thus throughout the world. Every atom of the solid ice which sheets the frozen lakes of the North has been fixed according to this law. Nature 'lays her beams in music,' and it is the function of science to purify our organs, so as to enable us to hear the strain.

(128) There are two points connected with this experiment, of great minuteness, but of great interest. You see these flowers by transmitted light—by the light, that is, which has passed through both the flowers and the ice. But when you examine them by allowing a beam to be reflected from them to your eye, you find in the centre of each flower a spot which shines with the lustre of burnished silver. You might be disposed to think this spot a bubble of air; but you can, by immersing it in hot water, melt away the circumjacent ice; the moment the spot is thus laid bare, it collapses, and no trace of a bubble is to be seen. *The spot is a vacuum.* Observe how truly Nature works—how rigidly she carries her laws into all her operations. We

know that ice in melting contracts, and here we find the fact making its appearance. The water of these flowers cannot quite fill the space of the ice by the fusion of which they are produced; hence a vacuum necessarily accompanies the formation of every liquid flower.

(129) When these beautiful figures were first observed, and at the moment when the central spot appeared, like a point of light suddenly formed within the ice, I thought I heard a clink, as if the ice had split asunder when the spot was formed. At first I suspected that it was my imagination which associated sound with the appearance of the spot, as it is said that people who see meteors often imagine a rushing noise when they really hear none. The clink, however, was a reality; and if you allow me, I will now conduct you from this trivial fact through a series of interesting phenomena to a far-distant question of practical science.

(130) All water holds a quantity of air within it in solution; by boiling you may liberate this imprisoned air. On heating a flask of water, air bubbles are seen crowding on its sides, long before it boils, rising through the liquid without condensation, and often floating on the top. The presence of this air in the water promotes the ebullition of the liquid. It acts as a kind of elastic spring, pushing the molecules apart, and thus helping them to take the gaseous form.

(131) When this antagonist to their intimate union is removed, the molecules lock themselves together in a far tighter embrace. The cohesion of the water is vastly augmented by the removal of the air. Here is a glass vessel which contains water purged of air. One effect of the withdrawal of the elastic buffer is, that the water falls with the sound of a solid body, and hence this instrument is called the water hammer. You hear how the liquid rings against the end of the tube, when it is turned upside

down. This other tube, A B C (fig. 35), bent into the form of a V, is intended to show how the cohesion of the water is affected by long-continued boiling. The water which partially fills the bent tube is first brought into one arm of the V. And now I tap the end of this arm against the table. You hear, at first, a loose and jingling sound. As long as you hear that jingle the water is not

FIG. 35.

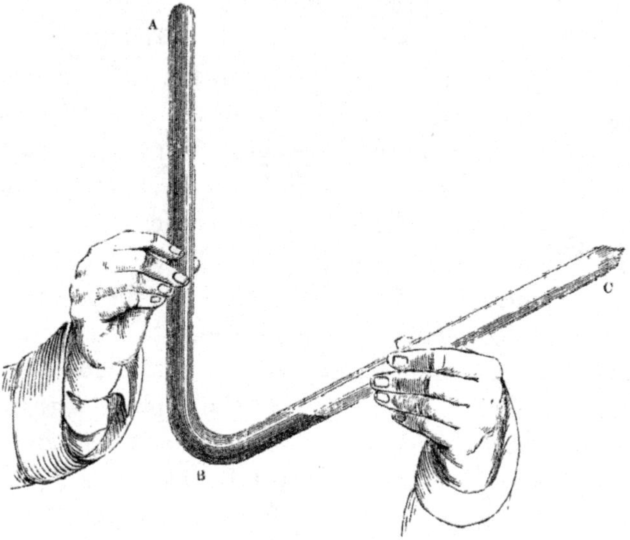

in true contact with the interior surface of the tube. As the tapping continues you notice an alteration in the sound; the jingling has now disappeared, the impact being hard, like that of solid against solid. I now raise the tube, and turn the column of water upside down, but there it stands in the arm A B. Its particles cling so tenaciously to the sides of the tube, and lock themselves so firmly together, that it refuses to behave like a liquid body; it declines to obey the law of gravity.

(132) So much for the augmentation of cohesion; but this very cohesion enables the liquid to resist ebullition.

Water thus freed of its air can be raised to a temperature 60° or 80° Fahr., above its ordinary boiling point, without ebullition. But mark what takes place when the liquid does boil. It has an enormous excess of heat stored up; the locked atoms finally part company, but they do so with the violence of a spring which suddenly breaks under strong tension, and ebullition is converted into explosion. To M. Donny, of Ghent, we are indebted for the discovery of this interesting property of water.

(133) Turn we now to our ice:—Water in freezing completely excludes the air from its crystalline architecture. All foreign bodies are squeezed out of it, and ice holds no air in solution. Supposing, then, that we melt a piece of pure ice, under conditions where air cannot approach it, we should have water in its most highly cohesive condition; and such water ought, if heated, to show the effects mentioned. That it does so has been proved by Faraday. He melted ice under spirit of turpentine, and found that the liquid thus formed could be heated far beyond its boiling point, and that the rupture of the liquid, by heating, took place with almost explosive violence. Let us apply these facts to the six-petaled ice-flowers, and their little central star. They are formed in a place where no air can come. Imagine the flower forming, and gradually augmenting in size. The cohesion of the liquid is so great, that it will pull the walls of its chamber together, or even expand its own volume, sooner than give way. But, as its size augments, the space which it tries to occupy becomes too large for it, until finally the liquid snaps with an audible clink, and a vacuum is formed.

(134) Let us take our final glance at this web of relations. It is very remarkable that a great number of locomotives have exploded on quitting the shed where they had remained for a time quiescent, and just as the

engineer turned on the steam. Now, if a locomotive has been boiling sufficiently long to expel the air contained in its water, *that* liquid will possess, in a greater or less degree, the high cohesive quality to which I have drawn your attention. It is at least conceivable, that while resting, previous to starting, an excess of heat might be thus stored up in the boiler, and, if stored up, the certain result would be, that the mechanical act of turning on the steam would produce the rupture of the cohesion, and steam of explosive force would instantly be generated. I do not say this *is* the case; but who can say it is *not* the case? We have been dealing throughout with a real agency, which is certainly competent, if its power be invoked, to produce the effects which have been ascribed to it.

(135) As you add heat, or, in other words, motion, to water, the particles from its free surface fly off in augmented numbers. You at length approach what is called the *boiling point* of the liquid, where the conversion into vapour is not confined to the free surface, but is most copious at the bottom of the vessel where the heat is applied. When water boils in a glass beaker, the steam is seen rising from the bottom to the top, where it often floats for a time, enclosed above by a dome-shaped liquid film. To produce these bubbles certain resistances must be overcome. First, we have the adhesion of the water to the vessel which contains it, and this force varies with the substance of the vessel. In the case of a glass vessel, for example, the boiling point may be raised two or three degrees by adhesion; while in metal vessels this is impossible. The adhesion is often overcome by fits and starts, which may be so augmented by the introduction of certain salts into the liquid, that a loud bumping sound accompanies the ebullition; the detachment is in some cases so sudden

and violent, as to cause the liquid to leap bodily out of the vessel.

(136) A second antagonism to the boiling of the liquid is the attraction of the liquid particles for each other; a force which, as we have seen, may become very powerful when the liquid is purged of air. This is not only true of water, but of other liquids—of all ethers and alcohols, for example. If we connect a small flask containing ether or alcohol with an air-pump, a violent ebullition occurs in the liquid when the pump is first worked; but after all the air has been removed from the liquid, we may, in many cases, continue to work the pump without producing any sensible ebullition; the free surface alone of the liquid yielding vapour.

(137) But in order that steam should exist in bubbles, in the interior of a mass of liquid, it must be able to resist two other things—the weight of the water above it, and the weight of the atmosphere above the water. What the atmosphere is competent to do may be thus illustrated. This tin vessel contains a little water, which is kept boiling by this small lamp. At the present moment all the space above the water is filled with steam, which issues from this stopcock. I shut the cock, withdraw the lamp, and pour cold water upon the tin vessel. The steam within it is condensed, the elastic cushion which pushed the sides outwards in opposition to the pressure of the atmosphere is withdrawn, and observe the consequence. The sides of the vessel are crushed and crumpled up by the atmospheric pressure. This pressure amounts to 15 lbs. on every square inch: how then can a thing so frail as a bubble of steam exist on the surface of boiling water? Simply because the elastic force of the steam within is exactly equal to that of the atmosphere without; the liquid film is pressed between two elastic cushions which exactly neutralise each other. If the steam were

predominant, the bubble would burst from within outwards; if the air were predominant, the bubble would be crushed inwards. Here, then, we have the true definition of the boiling point of a liquid. It is that temperature at which the tension of its vapour exactly balances the pressure of the atmosphere.

(138) As we ascend a mountain, the pressure of the atmosphere above us diminishes, and the boiling point is correspondingly lowered. On an August morning in 1859 the temperature of boiling water on the summit of Mont Blanc was found to be 184·95° F.; that is, about twenty-seven degrees lower than the boiling point at the sea level. On August 3, 1858, the temperature of boiling water on the summit of the Finsteraarhorn was 187° F. On August 10, 1858, the boiling point on the summit of Monte Rosa was 184·92° F. The boiling point on Monte Rosa is shown by these observations to be almost the same as it was found to be on Mont Blanc, though the latter exceeds the former in height by 500 feet. The fluctuations of the barometer are, however, quite sufficient to account for this anomaly. The lowering of the boiling point is about 1° F. for every 590 feet of elevation; and from the temperature at which water boils, we may approximately infer the height. It is said that to make good tea, in London, boiling water is essential; if this be so, it is evident that the beverage cannot be procured, in all its excellence, at the higher stations of the Alps.

(139) Our next experiment will illustrate the dependence of the boiling point on external pressure. This flask, F (fig. 36), contains water; while from this second and much larger one, G, the air has been removed by an air-pump. The two flasks are connected together by a system of cocks, which enables me to establish a communication between them. The water in the small flask has been kept boiling for some time, the steam

generated escaping from the cock *y*. I now remove the spirit-lamp, and turn this cock, so as to shut out the air. The water ceases to boil, and pure steam now fills the flask above it. We will give the water time to cool a little. At intervals you see a bubble of steam rising, because the pressure of the vapour above is gra-

FIG. 36.

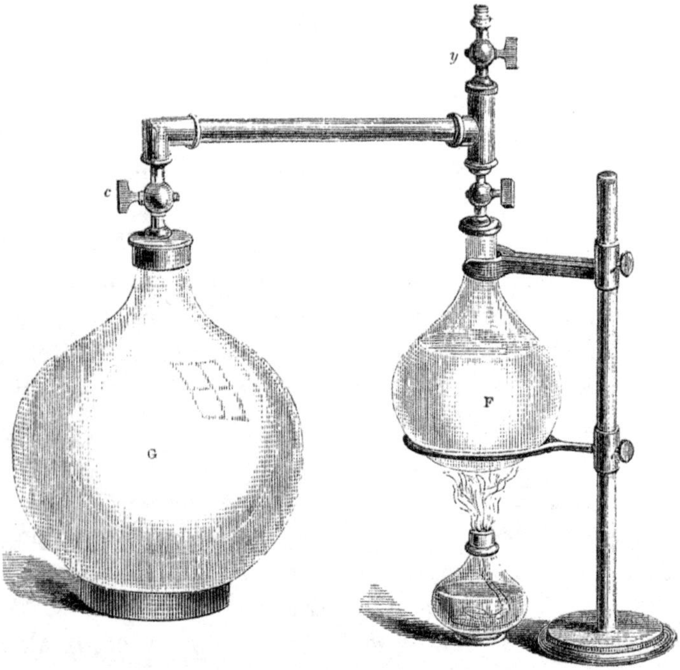

dually becoming less through its own slow condensation. I hasten the condensation by pouring cold water on the flask; the bubbles are more copiously generated. By plunging the flask bodily into cold water we might cause it to boil violently. The water in F is now at rest, and some degrees below its ordinary boiling point. I turn the cock *c*, which opens a way for the escape of the vapour into the exhausted vessel G; the moment the pressure is diminished ebullition begins in F; and observe how the condensed steam showers in a kind of rain against the sides of the vessel G. By keeping the vessel G cool, and thereby preventing the vapour in it from reacting upon the surface of the water in F, we can keep the small flask bubbling and boiling for a considerable time.

(140) Through high heating, the elastic force of steam may be rendered enormous. The Marquis of Worcester burst cannon with it, and our calamitous boiler explosions are so many illustrations of its power. By the skill of man this mighty agent has been controlled: by it Denis Papin raised a piston, which was pressed down again by the atmosphere, when the steam was condensed; Savery and Newcomen turned it to practical account, and James Watt completed this grand application of the moving power of heat. Pushing the piston up by steam, while the space above the piston is in communication with a condenser or with the free air, and again pushing down the piston, while the space below it is in communication with a condenser or with the air, we obtain a simple to-and-fro motion, which, by mechanical arrangements, may be made to take any form we please.

(140 *a*) But the principle of the conservation of force is illustrated here as elsewhere. For every stroke of work done by the steam-engine, for every weight that it lifts, and for every wheel that it sets in motion, an equivalent quantity of heat disappears. A ton of coal furnishes by its combustion a certain definite amount of heat. Let this quantity of coal be applied to a working steam-engine; and let all the heat communicated to the machine and the condenser, and all the heat lost by radiation and by contact with the air, be collected; it will fall short of the quantity produced by the simple combustion of the ton of coal, by an amount exactly equivalent to the work performed. Suppose that work to consist in lifting a

weight of 7,720 lbs. a foot high; the heat produced by the coal would fall short of its maximum by a quantity just sufficient to warm a pound of water 10° F. In an elaborate series of experiments, executed with extraordinary assiduity and on a grand scale by M. Hirn, the civil engineer at Colmar, this theoretic deduction has been reduced to fact.

(140 *b*) In the steam-engine employed by M. Hirn the steam left the boiler and entered the cylinder at a temperature of 146° C. The temperature of his condenser was 34° C. The steam was worked expansively; that is to say, it was permitted to enter the cylinder and exert its full pressure until the piston was raised through a certain fraction of its range. The steam was then cut off, the piston being urged through the remainder of its course by the expansive force of the steam already in the cylinder.

(140 *c*) In this case the space above the piston was connected with the condenser; and if the expansion were perfect the vapour underneath the piston at the moment it reached the highest point of its course would have the pressure due to the temperature of the condenser. Now supposing the expansion to be thus perfect, and that the expanded vapour all remains in the state of vapour; the experiments of M. Regnault enable us to calculate the fraction of the total heat which is converted into work. We find according to these calculations, and not without a feeling of astonishment at its smallness, that in the experiments of M. Hirn, the heat converted into work ought not to amount to $\frac{1}{19}$th of the whole.

(140 *d*) But as a matter of fact M. Hirn found that $\frac{1}{8}$th of the heat borrowed from the steam in the boiler was converted into mechanical effect. Thus actual experiment was at variance with calculation. A theoretic conclusion arrived at independently first by Mr. Rankine, and im-

mediately afterwards by M. Clausius, two of the founders of the mechanical theory of heat, reveals the cause of this discrepancy. In calculating the heat possessed by the vapour as it enters the condenser, it was assumed that the whole of the vapour borrowed from the boiler remained during its expansion in the vaporous condition. This Mr. Rankine and M. Clausius proved that it could not do. They showed that when saturated steam expands, as in M. Hirn's experiments, it is in part precipitated, thus yielding up a portion of the heat of vaporisation. Indeed, before anything correct was known about its cause, mechanical engineers met the nuisance arising from the water of condensation by surrounding the cylinder with a jacket of hot steam from the boiler. The mixture of vapour and liquid entering the condenser after the expansion would thus possess less heat than if it were all vapour; and hence it is that a greater amount of heat than that given by calculation is converted into work. I may add that the precipitation of the steam during its expansion was demonstrated experimentally by M. Hirn.

(140 *e*) But even $12\frac{1}{2}$ per cent. of the total heat implies enormous loss. Nor is this loss to be avoided in the steam engine. For the amount of heat converted into work depends upon two things, the temperature of the steam as it enters the cylinder and its temperature as it enters the condenser. The further the initial and the final temperatures are apart, the greater is the amount of heat converted into work; but to convert *all* the heat into work a condenser kept at the absolute zero of temperature would be required.*

* The *absolute temperature* of a body is its temperature reckoned from the absolute zero (explained in § 96). Thus the temperature of melting ice reckoned from this point is 273° C. Let T represent the initial temperature of the steam, and t its final temperature, both reckoned from the absolute zero; then the proportion of the total heat converted into work cannot under the most favourable conditions exceed the fraction $\frac{T-t}{T}$.

(141) My object however at present is to deal with nature rather than art, and I am compelled to pass quickly over the triumphs of man's skill in the application of steam to the purposes of life. Those who have walked through the workshops of Woolwich, or through any of our great factories where machinery is extensively employed, will have been sufficiently impressed with the aid which the mighty power of heat renders to man. Let it be remembered, that every wheel which revolves, every chisel, and plane, and punch, which passes through solid iron as if it were so much cheese, derives its moving energy from the clashing atoms in the furnace. The motion of these atoms is communicated to the boiler, thence to the water, whose particles are shaken asunder, and fly from each other with a repellent energy commensurate with the heat communicated. The steam is simply the apparatus, through the intermediation of which, the atomic motion is converted into the mechanical motion. And the motion thus generated can reproduce its parent. Look at the planing tools; look at the boring instruments—streams of water gush over them to keep them cool. Take up the curled iron shavings which the planing tool has pared off; you cannot hold them in your hand, they are so hot. Here the moving force is restored to its first form; the energy of the machine has been consumed in reproducing the power from which that energy was derived.

(142) I must now direct your attention to a natural steam engine, which long held a place among the wonders of the world;—the Great Geyser of Iceland. The surface of Iceland gradually rises from the coast towards the centre, where the general level is about 2,000 feet above the sea. On this, as on a pedestal, are planted the Jökull, or icy mountains of the island, which extend both ways in a north-easterly direction. Along this chain also occur the active volcanoes of Iceland, and the thermal

springs follow the same general direction. From the ridges and chasms which diverge from the mountains, enormous masses of steam issue at intervals, hissing and roaring; and when the escape occurs at the mouth of a cavern, the résonance of the cave often raises the sound to the loudness of thunder. Lower down, in the more porous strata, we have smoking mud pools, where a repulsive blue-black aluminous paste is boiled, rising at times in huge bubbles, which, on bursting, scatter their slimy spray to a height of fifteen or twenty feet. From the base of the hills upwards extend the glaciers, and above these are the snow-fields which crown the summits. From the arches and fissures of the glaciers, vast masses of water issue, falling at times in cascades over walls of ice, and spreading for miles over the country before they find definite outlet. Extensive morasses are thus formed. Intercepted by the cracks and fissures of the land, a portion of the water finds its way to the heated rocks beneath; and here, meeting with the volcanic gases which traverse these underground regions, both travel on together, to issue, at the first convenient opportunity, either as an eruption of steam or a boiling spring.

(143) The most famous of these springs is the Great Geyser. It consists of a tube, seventy-four feet deep, and ten feet in diameter. The tube is surmounted by a basin, which measures from north to south fifty-two feet across, and from east to west sixty feet. The interior of the tube and basin is coated with a beautiful smooth siliceous plaster, so hard as to resist the blows of a hammer; and the first question is, How was this wonderful tube constructed—how was this perfect plaster laid on? Chemical analysis shows that the water holds silica in solution, and it might therefore be conjectured that the water had deposited this silica against the sides of the tube and basin. But such is not the case: the water deposits no sediment;

no matter how long it may be kept, no solid substance is separated from it. It may be bottled up and preserved for years, as clear as crystal, without showing the slightest tendency to form a precipitate. To answer the question in this way would moreover assume that the shaft was formed by some foreign agency, the water merely lining it. The geyser-basin, however, rests upon the summit of a mound about forty feet high, and it is evident, from mere inspection, that the mound has been deposited by the geyser. But in building up this mound the spring must have formed the tube which perforates the mound, and hence the conclusion that the geyser is the architect of its own tube.

(144) If we place a quantity of the geyser water in an evaporating basin, the following takes place: In the centre of the basin the liquid deposits nothing, but at the sides, where it is drawn up by capillary attraction, and thus subjected to speedy evaporation, we find silica deposited. Round the edge a ring of silica is laid on, and not until the evaporation has continued a considerable time do we find the slightest turbidity in the middle of the water. This experiment is the microscopic representant of what occurs in Iceland. Imagine the case of a simple thermal siliceous spring, whose waters trickle down a gentle incline; the water thus exposed evaporates, and silica is deposited. This deposit gradually elevates the side over which the water passes, until, finally, the latter has to take another course. The same takes place here, the ground is elevated as before, and the spring has to move forward. Thus it is compelled to travel round and round, depositing its silica and deepening the shaft in which it dwells, until finally, in the course of ages, the simple spring has produced that wonderful apparatus which has so long puzzled and astonished both the tourist and the philosopher

(145) Previous to an eruption, both the tube and basin are filled with hot water; detonations which shake the ground are heard at intervals, and each is succeeded by a violent agitation of the water in the basin. The water in the pipe is lifted up, forms an eminence in the middle of the basin, and an overflow is the consequence. These detonations are evidently due to the production of steam in the ducts which feed the geyser tube, which steam, escaping into the cooler water of the tube, is there suddenly condensed, and produces the explosions. Professor Bunsen succeeded in determining the temperature of the geyser tube, from top to bottom, a few minutes before a great eruption; and these observations revealed the extraordinary fact, that at no part of the tube did the water reach its boiling point. In the annexed sketch (fig. 37) I have given, on one side, the temperatures actually observed, and on the other side the temperatures at which water would boil, taking into account both the pressure of the atmosphere and of the superincumbent column of water. The nearest approach to the boiling point is at A, thirty feet from the bottom: but even here the water is 2° Centigrade, or more than 3½° Fahr., below the temperature at which it could boil. How then is it possible that an eruption could occur under such circumstances?

(146) Fix your attention upon the water at the point A, where the temperature is within 2° C. of the boiling point. Call to mind the lifting of the column when the detonations are heard. Let us suppose that by the entrance of steam from the ducts near the bottom of the tube, the geyser column is elevated six feet, a height quite within the limits of actual observation; the water at A is thereby transferred to B. Its boiling point at A is 123·8°, and its actual temperature 121·8°; but at B its boiling point is only 120·8°; hence, when transferred from A to B, the heat

which it possesses is in excess of that necessary to make it boil. This excess of heat is instantly applied to the generation of steam: the column is lifted higher, and the water

FIG. 37.

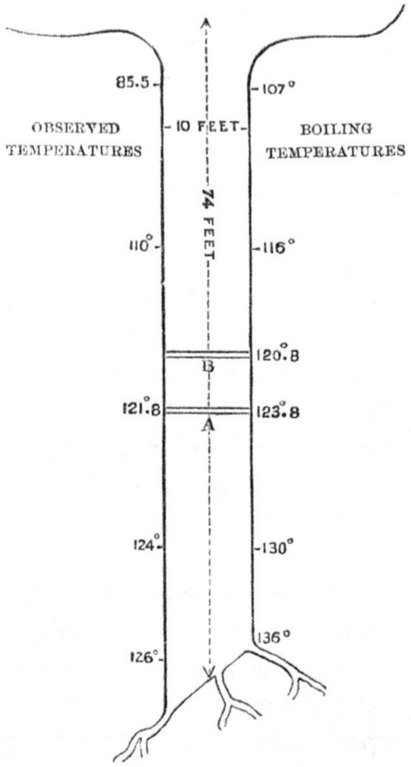

below is further relieved. More steam is generated; from the middle downwards the mass suddenly bursts into ebullition; the water above, mixed with steam-clouds, is projected into the atmosphere, and we have the geyser eruption in all its grandeur.

(147) By its contact with the air the water is cooled, falls back into the basin, partially refills the tube, in which

FIG. 38.

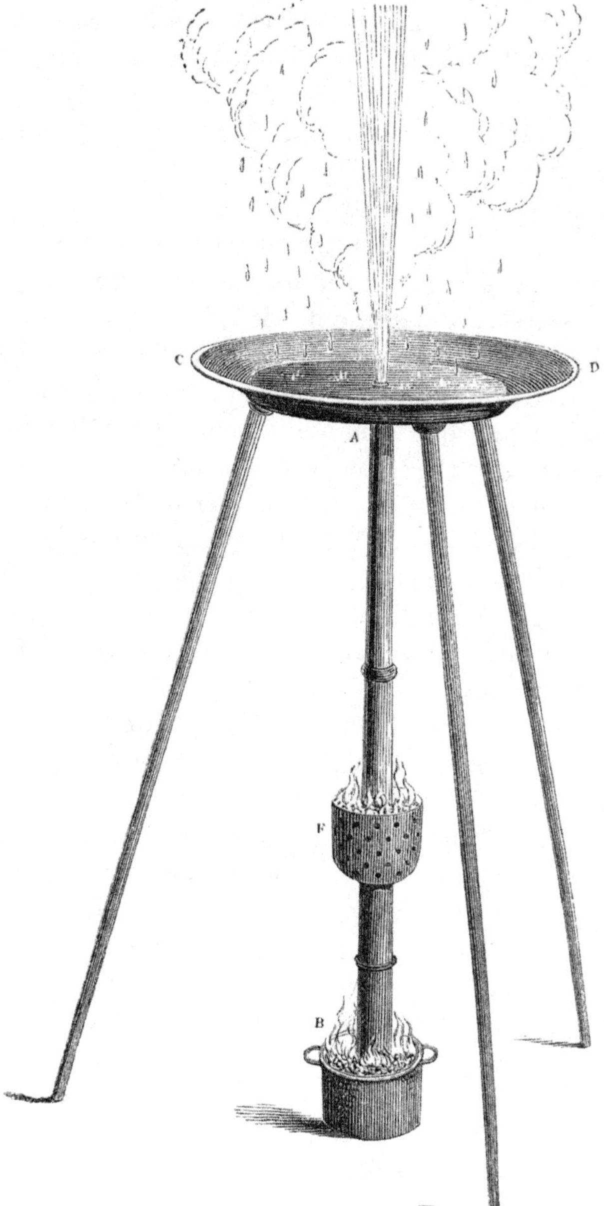

it gradually rises, and finally fills the basin as before. Detonations are heard at intervals, and risings of the water in the basin. These are so many futile attempts at an eruption, for not until the water in the tube comes sufficiently near its boiling temperature to make the lifting of the column effective, can we have a true eruption.

(148) To the celebrated Bunsen we owe this beautiful theory, and now let us try to justify it by experiment. Here is a tube of galvanised iron, six feet long, A B (fig. 38), surmounted by a basin, C D. It is heated by a fire underneath; and, to imitate as far as possible the condition of the geyser, the tube is encircled by a second fire, F, at a height of two feet from the bottom. Doubtless the high temperature of the water, at the corresponding part of the geyser tube, is due to a local action of the heated rocks. The tube is filled with water, which gradually becomes heated; and regularly, every five minutes, the liquid is ejected into the atmosphere.

(149) There is another famous spring in Iceland, called the Strokkur, which is usually forced to explode by stopping its mouth with clods. We can imitate the action of this spring by thus stopping the mouth of our tube A B with a cork. And now the heating progresses. The steam below will finally attain sufficient tension to eject the cork, and the water, suddenly relieved from the pressure, will burst forth in the atmosphere. The ceiling of this room is nearly thirty feet from the floor, but the eruption has reached the ceiling, from which the water now drips plentifully. In fig. 39 is given a section of the Strokkur.

(150) By stopping our model geyser-tube with corks, through which glass tubes of various lengths and diameters pass, the action of many of the other eruptive springs may be accurately imitated. We can readily, for example, produce an intermittent action; discharges of water, and

impetuous steam-gushes follow each other in quick succession, the water being squirted in jets fifteen or twenty feet high. These experiments prove that the geyser tube itself is the sufficient cause of the eruptions, and we are relieved from the necessity of imagining underground caverns filled with water and steam, which were formerly regarded as necessary to the production of these wonderful phenomena.

FIG. 39.

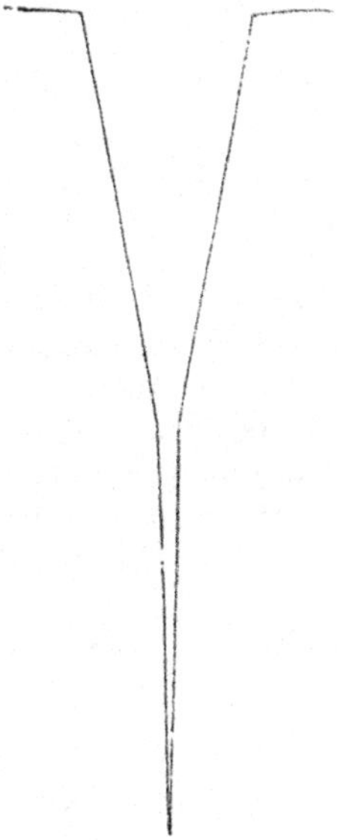

(151) A moment's reflection will suggest to you that there must be a limit to the operations of the geyser. When the tube has reached such an altitude that the water in the depths below, owing to the increased pressure, cannot attain its boiling point, the eruptions of necessity cease. The spring, however, continues to deposit its silica, and often forms a *Laug*, or cistern. Some of those in Iceland are forty feet deep. Their beauty, according to Bunsen, is indescribable, over the surface curls a light vapour, the water is of the purest azure, and tints with its own hue the fantastic incrustations on the cistern walls; while, at the bottom, is often seen the mouth of the once mighty geyser. There are also in Iceland traces of vast, but now extinct, geyser operations. Mounds are observed, whose shafts are filled with rubbish, the water having forced a passage underneath and retired to other scenes of action. We have, in fact, the geyser in its youth, manhood, old age, and death, here presented to us. In its youth, as a simple thermal spring; in its manhood, as the eruptive column; in its old age, as the tranquil *Laug*; while its death is recorded by the ruined shaft and forsaken mound, which testify the fact of its once active existence.

CHAPTER V.

APPLICATION OF THE DYNAMICAL THEORY TO THE PHENOMENA OF SPECIFIC AND LATENT HEAT—DEFINITION OF ENERGY: POTENTIAL AND DYNAMIC ENERGY—ENERGY OF MOLECULAR FORCES—EXPERIMENTAL ILLUSTRATIONS OF SPECIFIC AND LATENT HEAT—MECHANICAL VALUES OF THE ACTS OF COMBINATION, CONDENSATION, AND CONGELATION IN THE CASE OF WATER—SOLID CARBONIC ACID—THE SPHEROIDAL STATE OF LIQUIDS—FLOATING OF A SPHEROID ON ITS OWN VAPOUR—FREEZING OF WATER AND MERCURY IN A RED-HOT CRUCIBLE.

(152) WHENEVER a difficult expedition is undertaken in the Alps, the experienced mountaineer commences the day at a slow pace, so that when the real hour of trial arrives, he may find himself hardened instead of exhausted by his previous work. We, to-day, are about to enter on a difficult enterprise. Let us commence it in the same spirit; not with a flush of enthusiasm, which labour will extinguish, but with patient and determined hearts, which will not recoil should a difficulty arise.

(153) This lead weight, you observe, is attached to a string which passes over a pulley at the top of the room. We know that the earth and the weight are mutually attractive; the weight now rests upon the earth, and exerts a certain pressure upon its surface. The earth and the weight here *touch each other*; their mutual attractions are as far as possible satisfied, and *motion*, by their mutual approach, is no longer possible. As far as the attraction of gravity is concerned, the possibility of producing motion ceases, as soon as the two attracting bodies are actually in contact.

(154) I draw up this weight. It is now suspended at a height of sixteen feet above the floor, where it remains just as motionless as when it rested on the floor; but, by introducing a space between the floor and it, I entirely change the condition of the weight. By raising it, I have conferred upon it a motion-producing power. There is now an action possible to the weight, which was not possible when it rested upon the earth; *it can fall*, and, in its descent, can turn a machine, or perform other work. Let us employ the useful and appropriate term *energy* to denote the power of performing work; we might then fairly use the term *possible energy*, to express the power of motion, which our drawn-up weight possesses, but which has not yet been exercised by falling; or we might call it 'potential energy,' as some eminent men have already done. This potential energy is derived, in the case before us, from the pull of gravity, which pull, however, has not yet resulted in motion. But I now let the string go; the weight falls, and reaches the earth's surface with a velocity of thirty-two feet a second. At every moment of its descent it was pulled down by gravity, and its final moving force is the summation of the pulls. While in the act of falling, the energy of the weight is active. It may be called *actual* energy, in antithesis to *possible*; or it may be called *dynamic* energy, in antithesis to *potential*, or we might call the energy with which the weight descends *moving force*. The great thing, now, is to be able to distinguish energy *in store* from energy *in action*. Once for all, then, let us take the terms of Mr. Rankine, and call the energy in store 'potential,' and the energy in action 'actual.'* If, after this, I should use

* Helmholtz in his admirable memoir on 'Die Erhaltung der Kraft' (1847), divided all energy into *tension* and *vis viva*. (Spannkräfte und Lebendige Kräfte.)

the terms 'possible energy,' or 'dynamic energy,' or 'moving force,' you will have no difficulty in affixing the exact idea to these terms.

(155) Our weight started from a height of sixteen feet; let us fix our attention upon it, after it has accomplished the first foot of its fall. The total pull, if I may use the term, to be expended on it, has been then diminished by the amount expended in its passing through the first foot. At the height of fifteen feet it has one foot less of potential energy than it possessed at the height of sixteen feet, but at the height of fifteen feet it has an equivalent amount of dynamic or actual energy, which, if reversed in direction, would raise it again to its primitive height. Hence, as potential energy disappears, actual energy comes into play. *Throughout the universe, the sum of these two energies is constant.* To create or annihilate energy is as impossible as to create or annihilate matter; and all the phenomena of the material universe consist in transformations of energy alone. The principle here enunciated is called the law of the *conservation of energy*.

(156) It is as yet too early to refer to organic processes, but could we have observed the molecular condition of my arm as I drew up that weight, it would have been seen that in accomplishing this mechanical act, an equivalent amount of some other form of energy was consumed. That energy, we shall afterwards learn, is *heat*. If the weight were raised by a steam-engine, a portion of heat would also disappear, exactly equivalent to the work done. The weight is about a pound, and to raise it sixteen feet would consume as much heat as would raise the temperature of a cubic foot of air about 1° F. Conversely, this quantity of heat would be generated by the falling of the weight from a height of sixteen feet.

(157) It is easy to see that, if the force of gravity

were immensely greater than it is, an immensely greater amount of heat would have to be expended, in raising the weight. The greater the attraction, the greater would be the amount of heat necessary to overcome it; but conversely, the greater would be the amount of heat which a falling body would then develope by its collision with the earth.

(158) We must turn these conceptions, regarding sensible masses, to account, in forming conceptions regarding insensible masses. As an intellectual act, it is quite as easy to conceive the separation of two mutually attracting *atoms*, as to conceive the separation of the earth and weight. We have already had occasion to refer to the intensity of molecular forces, and here we must return to the subject. Closely locked together as they are, the atoms of bodies, though we cannot suppose them to be in contact, exert enormous attractions. It would require an almost incredible amount of ordinary mechanical force to widen the distances intervening between the atoms of any solid or liquid, so as to increase its volume in any sensible degree. It would also require a force of great magnitude to squeeze the particles of a liquid or solid together, so as to make the body sensibly less in size. I have vainly tried to augment permanently the density of a soft metal by pressure. Water, which yields so freely to the hand plunged in it, was for a long time regarded as absolutely incompressible. Great force was brought to bear upon it; but sooner than shrink, it oozed through the pores of the metal vessel which contained it, and spread like a dew on the surface.* By refined and powerful means we can

* I have to thank my friend, Mr. Spedding, for the following extract in reference to this experiment:—

'Now it is certain that rarer bodies (such as air) allow a considerable degree of contraction, as has been stated; but that tangible bodies (such as water) suffer compression with much greater difficulty and to a less extent. How far they do suffer it, I have investigated, in the following experiment:

now compress water, but the force necessary to accomplish this is very great. When, therefore, we wish to overcome molecular forces, we must attack them by their peers. Heat accomplishes what mechanical energy, as generally wielded, is incompetent to perform. Bodies, when heated, expand, and to effect this expansion their molecular attractions must be overcome; and where the attractions to be surmounted are so vast, we may infer that the quantity of heat necessary to overpower them will be commensurate.

(159) And now I must ask your entire attention. Suppose a certain amount of heat to be imparted to this lump of lead, how is that heat disposed of within the substance? It is applied to two distinct purposes —it performs two different kinds of work. One portion of it excites that species of motion which augments the temperature of the lead, and which is sensible to the thermometer; but another portion of it goes to force

I had a hollow globe of lead made capable of holding about two pints, and sufficiently thick to bear considerable force; having made a hole in it, I filled it with water, and then stopped up the hole with melted lead, so that the globe became quite solid. I then flattened the two opposite sides of the globe with a heavy hammer, by which the water was necessarily contracted into less space, a sphere being the figure of largest capacity; and when the hammering had no more effect in making the water shrink, I made use of a mill or press; till the water, impatient of further pressure, exuded through the solid lead like a fine dew. I then computed the space lost by the compression, and concluded that this was the extent of compression which the water had suffered, but only when constrained by great violence.' (Bacon's 'Novum Organum,' published in 1620: vol. iv. p. 209 of the translation.) Note by R. Leslie Ellis, vol. i. p. 324:—'This is perhaps the most remarkable of Bacon's experiments, and it is singular that it was so little spoken of by subsequent writers. Nearly fifty years after the production of the "Novum Organum," an account of a similar experiment was published by Megalotti, who was secretary of the Academia del Cimento at Florence; and it has since been familiarly known as the Florentine experiment.'

It is to be remembered that Leibnitz (Nouveaux Essais), in mentioning the Florentine experiment, says that the globe was of gold (p. 229, Erdmann), whereas the Florentine academicians expressly say why they preferred silver to either gold or lead.

the atoms of lead into new positions, and this portion *is lost as heat.* The pushing asunder of the atoms of the lead in this case, in opposition to their mutual attractions, is exactly analogous to the raising of our weight in opposition to the force of gravity, a loss of heat, in both cases, being the result. Let me try to make the comparison between the two actions still more strict. Suppose that a definite amount of force is to be expended upon our weight, and that this force is divided into two portions, one of which is devoted to the actual raising of the weight, while the other is employed to cause the weight, as it ascends, to oscillate like a pendulum, and to oscillate, moreover, with gradually augmented width and rapidity: we have, then, the analogue of that which occurs when heat is imparted to the lead. The atoms are pushed apart, but, during their recession, they vibrate with gradually augmented intensity. Thus, the heat communicated to the lead, resolves itself, in part, into atomic potential energy, and in part into actual energy, which may be regarded as a kind of atomic music, the musical part alone being competent to act upon our thermometers or to affect our nerves.

(160) In this case, then, the heat not only imparts actual energy to the vibrating atoms, but also accomplishes what we may call *interior work*;* it performs work within the body heated, by forcing its particles to take up new positions. When the body cools, the forces which were overcome in the process of heating come into play; the heat which was consumed in the recession of the atoms being restored upon their approach.

(161) Chemists have determined the relative weights of the atoms of different substances. Calling the weight of a hydrogen atom 1, the weight of an oxygen atom is 16. Hence, to make up a pound weight of hydrogen, sixteen times the number of atoms contained in a pound of oxygen

* See the excellent memoirs of Clausius in the Philosophical Magazine.

would be necessary. The number of atoms required to make up a pound is, evidently, inversely proportional to the atomic weight. We here approach a very delicate and important point. The experiments of Dulong and Pétit, and of MM. Regnault and Neumann, render it extremely probable, that all elementary atoms, great or small, light or heavy, when at the same temperature, possess the same amount of the energy we call heat, the lighter atoms making good by velocity what they want in mass. Thus, each atom of hydrogen has the same moving energy as an atom of oxygen, at the same temperature. But, inasmuch as a pound weight of hydrogen contains sixteen times the number of atoms, it must also contain sixteen times the amount of heat possessed by a pound of oxygen, at the same temperature.

(162) From this it follows, that to raise a pound of hydrogen a certain number of degrees in temperature—say from 50° to 60°—would require sixteen times the amount of heat, needed by a pound of oxygen, under the same circumstances. Conversely, a pound of hydrogen, in falling through 10°, would yield sixteen times the amount of heat yielded by a pound of oxygen in falling through the same number of degrees. The atomic weight of nitrogen being 14, the same reasoning leads to the conclusion, that a pound of hydrogen contains fourteen times the amount of heat contained by a pound of nitrogen. This conclusion, as we shall immediately learn, is verified by experiment.

(163) In oxygen and hydrogen we have no sensible 'interior work' to be performed; there are no molecular attractions of sensible magnitude to be overcome. But in solid and liquid bodies, besides the differences due to the number of atoms present in the unit of weight, we have also differences, due to the consumption of heat in interior work. Hence, it is clear that the absolute amounts

of heat, which different bodies possess, are not at all declared by their temperatures. To raise a pound of water, for example, one degree, would require thirty times the amount of heat necessary to raise a pound of mercury one degree. Conversely, the pound of water, in falling through one degree, would yield thirty times the amount of heat yielded by the pound of mercury.

(164) Let me illustrate, by a simple experiment, the differences which exist between bodies, as to the quantity of heat which they contain. Here is a cake of bees-wax, six inches in diameter and half an inch thick. Here also is a vessel containing oil, which is now at a temperature of 180° C. In the hot oil are immersed a number of balls of different metals—iron, lead, bismuth, tin, and copper. At present they all possess the same temperature, namely, that of the oil. I lift them out of the oil, and place them upon this cake of wax C D (fig. 40), which is supported by the ring of a retort-stand; they melt the wax underneath, and sink in it. But they are sinking with different velocities. The iron and the copper are working themselves much more vigorously into the fusible mass than the others; the tin comes next, while the lead and the bismuth lag entirely behind. And now the iron has gone clean through; the copper follows; the bottom of the tin ball just protrudes from the lower surface of the cake, but it cannot go farther; while the lead and bismuth have made but little way, being unable to sink to much more than half the depth of the wax.

FIG. 40.

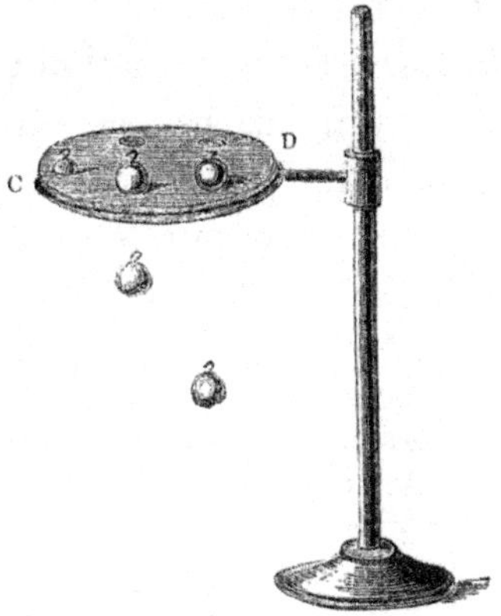

(165) If, then, equal weights of different substances were

all heated, say to 100°, and if the exact amount of heat which each of them gives out in cooling from 100° to 0° were determined, we should find very different amounts of heat for the different substances. Eminent men have solved this problem, by observing the *time* which a body requires to cool. Of course, the greater the amount of heat possessed or generated by its atoms, the longer would the body take to cool. The relative quantities of heat yielded up by different bodies have also been determined by plunging them, when heated, into cold water, and observing the gain of heat on the one hand and the loss on the other. The problem has also been solved, by observing the quantities of ice which different bodies can liquefy, in falling from 100° C. to 0°, or from 212° Fahr. to 32°. These different methods have given concordant results. According to the celebrated French experimenter Regnault, the following numbers express the relative amounts of heat given out by a unit of weight of each of the substances named in the table, in cooling from 98° C. to 15° C.

Aluminium	0·2143	Nickel	0·1086
Antimony	0·0508	Osmium	0·0311
Arsenic	0·0814	Palladium	0·0593
Bismuth	0·0308	Phosphorus (solid)	0·1887
Boron	0·2352	„ (amorphous)	0·1700
Bromine	0·1129	Platinum	0·0329
Cadmium	0·0567	Potassium	0·1696
Carbon	0·2414	Rhodium	0·0580
Cobalt	0·1067	Selenium	0·0827
Copper	0·0952	Silicon	0·1774
Diamond	0·1469	Silver	0·0570
Gold	0·0324	Sodium	0·2934
Iodine	0·0541	Sulphur (native)	0·1776
Iridium	0·0326	„ (recently melted)	0·2026
Iron	0·1138	Tellurium	0·0474
Lead	0·0314	Thallium	0·0336
Lithium	0·9408	Tin	0·0562
Magnesium	0·2499	Tungsten	0·0334
Manganese	0·1217	Water	1·0080
Mercury	0·0333	Zinc	0·0955

A moment's inspection of this table explains the reason why the iron and copper balls melted through the wax; it was in consequence of their high specific heat, while the lead and bismuth balls were incompetent to do so; it will also be seen that tin here occupies the position which our experiment with the cake of wax assigns to it; water, we see, yields more heat than any other substance in the list.

(166) Each of these numbers denotes what has been hitherto called the 'specific heat,' or the 'capacity for heat,' of the substance to which it is attached. As stated on a former occasion, those who considered heat to be a fluid, explained these differences by saying that some substances had a greater store of this fluid than others. We may, without harm, continue to use the term 'specific heat' or 'capacity for heat,' now that we know the true nature of the actions denoted by the term. It is a noteworthy fact, that as the specific heat increases, the *atomic weight* diminishes, and *vice versa*; so that the *product* of the atomic weight and specific heat is, in almost all cases, a sensibly constant quantity. This illustrates a remark already made, that the lighter atoms make good by velocity what they want in mass.

(167) The magnitude of the forces engaged in this atomic motion, and interior work, as measured by any ordinary mechanical standard, is enormous. This pound of iron, on being heated from 0° C. to 100° C., expands by about $\frac{1}{800}$th of the volume which it possesses at 0°. Its augmentation of volume would certainly escape the most acute eye; still, to give its atoms the motion corresponding to this augmentation of temperature, and to shift them through the small space indicated, an amount of heat is requisite which would raise about eight tons one foot high. The force of gravity almost vanishes in comparison with these molecular forces; the pull of

the earth upon the pound weight, as a mass, is as nothing compared with the mutual pull of its own molecules. Water furnishes a still subtler example. Water expands on both sides of 4° C. or 39° F.; at 4° C. it has its maximum density. Suppose a pound of water to be heated from $3\frac{1}{2}$° C. to $4\frac{1}{2}$° C.—that is, one degree—its volume at both temperatures is the same; there has been no forcing asunder whatever of the atomic centres, and still, though the volume is unchanged, an amount of heat has been imparted to the water, sufficient, if mechanically applied, to raise a weight of 1,390 lbs. a foot high. The interior work, done here by the heat, is simply that of causing the atoms of water to rotate. It separates their attracting poles by a tangential movement, but leaves their centres at the same distance asunder, first and last. The conceptions here dealt with may not be easy to those unaccustomed to such studies, but they can be realised, with perfect clearness, by all who have the patience to dwell upon them for a sufficient length of time.

(168) We thus see that there are descriptions of interior work, different from that of pushing the atoms more widely apart. An enormous quantity of interior work may be accomplished, while the atomic centres, instead of being pushed apart, approach each other. Polar forces—forces emanating from distant atomic points, and acting in distinct directions, give to crystals their symmetry, and the overcoming of these forces, while it necessitates a consumption of heat, may also be accompanied by a diminution of volume. This is illustrated by the deportment of both ice and bismuth in liquefying.

(169) The most important experiments on the specific heat of elastic fluids we owe to M. Regnault. He determined the quantities of heat necessary to raise equal weights of gases and vapours, and also the quantities necessary to raise equal volumes of them, through the same number of

degrees. Calling the specific heat of water 1, here are some of the results of this invaluable investigation—

SIMPLE GASES.

	Specific heats	
	Equal weights	Equal volumes
Air	0·237	
Oxygen	0·218	0·240
Nitrogen	0·244	0·237
Hydrogen	3·409	0·236
Chlorine	0·121	0·296
Bromine	0·055	0·304

(170) We have already arrived at the conclusion that, for equal weights, hydrogen would be found to possess sixteen times the amount of heat possessed by oxygen, and fourteen times that of nitrogen, because the hydrogen contains sixteen times the number of atoms, in the one case, and fourteen times the number, in the other. We here find this conclusion verified experimentally. Equal volumes, moreover, of all these gases contain the same number of atoms, and hence we should infer that the specific heats of equal volumes ought to be equal. They are very nearly so for oxygen, nitrogen, and hydrogen; but chlorine and bromine differ considerably from the other elementary gases. Now bromine is a *vapour*, and chlorine a gas, easily liquefied by pressure; hence, in both these cases, the mutual attraction of the atoms, which is insensible in oxygen, nitrogen, and hydrogen, requires a portion of heat to overcome it. The specific heats of chlorine and bromine at equal volumes are, therefore, higher.

(171) Certain simple gases unite to form compound ones, without any change of volume. Thus, one volume of chlorine combines with one volume of hydrogen, to form *two* volumes of hydrochloric acid. In other cases the act of combination is accompanied by a diminution of volume:

thus, two volumes of nitrogen combine with one of oxygen to form two volumes of the protoxide of nitrogen. By the act of combination, three volumes have, in this case, been condensed to two. M. Regnault finds that the compound gases formed without condensation, have, at equal volumes, the same specific heat as oxygen, nitrogen, and hydrogen; while with these which change the volume this is not the case.

COMPOUND GASES—WITHOUT CONDENSATION.

	Specific heats	
	Equal weights	Equal volumes
Nitric oxide . . .	0·232	0·241
Carbonic oxide . .	0·245	0·237
Hydrochloric acid . .	0·185	0·235

The specific heat of equal volumes of these compound gases is the same as that of the three simple gases already mentioned.

(172) COMPOUND GASES—3 VOLUMES CONDENSED TO 2.

	Specific heats	
	Equal weights	Equal volumes
Carbonic acid . .	0·217	0·331
Nitrous oxide . .	0·226	0·345
Aqueous vapour . .	0·480	0·299
Sulphurous acid . .	0·154	0·341
Sulphide of hydrogen .	0·243	0·286
Bisulphide of carbon .	0·157	0·412

(173) Here we find the specific heats of equal volumes neither equal to those of the elementary gases, nor equal to each other. It is worth bearing in mind that the specific heat of water is about double that of aqueous vapour, and also double that of ice.

(174) Comparing *equal weights*, the specific heat of water being 1, that of air is 0·237. Hence, a pound of

water, in losing one degree of temperature, would warm about 4·2 lbs. of air one degree. But water is 770 times heavier than air; hence, comparing *equal volumes*, a cubic foot of water, in losing one degree of temperature, would raise $770 \times 4{\cdot}2 = 3{,}234$ cubic feet of air one degree.

(175) The vast influence which the ocean must exert, as a moderator of climate, here suggests itself. The heat of summer is stored up in the ocean, and slowly given out during the winter. Hence one cause of the absence of extrèmes in an island climate. The summer of the island can never attain the fervid heat of the continental summer, nor can the winter of the island be so severe as the continental winter. In various parts of the Continent fruits grow which our summers cannot ripen; but in these same parts our evergreens are unknown; they cannot live through the winter cold. Winter in Iceland is, as a general rule, milder than in Lombardy.

(176) We have hitherto confined our attention to the heat consumed in the molecular changes of solid and liquid bodies, while these bodies continue solid and liquid. We shall now direct our attention to the phenomena which accompany *changes of the state of aggregation*. When sufficiently heated, a solid becomes a liquid; and when sufficiently heated, a liquid assumes the form of gas. Let us take the case of ice, and trace it through the entire cycle. The block of ice before you has now a temperature of 10° C. below zero. Let us warm it; a thermometer fixed in it rises to 0°, and at this point the ice begins to melt; the thermometric column, which rose previously, *is now arrested in its march, and becomes perfectly stationary*. The warmth is still applied, but there is no augmentation of temperature; and, not until the last film of ice has been removed from the bulb of the thermometer, does the mercury resume its motion. It is now again ascending; it

reaches 30°, 60°, 100°: here steam-bubbles appear in the liquid; it boils, and from this point, onwards, *the thermometer remains stationary at* 100°.

(177) But during the melting of the ice, and during the evaporation of the water, heat is incessantly communicated: to simply liquefy the ice, as much heat is imparted as would raise the same weight of water 79·4° C., or 79·4 times that weight one degree in temperature; and to convert a pound of water at 100° C. into a pound of steam, at the same temperature, 537·2 times as much heat is required, as would raise a pound of water one degree in temperature. The former number, 79·4° C. (or 143° F), represents what has been hitherto called the *latent heat* of water; and the latter number, 537·2° C. (or 967° F.) represents the latent heat of steam. It was manifest to those who first used these terms, that throughout the entire time of melting, and throughout the entire time of boiling, heat was communicated; but inasmuch as this heat was not revealed by the thermometer, the fiction was invented that it was rendered *latent*. The fluid of heat was supposed to hide itself, in some unknown way, in the interstitial spaces of the water and the steam. According to our present theory, the heat expended in melting is consumed in conferring potential energy upon the atoms. It is, virtually, the lifting of a weight. So, likewise, as regards steam, the heat is consumed in pulling the liquid molecules asunder, conferring upon them a still greater amount of potential energy. When the heat is withdrawn, the vapour condenses, the molecules again clash with a dynamic energy, equal to that which was employed to separate them, and the precise quantity of heat then consumed reappears.

(178) The act of liquefaction consists of interior work—of work expended in moving the atoms into new positions. The act of vaporisation is also, for the most part, interior work; to which however must be added the exterior work

of forcing back the atmosphere, when the liquid becomes vapour.

(179) We are indebted to an eminent man to whom I have already often referred, for the first accurate determinations of the calorific power of fuel. 'Rumford estimated the calorific power of a body by the number of parts, by weight, of water, which one part, by weight, of the body would, on perfect combustion, raise one degree in temperature. Thus, one pound of charcoal, in combining with $2\frac{2}{3}$ lbs. of oxygen, to form carbonic acid, evolves heat sufficient to raise the temperature of about 8,000 lbs. of water 1° C. Similarly, one pound of hydrogen, in combining with eight pounds of oxygen, to form water, generates an amount of heat sufficient to raise 34,000 lbs. of water 1° C. The calorific powers, therefore, of carbon and hydrogen are as 8 : 34.'* The recent refined researches of Favre and Silbermann entirely confirm the determinations of Rumford.

(180) Let us, then, fix our attention upon this wonderful substance, water, and trace it through the various stages of its existence. First, we have its constituents as free atoms of oxygen and hydrogen, which attract each other and clash together. The mechanical value of this atomic act is easily determined. The heating of 1 lb. of water 1° C. is equivalent to 1,390 foot-pounds; hence the heating of 34,000 lbs. of water 1° C. is equivalent to $34{,}000 \times 1{,}390$ foot-pounds. We thus find that the concussion of our 1 lb. of hydrogen with 8lbs. of oxygen is equal, in mechanical value, to the raising of forty-seven million pounds one foot high. It was no overstatement which affirmed that the force of gravity, as exerted near the earth, is almost a vanishing quantity, in comparison with these molecular forces. The distances which separate the atoms before combination are so small as to be utterly immea-

* Percy's Metallurgy, p. 53.

surable; still, it is in passing over these distances that the atoms acquire a velocity, sufficient to cause them to clash with the tremendous energy indicated above.

(181) After combination, the substance is in a state of vapour, which sinks to 100° C., and afterwards condenses to water. In the first instance, the atoms fall together to form the compound; in the next instant the molecules of the compound fall together, to form a liquid. The mechanical value of this act is also easily calculated: 9 lbs. of steam, in falling to water, generate an amount of heat sufficient to raise $537{\cdot}2 \times 9 = 4{,}835$ lbs. of water 1° C., or $967 \times 9 = 8{,}703$ lbs. 1° F. Multiplying the former number by 1,390, or the latter by 772, we have, in round numbers, a product of 6,720,000 foot-pounds, as the mechanical value of the mere act of condensation.* The next great fall is from the state of liquid to that of ice, and the mechanical value of this act is equal to 993,564 foot-pounds. Thus, our 9 lbs. of water, at its origin and during its progress, falls down three great precipices: the first fall is equivalent, in energy, to the descent of a ton weight down a precipice 22,320 feet high; the second fall is equal to that of a ton down a precipice 2,900 feet high; and the third is equal to the fall of a ton down a precipice 433 feet high. I have seen the wild stone-avalanches of the Alps, which smoke and thunder down the declivities, with a vehemence almost sufficient to stun the observer. I have also seen snow-flakes descending so softly, as not to hurt the fragile spangles of which they were composed; yet to produce, from aqueous vapour, a quantity, which a child could carry, of that tender material,

* In Rumford's experiments the heat of condensation was included in his estimate of calorific power; deducting the above number from that found for the chemical union of the hydrogen and oxygen, forty millions of foot-pounds would still remain as the mechanical value of the act of combination.

demands an exertion of energy competent to gather up the shattered blocks of the largest stone-avalanche I have ever seen, and pitch them to twice the height from which they fell.

(182) A few experimental illustrations of the calorific effects which accompany the change of aggregation will not be out of place here. I place the thermo-electric pile with its back upon the table, and on its naked face this thin silver basin, B (fig. 41), which contains a quantity of water slightly warmed; the needle of the galvanometer moves to 90°, and remains permanently deflected at 70°. I now put a little powdered nitre, not more than will cover a threepenny-piece, in the basin, and allow it to dissolve. The nitre was previously placed before the fire, so that not only was the liquid warm, but also the solid powder. Observe the effect of their

FIG. 41

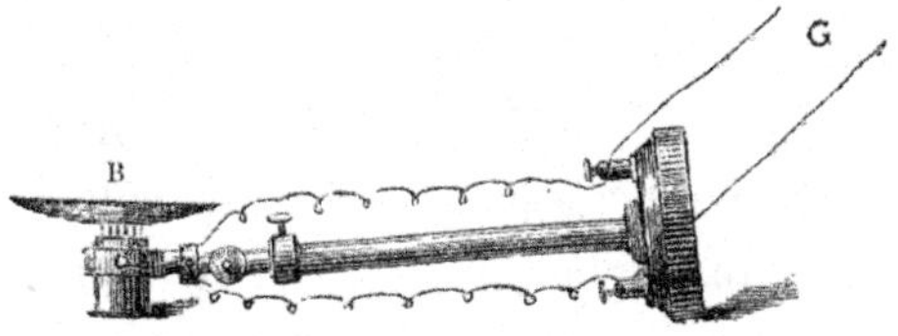

mixture. The nitre dissolves in the water; and to produce this change, all the heat which both the water and the nitre possess, in excess of the temperature of this room, is consumed, and, indeed, a great deal more. The needle, you see, not only sinks to zero, but moves strongly up at the other side, showing that the face of the pile is now powerfully chilled.

(183) Pouring out the chilled liquid, and replacing it with warm water, I introduce a pinch of common salt. The needle was at 70° when the salt was put in: it is

now sinking, it reaches zero, and moves up on the side which indicates cold. But the action is not at all so strong as in the case of saltpetre. As regards latent heat, then, we have differences similar to those which we have already illustrated as regards specific heat. Putting a little sugar, instead of salt, in the water, the amount of heat absorbed is sensible; the liquid is chilled, but the amount of chilling is much less than in either of the former cases. Thus, when you sweeten your hot tea, you cool it in the most philosophical manner; when you put salt in your soup, you do the same; and if you were concerned with the act of cooling alone, and careless of the flavour of your soup, you might hasten its refrigeration by adding to it saltpetre.

(184) On a former occasion a mixture of pounded ice and salt was employed to obtain intense cold. Both the salt and the ice, when they are thus mixed together, change their state of aggregation, and, as a consequence, the temperature of the mixture sinks many degrees below the freezing point of water. Here is a nest of watch-glasses wrapped in tin foil, and immersed in a mixture of ice and salt. Into each watch-glass was poured a little water, in which the next glass rested. They are now all frozen together to a solid cylinder, by the cold of this mixture of ice and salt.

(185) I will now reverse the process, and endeavour to show you the heat developed, in passing from the liquid to the solid state. But first let me show you that heat is rendered latent, when sulphate of soda is dissolved. Testing the substance as the nitre was tested, you see that as the crystals melt in the water the pile is chilled. And now for the complementary experiment. This large glass vessel, B (fig. 42), with its long neck, is filled with a solution of sulphate of soda. Yesterday, my assistant dissolved the substance in a pan over our laboratory fire, and filled this

bottle with the solution. He then covered the top carefully with a piece of bladder, and placed the bottle behind this table, where it has remained undisturbed throughout the night. The liquid is, at the present moment, supersaturated with sulphate of soda. When the water was hot, it melted more than it could melt when cold; but now the temperature has sunk much lower than that which corresponds to the point of saturation. This state of things is secured by keeping the solution perfectly still, and permitting nothing to fall into it. Water, kept thus still, may be cooled many degrees below its freezing point. Some of you may have noticed the water in your jugs, after a cold winter night, suddenly freeze, on being poured out in the morning. In cold climates this is not uncommon. The particles of sulphate of soda, in this solution, are on the brink of a precipice, and may be pushed over it, by simply dropping a small crystal of the substance, not larger than a grain of sand, into the solution. Observe what takes place—the bottle now contains a clear liquid; I drop the bit of crystal in, it does not sink; the molecules have closed round it, to form a solid, in which it is now embedded. The passage of the atoms from a state of freedom to a state of bondage goes on quite gradually; you see the solidification extending down the neck of the bottle. The naked face of the thermo-electric pile rests against the convex surface of the bottle, and the needle of the galvanometer points to zero. The process of crystallisation has not yet reached

FIG. 42.

the liquid in front of the pile, but you see it approaching. The salt is now solidified opposite the pile, and mark the effect. The atoms, in falling to the solid form, develope heat; this heat communicates itself to the glass envelope, which, in its turn, warms the pile, and the needle, as you see, flies to 90°. The quantity of heat thus rendered sensible by solidification is exactly equal to that which was rendered latent by liquefaction.

(186) We have, in these experiments, dealt with the latent heat of liquids; let me now direct your attention to a few experiments illustrative of what has been called the latent heat of vapours—in other words, the heat consumed in conferring potential energy, when a body passes from the liquid to the gaseous state. As before, the pile is laid upon its back, with its naked face upwards; on this face is placed the silver basin already used, and which now contains a small quantity of a volatile liquid, purposely warmed. The needle moves, indicating heat. But scarcely has it attained 90°, when it turns promptly, descends to 0°, and flies up with violence on the side of cold. The liquid here used is sulphuric ether; it is very volatile, and the speed of its evaporation is such, that it consumes rapidly the heat at first communicated to it, and then abstracts heat from the face of the pile. I remove the ether, and supply its place by alcohol, slightly warm; the needle, as before, ascends on the side of heat. By these small bellows we can promote the evaporation of the alcohol; you see the needle descending, and now it is up at 90° on the side of cold. Water is not nearly so volatile as alcohol; still, with this arrangement, the absorption of heat by the evaporation of water may be shown. We use sometimes unglazed pottery for holding water, which admits of a slight percolation of the liquid, and thus causes a dewiness on the external surface. From that surface evaporation goes on, and the heat necessary for this molecular work,

being drawn in great part from the water within, keeps it cool. Butter-coolers are made on the same principle.

(187) The extent to which refrigeration may be carried by the evaporation of water is illustrated by the fact that water may be frozen, through the simple abstraction of heat by its own vapour. The instrument which effects this is called a *cryophorus*, or ice-carrier, which was invented by Dr. Wollaston. It is made in this way—a little water is put into one of these bulbs, A (fig. 43); the other bulb, B, while softened by heat, had a tube drawn out from it, with a minute aperture at the end. The water was boiled in A, and steam was produced, until it had chased all the air away through the small aperture in the distant bulb. When the bulbs, and connecting tube,

FIG. 43.

were filled with pure steam, the small orifice was sealed with a blowpipe. Here, then, we have water and its vapour, with scarcely a trace of air. You hear how the liquid rings, exactly as it did in the case of the water-hammer.

(188) I turn all the liquid into one bulb, A, which is dipped into an empty glass, to protect it from air currents, The *empty* bulb, B, is plunged into a freezing mixture; thus, the vapour which escapes from the liquid in the bulb A, is condensed, by the cold, to water in B. This condensation permits of the formation of new quantities of vapour. As the evaporation continues, the water which supplies the

vapour becomes more and more chilled. In a quarter of an hour, or twenty minutes, it will be converted into a cake of ice. Here, in fact, is the opalescent solid, formed in a second instrument, which was set in action about half an hour ago. The whole process of refrigeration consists in the uncompensated transfer of atomic motion from the one bulb to the other.

(189) But the most striking example of the consumption of heat, in changing the state of aggregation, is furnished by a substance which is imprisoned in this strong iron bottle. This bottle contains carbonic acid, liquefied by enormous pressure. The substance, you know, is a gas under ordinary circumstances. This glass jar is full of the gas, which, though it manifests its nature by extinguishing a taper, is not to be distinguished, by the eye, from common air. When the cock attached to the iron bottle is turned, the pressure upon the acid is relieved; the liquid boils—flashes, as it were, suddenly into gas, which rushes from the orifice with impetuous force. You can trace this current through the air; mixed with it you see a white substance, which is blown to a distance of eight or ten feet. What is this white substance? It is carbonic acid *snow*. The cold produced, in passing from the liquid to the gaseous state, is so intense, that a portion of the carbonic acid is actually frozen to a solid, which mingles, in small flakes, with the issuing stream of gas. This snow may be collected in a suitable vessel. Here is a cylindrical box, with two hollow handles, through which the gas is allowed to issue. Right and left you see the turbid current, but a large portion of the frozen mass is retained in the box. On being opened, you see it filled with this perfectly white solid.

(190) The solid disappears very gradually; its conversion into vapour is slow, because it can only slowly collect, from surrounding substances, the heat necessary to vaporise

it. You can handle it freely, but not press it too much, lest it should burn you. It is cold enough to burn the hand. When a piece of it is plunged into water, and held there, you see bubbles rising through the water—these are pure carbonic acid gas. It possesses all the properties of the substance as commonly prepared. I put a bit of the acid into my mouth, taking care not to inhale, while it is there. Breathing against this candle, my breath extinguishes the flame. How it is possible to keep so cold a substance in the mouth without injury will be immediately explained. A piece of iron, of equal coldness, would do serious damage.

(191) Water will not melt this snow, but sulphuric ether will; and on pouring a quantity of this ether on the snow, a pasty mass is obtained, which has an enormous power of refrigeration. Here are some thick and irregular masses of glass—the feet, in fact, of drinking-glasses. I place a portion of the solid acid on them, and wet it with ether; you hear the glasses crack; they have been shattered by the contraction produced by the intense cold.

(192) In this basin is spread a little paper, and over the paper is poured a pound or two of mercury; on the mercury is placed some solid carbonic acid, and over the acid is poured a little ether. Mercury, you know, requires a very low temperature to freeze it; but here it is frozen. It is now before you as a solid mass; the solid can be hammered, and also cut with a knife. To enable me to lift the mercury out of the basin, a wire is frozen into it; by the wire I raise the mercury, and plunge it into a glass jar containing water. It liquefies, and showers downwards through the water; but every fillet of mercury freezes the water with which it comes into contact, and thus, round each fillet is formed a tube of ice, through which you can see the liquid metal descending. These experiments might

be multiplied almost indefinitely; but enough has been done to illustrate the chilling effect of vaporisation.

(193) I have now to direct your attention to another, and very singular class of phenomena, connected with the production of vapour. On the table is a broad porcelain basin, B (fig. 44), filled with hot water, and over this lamp is a light silver basin, heated to redness. If the silver basin be placed on the water, as at S, what will occur? You might naturally reply, that the basin will impart its heat instantly to the water, and be cooled down to the temperature of the latter. But nothing of this kind occurs. The silver for a time developes a sufficient amount of vapour underneath it, to lift it entirely out of contact with the water; or, in the language of the hypothesis developed on a former occasion, it is lifted by the discharge of molecular projectiles against its under surface. This will go on, until the temperature of the basin sinks, and it is no longer able to produce vapour of sufficient tension to support it. Then it comes into contact with the water, and the ordinary hissing of a hot metal, together with the cloud which forms overhead declares the fact.

FIG. 44.

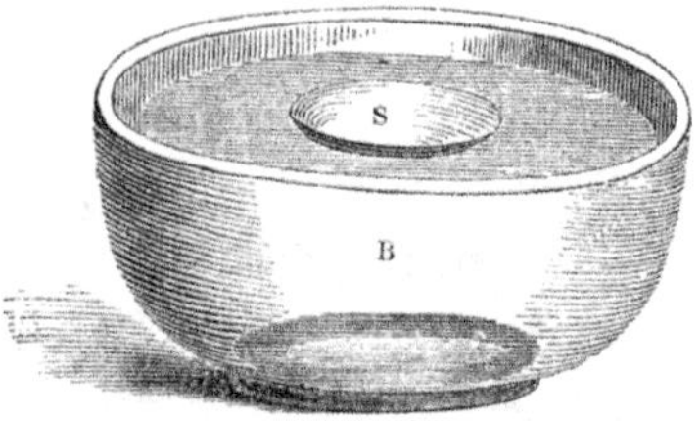

(194) Let us now reverse the experiment, and instead of placing the basin in water, place water in the basin—first of all, however, heating the latter to redness by a lamp. You hear no noise of ebullition, no hissing of the water, as it falls into the hot basin; the drop rolls about on its own vapour—that is to say, it is sustained by the recoil of the molecular projectiles, discharged from its under surface. I withdraw the lamp, and allow the basin to cool, until it is no longer able to produce vapour

strong enough to support the drop. The liquid then touches the metal; the instant it does so, violent ebullition sets in, and the cloud, which you now observe, forms above the basin.

(195) You cannot, from your present position, see this flattened spheroid rolling about in the hot basin; but it may be shown to you, and, if we are fortunate, you will see something very beautiful. here is, underneath the drop, an incessant development of vapour which, as incessantly, escapes from it laterally. If the drop rest upon a flattish surface, so that the lateral escape is very difficult, the vapour will burst up through the middle of the drop. But

FIG. 45.

matters are here so arranged, that the vapour shall issue laterally; and it sometimes happens that the escape is rhythmic; the vapour issues in regular pulses, and then we have our drop of water moulded to a most beautiful rosette. It is there now—a round mass of liquid, two inches in diameter, with a beautifully crimped border. Throwing the beam of the electric lamp upon this drop, so as to illuminate it, and holding this lens over it, I hope to cast its image on the ceiling, or on the screen. It is now perfectly defined, forming a figure (fig. 45), eighteen inches in diameter, with the vapour breaking, as if in music, from its edge. If a little ink be added, so as to darken the liquid, the definition of its outline is augmented, but the pearly lustre of its surface is lost. I withdraw the heat; the undulation continues for some time, diminishing gradually: the border finally becomes unindented. The drop is now perfectly motionless—a liquid spheroid; and now it suddenly spreads upon the surface, for contact has been established, and the 'spheroidal condition' is at an end.

(196) When this silver basin is placed, with its bottom upwards, in front of the electric lamp, by means of a lens in

FIG. 46.

front, the rounded outline of the basin may be brought to a focus on the screen; I dip this bit of sponge in alcohol and squeeze it over the cold basin, so that the drops fall upon the surface of the metal: you see their magnified images upon the screen, and you observe, that when they strike the inverted basin they spread out and trickle down along it. Let us now heat the basin by placing a lamp underneath it. Observe what occurs: when the sponge is squeezed the drops descend as before, but, when they come in contact with the basin, they no longer spread, but roll over the surface as liquid spheres (fig. 46). See how they bound and dance, as if they had fallen upon elastic springs; and so, in fact, they have. Every drop, as it strikes the hot surface, and rolls along it, developes vapour which lifts it out of contact, thus destroying all cohesion between the surface and the drop, and enabling the latter to preserve its spherical or spheroidal form.

(197) The arrangement now before you was suggested by

FIG. 47.

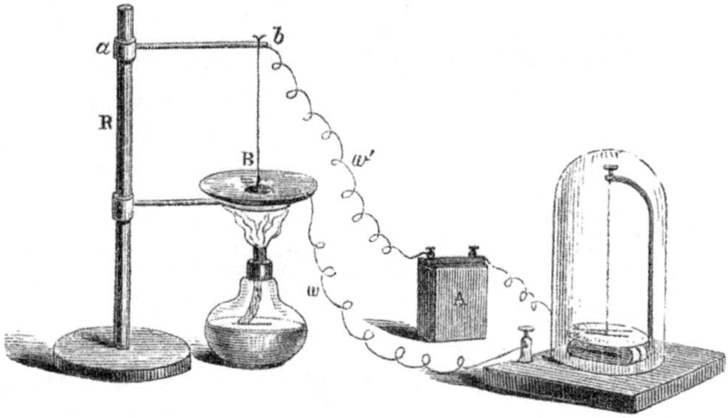

Professor Poggendorff, and shows, in a very ingenious manner, the interruption of contact between the spheroidal drop and its supporting surface. From this silver basin, B (fig. 47), intended to hold the drop, a wire, *w*, is carried round yonder magnetic needle; the other end of the galvanometer wire is attached to one end of this battery, A. From the opposite pole of the little battery a wire, *w′*, is carried to the moveable arm, *a b*, of this retort-stand, R. I heat the basin, pour in the water, and lower the wire till the end of it dips into the spheroidal mass:

you see no motion of the galvanometer needle; still, the only gap in the entire circuit is that which now exists underneath the drop. If the drop were in contact, the current would pass. This is proved by withdrawing the lamp; the spheroidal state will soon end; the liquid will touch the bottom. It now does so, and the needle instantly flies aside.

(198) You can actually *see* the interval between the drop and the hot surface upon which it rests. A private experiment may be made in this way: Let a flattish basin, B (fig. 48), be turned upside down, and let the bottom of

FIG. 48.

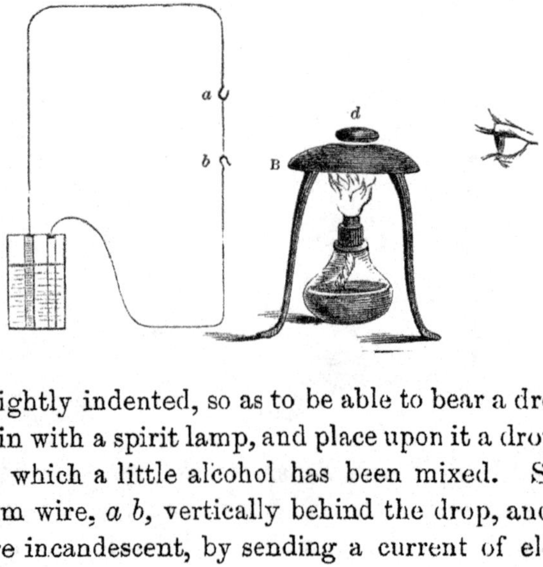

it be slightly indented, so as to be able to bear a drop; heat the basin with a spirit lamp, and place upon it a drop of ink, *d*, with which a little alcohol has been mixed. Stretch a platinum wire, *a b*, vertically behind the drop, and render the wire incandescent, by sending a current of electricity through it. Bring your eye to a level with the bottom of the drop, and you will be able to see the red-hot wire, through the interval between the drop and the surface which supports it. Let me show you this interval. I place a heated basin, B (fig. 49), as before, with its bottom upward, in front of the electric lamp; and bring carefully down upon it a drop, *d*, dependent from a pipette.

When it seems to rest upon the surface, and when the lens is brought to its proper position in front, you see be tween the drop and the silver, a bright line of light, indicating that the beam has passed, underneath the drop, to the screen.

FIG. 49.

(199) The spheroidal condition was first observed by Leidenfrost, and fifty other illustrations of it might be given. Liquids can be made to roll on liquids. If this red-hot copper ball be plunged into a vessel of hot water, a loud sputtering is produced, due to the escape of the vapour generated; still, the contact of the liquid and solid is not established: but we will let the ball remain until it cools, the liquid at length touches it, and then the ebullition is so violent as to project the water from the vessel on all sides.

(200) M. Boutigny has lent new interest to this subject by expanding the field of illustration, and applying it to the explanation of many extraordinary effects. If the hand be wet, it may be passed through a stream of molten metal without injury. I have seen M. Boutigny pass his wet hand through a stream of molten iron, and toss with his fingers the fused metal from a crucible: a blacksmith

will lick a white-hot iron without fear of burning his tongue. The tongue is effectually preserved from contact with the iron, by the vapour developed; and it was to the vapour, of the carbonic acid, which shielded me from its contact, that I owed my safety, when the substance was put into my mouth. To the same protective influence, many escapes from the fiery ordeal of ancient times have been attributed, by M. Boutigny. It may be added, that the explanation of the spheroidal condition, given by M. Boutigny, has not been accepted by scientific men. The foregoing experiments reduce its cause to ocular demonstration.

(201) Boiler explosions have also been ascribed to the water in the boiler assuming the spheroidal state; the sudden development of steam, by subsequent contact with the heated metal, causing the explosion. We are more ignorant of these things than we ought to be. Experimental science has brought a series of true causes to light, which may produce these terrible catastrophes, but practical science has not yet determined the extent to which

FIG. 50.

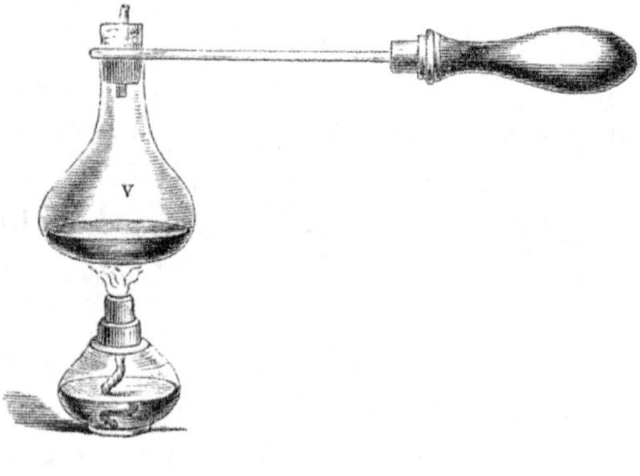

they actually come into operation. The effect of a sudden generation of steam has been illustrated by an experiment which may be made in your presence. Here is a copper vessel, V (fig. 50), with a neck stopped with a cork, through which half an inch of fine glass tubing passes. When the vessel is heated, a little water poured into it assumes the spheroidal state. The vessel is then corked, the small quantity of steam developed, while the water remains spheroidal, escaping through the glass tube. On removing the vessel from the lamp, and waiting for a minute or two, the water comes into contact with the copper. When it does so, the cork is driven, as if by the explosion of gunpowder, to a considerable height into the air.

(202) The spheroidal condition enables us to perform the extraordinary experiment of freezing a liquid in a red-hot vessel. M. Boutigny, by means of sulphurous acid, first froze water in a red-hot crucible; and Mr. Faraday subsequently froze mercury, by means of solid carbonic acid. This latter result may be reproduced; but first let me operate with water. Here is a hollow sphere of brass, about two inches in diameter, now accurately filled with water. Into the sphere is screwed a wire, which is to serve as a handle. This platinum crucible is heated to glowing redness, and in it are placed some lumps of solid carbonic acid. When ether is poured on the acid, neither of them comes into contact with the hot crucible; they are protected from contact by the elastic cushion of vapour which surrounds them. I lower the sphere of water down upon the mass, and carefully pile fragments of carbonic acid over it, adding also a little ether. The pasty substance, within the red-hot crucible, remains intensely cold; and now you hear a crack! You are thereby assured that the experiment has succeeded. The freezing water has burst the brass sphere, as it burst the iron bottles in a former experiment. Raising the sphere, and peeling off the shattered brass shell, you have a solid sphere of ice, extracted from the red-hot crucible.

(203) I place a quantity of mercury in a conical copper

spoon, and dip it into the crucible. The ether in the crucible has taken fire, which was not intended. The experiment ought to be so made, that the carbonic acid gas—the choke-damp of mines—ought to keep the ether from ignition. But the mercury will freeze notwithstanding. Out of the fire, and through the flame, I draw the spoon, and the frozen metal is now before you.

CHAPTER VI.

CONVECTION OF HEATED AIR—WINDS—THE UPPER AND LOWER 'TRADES.'—EFFECT OF THE EARTH'S ROTATION ON THE DIRECTION OF WIND—INFLUENCE OF AQUEOUS VAPOUR UPON CLIMATE—EUROPE THE CONDENSER OF THE WESTERN ATLANTIC—RAINFALL IN IRELAND—THE GULF STREAM—FORMATION OF SNOW—FORMATION OF ICE FROM SNOW—GLACIERS—PHENOMENA OF GLACIER MOTION—REGELATION—MOULDING OF ICE BY PRESSURE—ANCIENT GLACIERS.

(204) I PROPOSE devoting an hour to-day to the consideration of some of the thermal phenomena which occur, on a large scale, in Nature. And first, with regard to winds. You see those sunburners, intended to illuminate this room, when the daylight is intercepted, or gone. Not to give light alone were they placed there, but, in part, to promote ventilation. The air, heated by the gas flames, expands, and issues in a strong vertical current into the atmosphere. The air of the room is thereby incessantly drawn upon, and a fresh supply must be introduced to make good the loss. Our chimney draughts are so many vertical winds, due to the heating of the air by our fires.

(205) When this piece of brown paper is ignited, the flame ascends; and when the flame is blown out, the smouldering edges warm the air, and produce currents which carry the smoke upward. I dip the smoking paper into a large glass vessel, and stop the neck of the vessel to prevent the escape of the smoke; the smoke ascends with the light air in the middle, spreads out laterally above, is cooled, and falls, like a cascade of cloud, along

the sides of the vessel. When this heavy iron spatula, heated to dull redness, is held thus in the air, you cannot see the currents of air ascending from it. But they reveal themselves by their action on a strong light. Placing the spatula in the beam of the electric lamp, its shadow is thrown on the screen, and those waving lines of light and shade mark the streaming upwards of the heated air. If this fragment of sulphur, contained in an iron spoon, be heated until it ignites, and then plunged into a jar of oxygen, the combustion becomes brilliant and energetic, and the air of the jar is thrown into intense commotion. The fumes of the sulphur enable you to track the storms, which the heating of the air produces within the jar. I use the word 'storms' advisedly, for the hurricanes which desolate the earth are nothing more than large illustrations of the effect produced in this glass jar.

(206) From the heat of the sun our winds are all derived. We live at the bottom of an aërial ocean, in a remarkable degree permeable to the sun's rays, and but little disturbed by their direct action. But those rays, when they fall upon the earth, heat its surface, and, when they fall upon the ocean, they provoke evaporation. The air in contact with the surface shares its heat, is expanded, and ascends into the upper regions of the atmosphere, while the vapour from the ocean also ascends, because of its lightness, carrying, no doubt, air along with it. Where the rays fall vertically on the earth, that is to say, between the tropics, the heating of the surface is greatest. Here aërial currents ascend and flow laterally, north and south, towards the poles, the heavier air of the polar regions streaming in to supply the place vacated by the light and warm air. Thus, we have an incessant circulation. A few days ago, the following experiment was made in the hot room of a Turkish bath. I opened wide the door, and held a lighted taper in the doorway, midway between top and

bottom. The flame rose straight from the taper. When placed at the bottom, the flame was blown violently inwards; when placed at the top, it was blown violently outwards. Here we had two currents, or winds, sliding over each other, and moving in opposite directions. Thus, also, as regards a portion of our hemisphere, a current from the equator sets in towards the north, and flows in the higher regions of the atmosphere, while another flows towards the equator, in the lower regions of the atmosphere. These are the upper and the lower Trade Winds.

(207) Were the earth motionless, these two currents would run directly north and south, but the earth rotates from west to east round its axis, once in twenty-four hours. In virtue of this rotation, the air at the equator is carried round with a velocity of 1,000 miles an hour. You have observed what takes place when a person incautiously steps out of a carriage in motion. He shares the motion of the carriage, and when his feet touch the earth he is thrown forward in the direction of the motion. This is what renders leaping from a railway carriage, when the train is at full speed, almost always fatal. As we withdraw from the equator, the velocity due to the earth's rotation diminishes, and it becomes nothing at the poles. It is proportional to the radius of the parallel of latitude, and diminishes as these circles diminish in size. Imagine, then, an individual suddenly transferred from the equator to a place where the velocity, due to rotation, is only 900 miles an hour; on touching the earth here he would be thrown forward in an easterly direction, with a velocity of 100 miles an hour, this being the difference between the equatorial velocity with which he started, and the velocity of the earth's surface in his new locality.

(208) Similar considerations apply to the transfer of air from the equatorial to the northern regions, and vice versâ.

At the equator the air possesses the velocity of the earth's surface there, and, on quitting this position, it not only has its tendency northwards to obey, but also a tendency to the east, and it must take a resultant direction. The farther it goes north, the more it is deflected from its original course; the more it turns towards the east, and tends to become what we should call a westerly wind. The opposite holds good for the current proceeding *from* the north; this passes from places of slow motion to places of quick motion: it is met by the earth; hence, the wind which started as a north wind becomes a north-east wind, and, as it approaches the equator, it becomes more and more easterly.

(209) It is not by reasoning alone that we arrive at a knowledge of the existence of the upper atmospheric current, though reasoning is sufficient to show that compensation must take place somehow,—that a wind cannot blow in any direction without an equal displacement of air taking place, in the opposite direction. But clouds are sometimes seen in the tropics, high in the atmosphere, and moving in a direction opposed to that of the constant wind below. Could we discharge a light body with sufficient force to cause it to penetrate the lower current, and reach the higher, the direction of that body's motion would give us that of the wind above. Human strength cannot perform this experiment, but it has nevertheless been made. Ashes have been shot through the lower current by volcanoes, and, from the places where they have subsequently fallen, the direction of the wind which carried them has been inferred. Professor Dove, in his 'Witterungs Verhältnisse von Berlin,' cites the following instance: 'On the night of April 30th, explosions like those of heavy artillery were heard at Barbadoes, so that the garrison at Fort St. Anne remained all night under arms. On May 1, at daybreak, the eastern portion of the horizon appeared

clear, while the rest of the firmament was covered by a black cloud, which soon extended to the east, quenched the light there, and at length produced a darkness so dense that the windows in the rooms could not be discerned. A shower of ashes descended, under which the tree branches bent and broke. Whence came these ashes? From the direction of the wind, we should infer that they came from the Peak of the Azores; they came, however, from the volcano Morne Garou in St. Vincent, which lies about 100 miles west of Barbadoes. The ashes had been cast into the current of the upper trade. A second example of the same kind occurred on January 20, 1835. On the 24th and 25th the sun was darkened in Jamaica by a shower of fine ashes, which had been discharged from the mountain Coseguina, distant 800 miles. The people learned in this way that the explosions previously heard were not those of artillery. These ashes could only have been carried by the upper current, as Jamaica lies north-east from the mountain. The same eruption gives also a beautiful proof that the ascending air-current divides itself above, for ashes fell upon the ship Conway, in the Pacific, at a distance of 700 miles south-west of Coseguina.

(210) 'Even on the highest summits of the Andes,' continues Dove, 'no traveller has as yet reached the upper trade. From this some notion may be formed of the force of the explosions; they were indeed tremendous in both instances. The roaring of Coseguina was heard at San Salvador, a distance of 1,000 miles. Union, a seaport on the west coast of Conchagua, was in absolute darkness for forty-three hours; as light began to dawn, it was observed that the sea shore had advanced 800 feet upon the ocean, through the mass of ashes which had fallen. The eruption of Morne Garou forms the last link of a chain of vast volcanic actions. In June and July, 1811, near St. Miguel, one of the Azores, the island Sabrina rose, accompanied by

smoke and flame, from the bottom of a sea 150 feet deep, and attained a height of 300 feet, and a circumference of a mile. The small Antilles were afterwards shaken, and subsequently the valleys of the Mississippi, Arkansas, and Ohio. But the elastic forces found no vent; they sought one, then, on the north coast of Columbia. March 26 began as a day of extraordinary heat in Caraccas; the air was clear and the firmament cloudless. It was Green Thursday, and a regiment of troops of the line stood under arms in the barracks of the quarter San Carlos, ready to join in the procession. The people streamed to the churches. A loud subterranean thunder was heard, and immediately afterwards came an earthquake shock so violent, that the church of Alta Gracia, 150 feet in height, borne by pillars fifteen feet thick, formed a heap of crushed rubbish not more than six feet high. In the evening the almost full moon looked down with mild lustre upon the ruins of the town, under which lay the crushed bodies of upwards of 10,000 of its inhabitants. But even here there was no exit granted to the elastic forces underneath. Finally, on April 27, they succeeded in opening once more the crater of Morne Garou, which had been closed for a century; and the earth, for a distance equal to that from Vesuvius to Paris, rang with the thunder-shout of the liberated prisoner.'

(211) On this terrestrial globe, I trace with my hand two meridians. At the equator of the globe they are a foot apart, which would correspond to about 1,000 miles on the earth's surface. But these meridians, as they proceed northward, gradually approach each other, and meet at the north pole. It is manifest that the air which rises between these meridians, in the equatorial regions, must, if it went direct to the pole, squeeze itself into an ever-narrowing bed. Were the earth a cylinder, instead of a sphere, we might have a circulation from the middle of the cylinder quite to each end,

and a return current from each end to the middle. But this, in the case of the earth, is impossible, simply because the space around the poles is unable to embrace the air from the equator. The cooled equatorial air sinks, and the return current sets in, before the poles are attained, and this occurs more or less irregularly. The two currents, moreover, instead of flowing one over the other, often flow beside each other. They constitute rivers of air, with incessantly shifting beds.

(212) These are the great winds of our atmosphere, which, however, are materially modified by the irregular distribution of land and water. Winds of minor importance also occur, through the local action of heat, cold, and evaporation. There are winds produced by local action in the Alps, which sometimes rush with sudden and destructive violence down the gulleys of the mountains: gentler down-flows of air are produced by the presence of glaciers upon the heights. There are the land breeze and the sea breeze, due to the varying temperature of the seaboard soil, by day and night. The morning sun, heating the land, produces vertical displacement, and the air from the sea moves landward. In the evening the land is more chilled superficially, by radiation, than the sea, and the conditions are reversed; the heavy air of the land now flows seaward.

(213) Thus, then, a portion of the heat of the tropics is sent, by an aërial messenger, towards the poles, a more equable distribution of terrestrial warmth being thus secured. But in its flight northward the air is accompanied by another substance—by the vapour of water, which, you know, is perfectly transparent. Imagine the ocean of the tropics, giving forth its vapour, which promotes by its lightness the ascent of the associated air. Both expand, as they ascend: at a height of 16,000 feet the air and vapour occupy twice the volume which they

embraced at the sea level. To secure this space they must, by their elastic force, push away the air in all directions round them; they must perform work; and this work cannot be performed, save at the expense of the warmth with which they were, in the first instance, charged.

(214) The vapour, thus chilled, is no longer competent to retain the gaseous form. It is precipitated, as cloud: the cloud descends, as rain; and in the region of calms, or directly under the sun, where the air is first drained of its aqueous load, the descent of rain is enormous. The sun does not remain always vertically over the same parallel of latitude—he is sometimes north of the equator, sometimes south of it, the two tropics limiting his excursion. When he is south of the equator, the earth's surface, north of it, is no longer in the region of calms, but in one across which the aërial current from the north flows towards the region of calms. The moving air is but slightly charged with vapour, and, as it travels from north to south, it becomes ever warmer; it constitutes a dry wind, and its capacity to retain vapour is continually augmenting. It is plain, from these considerations, that each place between the tropics must have its dry season and rainy season; dry, when the sun is at the opposite side of the equator, and wet, when the sun is overhead.

(215) Gradually, however, as the upper stream, which rises from the equator, and flows towards the poles, becomes chilled and dense, it sinks towards the earth; at the Peak of Teneriffe it has already sunk below the summit of the mountain. With the contrary wind blowing at the base, the traveller finds the wind from the equator blowing strongly over the top. Farther north the equatorial wind sinks lower still, and, finally, quite reaches the surface of the earth. Europe, for the most part, is overflowed by this equatorial current. Here, in London, for eight or nine

months in the year, south-westerly winds prevail. But mark what an influence this must have upon our climate. The moisture of the equatorial ocean comes to us, endowed with potential energy; with its molecules separate, and therefore competent to clash and develope heat by their collision; it comes, if you will, charged with latent heat. In our northern atmosphere condensation takes place, and the heat generated is a main source of warmth to our climate. Were it not for the rotation of the earth, we should have over us the hot dry blasts of Africa; but owing to this rotation, the wind which starts northward from the Gulf of Mexico is deflected to Europe. Europe is, therefore, the recipient of those stores of latent heat which were amassed in the western Atlantic. The British Isles come in for the greatest share of this moisture and heat, and this circumstance adds itself to that already dwelt upon—the high specific heat of water—to preserve our climate from extremes. It is this condition of things which makes our fields so green, and which also gives the bloom to our maidens' cheeks.

(216) Another property of this wonderful substance, to which is probably due its main influence as a meteorological agent, shall be examined on a future occasion.*

(217) As we travel eastward in Europe, the amount of aqueous precipitation grows less and less; the air becomes more and more drained of its moisture. Even between the east and west coasts of our own islands, the difference is sensible; local circumstances, also, have a powerful influence on the amount of precipitation. Dr. Lloyd finds the mean yearly temperature of the western coast of Ireland to be about two degrees Fahr. higher than that of the eastern coast at the same elevation, and in the same parallel of latitude. The total amount of rain which fell in the year

* See Chapter XI.

1851, at various stations in the island, is given in the following table—

Station	Rain in Inches
Portarlington	21·2
Killough	23·2
Dublin	26·4
Athy	26·7
Donaghadee	27·9
Courtown	29·6
Kilrush	32·6
Armagh	33·1
Killybegs	33·2
Dunmore	33·5
Portrush	37·2
Burincrana	39·3
Markree	40·3
Casteltownsend	42·5
Westport	45·9
Cahirciveen	59·4

With reference to this table, Dr. Lloyd remarks—

(218) '1. That there is great diversity in the yearly amount of rain at the different stations, all of which (excepting four) are but a few feet above the sea level; the greatest rain (at Cahirciveen) being nearly three times as great as the least (at Portarlington).

(219) '2. That the stations of least rain are either inland or on the eastern coast, while those of the greatest rains are at or near the western coast.

(220) '3. That the amount of rain is greatly dependent on the proximity of a mountain chain or group, being always considerable in such neighbourhood, unless the station lie to the north-east of the same.

'Thus, Portarlington lies to the north-east of Slieve-bloom; Killough to the north-east of the Mourne range; Dublin, north-east of the Wicklow range, and so on. On the other hand, the stations of greatest rain, Cahirciveen, Castletownsend, Westport, &c., are in the vicinity of high mountains, but on a different side.'*

* The greatest rainfall recorded by Sir John Herschel in his table

(221) This distribution of heat by the transfer of masses of heated air from place to place, is called '*convection,*' in contradistinction to the process of conduction, which will be treated in its proper place. Heat is distributed in a similar manner through liquids. This glass cell, c (fig. 51), contains warm water. Placed in front of the electric lamp, by means of a converging lens, a magnified image of the cell is thrown upon the screen. I now introduce the end of this pipette into the water of the cell, and allow a little cold water gently to enter it. The difference of refraction between the two enables you to see the heavy cold water falling through the lighter warm water. The experiment succeeds still better when a fragment of ice is allowed to float upon the surface of the water. As

FIG. 51.

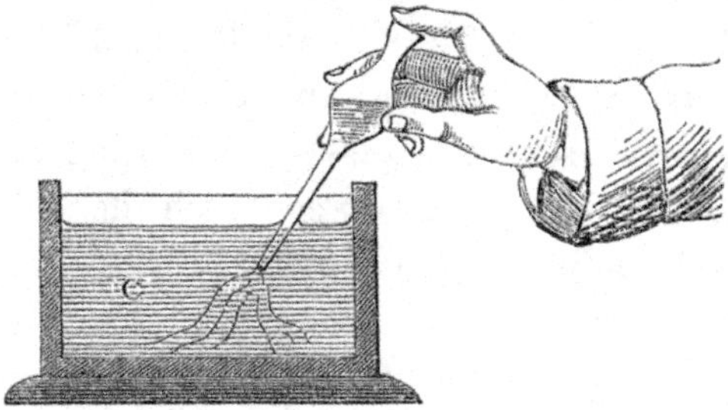

the ice melts, it sends long heavy striæ downwards to the bottom of the cell. I now reverse the experiment, placing cold water in the cell, and hot water in the pipette. Care is here necessary to allow the warm water to enter without any momentum, which would carry it mechanically down. You notice the effect. The point of the pipette is in the

(Meteorology, 110, &c.) occurs at Cherra Pungee, where the annual fall is 592 inches. It is not my object to enter far into the subject of meteorology; for the fullest and most accurate information the reader will refer to the excellent works of Sir John Herschel and Professor Dove.

middle of the cell, and as the warm water enters, it speedily turns upwards (fig. 52) and spreads out at the top, almost as oil would do, under the same circumstances.

(222) When a vessel, containing water, is heated at the bottom, the warmth communicated is diffused by convection. You may see the direction of the ascending warm

FIG. 52.

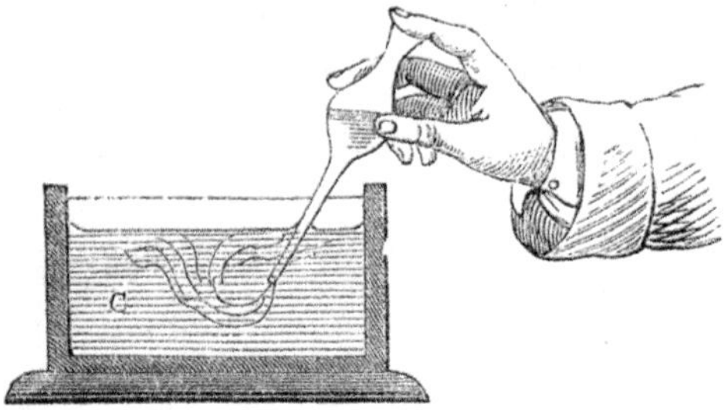

currents by means of the electric lamp, and also that of the currents which descend to occupy the place of the lighter water. Here is a vessel, containing cochineal, the fragments of which, being not much heavier than the water, freely follow the direction of its currents. The pieces of cochineal break loose from the heated bottom, ascending along the middle of the jar, and descending again by the sides. In the Geyser of Iceland this convection occurs on a grand scale. A fragment of paper thrown upon the centre of the water which fills the pipe, is instantly drawn towards the side, and there sucked down by the descending current.

(223) Partly to this cause, and partly, perhaps, to the action of winds, currents establish themselves in the ocean, and powerfully influence climate, by the heat which they distribute. The most remarkable of these currents, and by far the most important for us, is the Gulf Stream, which sweeps across the Atlantic, from the equatorial regions, through the Gulf of Mexico, whence it derives

its name. As it quits the Straits of Florida it has a temperature of 83° Fahr., thence it follows the coast of America as far as Cape Fear, whence it starts across the Atlantic, taking a north-easterly course, and, finally, washing the coast of Ireland, and the north-western shores of Europe generally. As might be expected, the influence of this body of warm water makes itself most evident during our winter. It then entirely abolishes the difference of temperature, due to the difference of latitude of north and south Britain; if we walk from the Channel to the Shetland Isles, in January, we encounter everywhere the same temperature. The isothermal line runs, then, north and south. The presence of this water renders the climate of western Europe totally different from that of the opposite coast of America. The river Hudson, for example, in the latitude of Rome, is frozen for three months in the year. Starting from Boston in January, and proceeding round St. John's, and thence to Iceland, we meet everywhere the same temperature. The harbour of Hammerfest derives great value from the fact, that it is clear of ice all the year round. This is due to the Gulf Stream, which sweeps round the North Cape, and so modifies the climate there, that at some places, by proceeding northward, you enter a warmer region. The contrast between northern Europe and the east coast of America caused Halley to surmise, that the north pole of the earth had shifted; that it was formerly situate somewhere near Behring's Straits, and that the intense cold, observed in these regions, is really the cold of the ancient pole, which had not been entirely subdued since the axis changed its direction. But now we know that the Gulf Stream, and the diffusion of heat by winds and vapours, are the real causes of European mildness. On the western coast of America, between the Rocky Mountains and the ocean, we find a European climate.

(224) Europe, then, is the condenser of the Atlantic; and the mountains are the chief condensers in Europe. On them, moreover, when they are sufficiently high, the condensed vapour descends, not in a liquid, but a solid form. Let us look to this water in its birthplace, and follow it through its subsequent course. Clouds float in the air, and hence has arisen the surmise that they are composed of vesicles of bladders or water, thus forming *shells* instead of *spheres*. It is certain, however, that if the particles of water be sufficiently small they will float for an indefinite period without being vesicular. It is also certain that water-particles at high elevations possess on or after precipitation, the power of building themselves into crystalline forms; they thus bring forces into play which we have hitherto been accustomed to regard as molecular, and which could not be ascribed to the aggregates necessary to form vesicles.

(225) Snow, perfectly formed, is not an irregular aggregate of ice-particles; in a calm atmosphere, the atoms arrange themselves, so as to form the most exquisite figures. You have seen those six-petalled flowers which show themselves within a block of ice, when a beam of heat is sent through it. The snow-crystals, formed in a calm atmosphere, are built upon the same type: the molecules arrange themselves to form hexagonal stars. From a central nucleus shoot six spiculæ, every two of which are separated by an angle of 60°. From these central ribs smaller spiculæ shoot right and left, with unerring fidelity to the angle 60°, and from these again other smaller ones diverge at the same angle. The six-leaved blossoms assume the most wonderful variety of form; their tracery is of the finest frozen gauze; and round about their corners other rosettes of smaller dimensions often cling. Beauty is superposed upon beauty, as if Nature, once committed to her task, took delight in showing

even within the narrowest limits, the wealth of her resources.*

(226) These frozen blossoms constitute our mountain snows; they load the Alpine heights, where their frail architecture is soon destroyed by the weather. Every winter they fall, and every summer they disappear, but this rhythmic action does not perfectly compensate itself. Below a certain line, warmth is predominant, and the quantity which falls every winter is entirely swept away; above this line, cold is predominant; the quantity which falls is in excess of the quantity melted, and an annual residue remains. In winter the snows reach to the plains; in summer they retreat to the *snow-line*,—to that particular line where the snow-fall of every year is exactly balanced by the consumption, and above which is the region of eternal snows. But, if a residue remains annually above the snow-line, the mountains must be loaded with a burden which increases every year. Supposing, at a particular point above the line referred to, a layer of three feet a year to be added annually to the mass; this deposit, accumulating even through the brief period of the Christian era, would produce an elevation of 5,580 feet. And did such accumulations continue throughout geologic, instead of historic ages, we cannot estimate the height to which the snows would pile themselves. It is manifest that no accumulation of this kind takes place; the quantity of snow on the mountains is not augmenting in this way. By some means or other the sun is prevented from lifting the ocean out of its basins, and piling its waters permanently upon the hills.

(227) How then is this annually augmenting load taken off the shoulders of the mountains? The snows sometimes detach themselves, and rush down the slopes in avalanches,

* See fig. 53, in which are copied some of the beautiful drawings of Mr. Glaisher.

FIG. 53.

melting to water in the warmer air below. But the violent rush of the avalanche is not their only motion; they also creep, by almost insensible degrees, down the slopes. As layer, moreover, heaps itself upon layer, the deeper portions of the mass become squeezed and consolidated; the air, first entrapped in the meshes of the snow, is squeezed out, and the compressed mass approximates more and more to the character of ice. You know how the granules of a snowball will adhere; and you know how hard you can make the ball if mischievously inclined. The snowball is incipient ice; augment your pressure, and you actually convert it into ice. But even after it has attained a compactness which would entitle it to be called ice, it is still capable of yielding more or less, as the snow yields, to pressure. When, therefore, a sufficient depth of the substance collects upon the earth's surface, the lower portions are squeezed out by the pressure of the upper ones, and if the snow rests upon a slope, it will yield principally in the direction of the slope, and move downwards.

(228) This motion is incessantly going on along the slopes of every snow-laden mountain; in the Himalayas, in the Andes, in the Alps; but in addition to this motion, which depends upon the power of the substance itself to yield to pressure, there is also a sliding motion, over the inclined bed. The consolidated snow moves bodily over the mountain slope, grinding off the asperities of the rocks, and polishing their hard surfaces. The under surface of the mighty polisher is also scarred and furrowed, by the rocks over which it has passed; but as the compacted snow descends, it enters a warmer region, is more copiously melted, and sometimes, before the base of its slope is reached, it is wholly cut off by fusion. Sometimes, however, large and deep valleys receive the gelid masses thus sent down; in these valleys it is further consolidated, and through them it moves, at a slow but measurable pace,

imitating in all its motions those of a river. The ice is thus carried far beyond the limits of perpetual snow, until, at length, the consumption below equals the supply above, and at this point the glacier ceases. From the snow-line downwards in summer, we have *ice*; above the snow-line, both summer and winter, we have, on the surface, *snow*. The portion below the snow-line is called a *glacier*, that above the snow-line is called the *névé*. The névé, then, is the feeder of the glacier.

(229) Several valleys, thus filled, may unite in a single valley, the tributary glaciers welding themselves together to form a trunk glacier. Both the main valley, and its tributaries, are often sinuous, and the tributaries must change their direction, to form the trunk. The width of the valley, also, often changes: the glacier is forced through narrow gorges, widening after it has passed them; the centre of the glacier moves more quickly than the sides, and the surface more quickly than the bottom. The point of swiftest motion follows the same law as that observed in the flow of rivers, changing from one side of the centre to the other, as the flexure of the valley changes. Most of the great glaciers in the Alps have, in summer, a central velocity of two feet a day. There are points on the Mer-de-Glace, opposite the Montanvert, which have a daily motion of thirty inches in summer, and, in winter, have been found to move at half this rate.

(230) The power of accommodating itself to the channel through which it moves, has led eminent men to assume that ice is viscous; and the phenomena at first sight seem to enforce this assumption. The glacier widens, bends, and narrows, and its centre moves more quickly than its sides; a viscous mass would undoubtedly do the same. But the most delicate experiments on the capacity of ice to yield to strain, to stretch out like treacle, honey, or tar, have failed to detect this stretching power. Is

there, then, any other physical quality to which the power of accommodation, possessed by glacier ice, may be referred?

(231) Let us approach this subject gradually. We know that vapour is continually escaping from the free surface of a liquid; that the particles at the surface attain their gaseous liberty sooner than the particles within the liquid; it is natural to expect a similar state of things with regard to ice; that when the temperature of a mass of ice is uniformly augmented, the first particles to attain liquid liberty will be those at the surface; for here they are entirely free, on one side, from the controlling action of the surrounding particles. Supposing, then, two pieces of ice, raised throughout to 32°, and melting, at this temperature, at their surfaces; what may be expected to take place if we place the liquefying surfaces close together? We thereby virtually transfer these surfaces to the centre of the ice, where the motion of each molecule is controlled, all round, by its neighbours. As might reasonably be expected, the liberty of liquidity, at each point where the surfaces touch each other, is arrested, and the two pieces freeze together at these points. Let us make the experiment: Here are two masses just cut asunder with a saw; I place their flat surfaces together; a second's contact will suffice; they are now frozen together, and by taking hold of one of them I thus lift them both.

(232) This is the effect to which attention was first directed by Mr. Faraday, in June 1850, and which is now known under the name of *Regelation*.* On a hot summer's day, I have gone into a shop in the Strand, where fragments of ice were exposed in a basin in the window; and, with the shopman's permission, have laid hold of the topmost piece of ice, and, by means of it, have lifted the whole of

* A term suggested by Dr. Hooker to Mr. Huxley and myself, on the publication of our first paper upon glaciers.

the pieces bodily out of the dish. Though the thermometer at the time stood at 80°, the pieces of ice had frozen together at their points of junction. Even under hot water this effect takes place. This basin contains water as hot as my hand can bear; I plunge into it these two pieces of ice, and hold them together for a moment: they are now frozen together, notwithstanding the presence of the heated liquid. A pretty experiment of Mr. Faraday's consists in placing a number of small fragments of ice in a dish of water deep enough to float them. When one piece touches the other, even at a single point, regelation instantly sets in. Thus, a train of pieces may be caused to touch each other, and, after they have once so touched, you may take the terminal piece of the train and, by means of it, draw all the others after it. When we seek to bend two pieces, thus united at their point of junction, the frozen points suddenly separate by fracture, but, at the same moment, other points come into contact, and regelation sets in between them. Thus a wheel of ice might be caused to roll on an ice surface, the contacts being incessantly ruptured, with a crackling noise, and others as quickly established by regelation. In virtue of this property of regelation, ice is able to reproduce many of the phenomena which are usually ascribed to viscous bodies.

(233) Here, for example, is a straight bar of ice: by passing it successively through a series of moulds, each more curved than the last, it is finally turned out as a semi-ring. The straight bar, on being squeezed into the curved mould, breaks; but by continuing the pressure, new surfaces come into contact, and the continuity of the mass is restored. A handful of those small ice fragments when squeezed together, freeze at their points of contact, and form one aggregate. The making of a snowball, as remarked by Mr. Faraday, illustrates the same

principle. In order that this freezing shall take place, the snow ought to be at 32°, and moist. When below 32°, and dry, on being squeezed it behaves like salt. The crossing of snow-bridges, in the upper regions of the Swiss glaciers, is often rendered possible solely by the regelation of the snow granules. The climber treads down the mass carefully, and causes its granules to regelate: he thus obtains an amount of rigidity which, without the act of regelation, would be quite unattainable. To those unaccustomed to such work, the crossing of snow bridges, spanning, as they often do, fissures 100 feet, and more, in depth, must appear quite appalling.

(234) When this mass of ice fragments is still further squeezed, they are brought into closer proximity. My hand, however, is incompetent to squeeze them very closely together. Placing them in this boxwood mould, which is a shallow cylinder, and inserting a flat piece of boxwood overhead, I introduce both between the plates of a small hydraulic press, and squeeze the mass forcibly into the mould. The substance is converted by the pressure into a coherent cake of ice. We can place it in this lenticular cavity and again squeeze it. It is crushed by the pressure, of course, but new contacts are established, and now the mass is turned into a lens of ice. Let us now transfer the lens to this hemispherical cavity, H (fig. 54), bring down upon it a hemispherical protuberance, P, which is not quite able to fill the cavity, and squeeze the mass: the ice, which a moment ago was a lens, is now pressed into the space between the two spherical surfaces: on removing the protuberance, you see the interior surface of a cup of glassy ice. When detached from the mould, it is a hemispherical cup, which may be filled with cold wine, without the escape of a drop. I scrape, with a chisel, a quantity of ice from this block, and, placing the spongy mass within this spherical cavity, C (fig. 55),

squeeze it and add to it, till, finally, by bringing down upon it another spherical cavity, D, it is enclosed as a sphere between both. As the press is worked, the substance becomes more and more compact. I add more material,

FIG. 54.

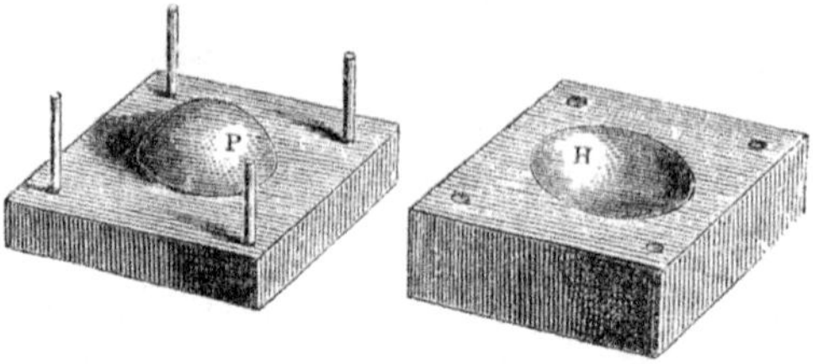

and again squeeze; by every such act the mass is made harder, and now you have a snowball before you such as you never saw before. It is a sphere of hard translucent ice, B. Thus, you see, broken ice can be compacted together by pressure, and in virtue of the property of regelation, which cements its touching surfaces, the substance may be made to take any shape we please. Were the experi-

FIG. 55.

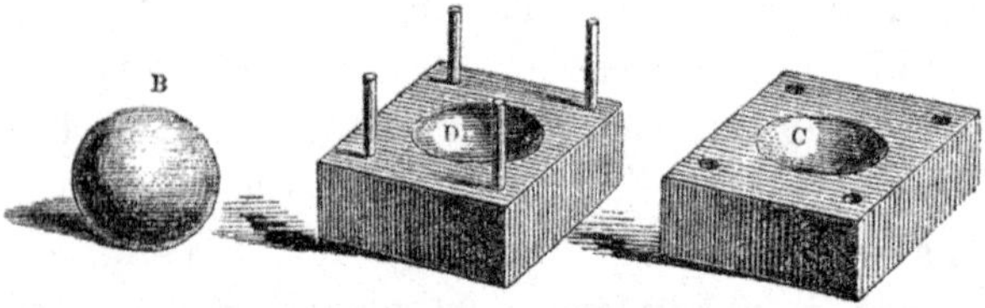

ment worth the trouble, *a rope of ice* might be formed from this block, and afterwards coiled into a knot. Nothing, of course, can be easier than to produce statuettes of the substance from suitable moulds.

(235) It is easy to understand, how a substance so endowed can be squeezed through the gorges of the Alps—can bend so as to accommodate itself to the flexures of the Alpine valleys, and can permit of a differential motion of its parts, without, at the same time, possessing a sensible trace of viscosity. The hypothesis of viscosity, first started by

Rendu, and worked out with such ability by Professor Forbes, accounts, certainly, for half the facts. Where pressure comes into play, the deportment of ice is, apparently, that of a viscous body; where tension comes into play, the analogy with a viscous body ceases.*

(236) I have thus briefly sketched the phenomena of existing glaciers, as far as they are related to our present subject; but the scientific explorer of mountain regions soon meets with appearances, which carry his mind back to a state of things very different from that of the present day. The unmistakable traces which they have left behind them show that vast glaciers once existed, in places from which they have for ages disappeared. Go, for example, to the glacier of the Aar in the Bernese Alps, and observe its present performances; look to the rocks upon its flanks as they are at this moment, rounded, polished, and scarred by the moving ice. And, having by patient and varied exercise educated your eye and judgment, in these matters, walk down the glacier towards its end, keeping always in view the evidences of glacial action. After quitting the ice, continue your walk down the valley towards the Grimsel: you see everywhere the same unmistakable record. The rocks which rise from the bed of the valley are rounded like hogs' backs; these are the 'roches moutonnés' of Charpentier and Agassiz; you observe upon them the larger flutings of the ice, and also the smaller scars, scratched by pebbles, which the glacier held as a kind of emery on its under surface. All the rocks of the Grimsel have been thus planed down. Walk down the valley of Hasli and examine the mountain sides right and left; without the key, which I now suppose you to possess, you would be in a land of enigmas; but with this key all is plain—you see everywhere the well-known scars and flutings and

* For further information regarding glacier phenomena, I must refer the reader to the 'Glaciers of the Alps,' Murray, London.

furrowings. In the bottom of the valley you have the rocks filed down, in some places, to dome-shaped masses, and, in others, polished so smoothly that to pass over them, even when the inclination is moderate, steps must be hewn. All the way down to Meyringen, and beyond it, if you wish to pursue the enquiry, these evidences abound. For a preliminary lesson in recognising the traces of ancient glaciers, no better ground than this can be chosen.

(237) Similar evidences are found in the valley of the Rhone; you may track them through the valley for eighty miles, and lose them at length in the Lake of Geneva. But on the flanks of the Jura, at the opposite side of the Canton de Vaud, the evidences reappear. All along these limestone slopes are strewn the granite boulders of Mont Blanc. Right and left, also, from the great Rhone valley, the lateral valleys show that they were once filled with ice. On the Italian side of the Alps the remains are, if possible, more stupendous than those on the northern side. Grand as the present glaciers seem to those who explore them to their full extent, they are mere pigmies in comparison with their predecessors.

(238) Not in Switzerland alone—not alone in proximity with existing glaciers—are these well-known vestiges of the ancient ice discernible; on the hills of Cumberland they are almost as clear as among the Alps. Where the bare rock has been exposed for ages to the action of weather, the finer marks have, in most cases, disappeared; and the mammillated forms of the rocks are the only evidences. But the removal of the soil which has protected them often discloses rock surfaces, scarred as sharply, and polished as cleanly, as those which are now being scratched and polished by the glaciers of the Alps. Round about Scawfell, the traces of ancient ice appear, both in *roches moutonnés* and *blocs perchés*; and there are ample facts to show that Borrodale was once occupied by glacier ice.

In North Wales, also, the ancient glaciers have placed their stamp so firmly upon the rocks, that the ages which have since elapsed have failed to obliterate even their superficial marks. All round Snowdon these evidences abound. On the south-west coast of Ireland rise the Reeks of Macgillicuddy, which tilt upwards, and catch upon their cold crests the moist winds of the Atlantic; precipitation is copious, and rain at Killarney seems the order of Nature. In this moist region every crag is covered with rich vegetation; but the vapours, which now descend as mild and fertilising rain, once fell as snow, which formed the material for noble glaciers. The Black Valley was once filled by ice, which planed down the sides of the Purple Mountain, as it moved towards the Upper Lake. The ground occupied by this lake was entirely covered by the ancient ice, and every island that now emerges from its surface is a glacier-dome. The fantastic names, which many of the rocks have received, are suggested by the shapes into which they have been sculptured by the mighty moulding plane which once passed over them. North America is also thus glaciated. But the most notable observation, in connection with this subject, is one recently made by Dr. Hooker during a visit to Syria: he has found that the celebrated cedars of Lebanon grow upon ancient glacier moraines.

(239) To determine the condition, which permitted of the formation of those vast masses of ice, has long been a problem with philosophers, and a consideration of the solutions which have been offered, from time to time, will not be uninstructive. I have no new hypothesis to offer, but it seems possible to give a truer direction and more definite aim to our enquiries than they can at present boast of. The aim of all the writers on this subject, with whom I am acquainted, has been the attainment of *cold*. Some eminent men have thought, and some still think, that the reduction of temperature, during the glacier

epoch, was due to a temporary diminution of solar radiation; others have thought that, in its motion through space, our system may have traversed regions of low temperature, and that, during its passage through these regions, the ancient glaciers were produced. Others have sought to lower the temperature, by a redistribution of land and water. If I understand the writings of the eminent men who have propounded and advocated the above hypotheses, all of them seem to have overlooked the fact, that the enormous extension of glaciers in bygone ages demonstrates, just as rigidly, the operation of heat as the action of cold.

(240) Cold alone will not produce glaciers. You may have the bitterest north-east winds here in London throughout the winter, without a single flake of snow. Cold must have the fitting object to operate upon, and this object—the aqueous vapour of the air—is the direct product of heat. Let us put this glacier question in another form: the latent heat of aqueous vapour, at the temperature of its production in the tropics, is about 1,000° Fahr., for the latent heat augments, as the temperature of evaporation descends. A pound of water, then, vaporised at the equator, has absorbed 1,000 times the quantity of heat which would raise a pound of the liquid one degree in temperature. But the quantity of heat which would raise a pound of water one degree, would raise a pound of cast iron ten degrees: hence, simply to convert a pound of the water of the equatorial ocean into vapour, would require a quantity of heat, sufficient to impart to a pound of cast iron 10,000 degrees of temperature. But the fusing-point of cast iron is 2,000° Fahr.; therefore, for every pound of vapour produced, a quantity of heat has been expended by the sun, sufficient to raise 5 lbs. of cast iron to its melting point. Imagine, then, every one of those ancient glaciers with its mass of ice quintupled; and imagine the place of the mass

so augmented, to be taken by an equal weight of cast iron raised to the white heat of fusion, we shall then have the exact expression of the solar action, involved in the production of the ancient glaciers. Substitute the hot iron for the cold ice—our speculations would instantly be directed to account for the *high* temperature of the glacial epoch, and a complete reversal of some of the hypotheses above quoted would probably ensue.

(241) It is perfectly manifest, that by weakening the sun's action, either through a defect of emission, or by the steeping of the entire solar system in space of a low temperature, we should be cutting off the glaciers at their source. Vast masses of mountain ice indicate, infallibly, the existence of commensurate masses of atmospheric vapour, and a proportionately vast action on the part of the sun. In a distilling apparatus, if you required to augment the quantity distilled, you would not surely attempt to obtain the low temperature, necessary to condensation, by taking the fire from under your boiler; but this, if I understand them aright, is what has been done by those philosophers who have sought to produce the ancient glaciers by diminishing the sun's heat. It is quite manifest that the thing most needed to produce the glaciers is an *improved condenser*; we cannot afford to lose an iota of solar action; we need, if anything, more vapour, but we need a condenser so powerful, that this vapour, instead of falling in liquid showers to the earth, shall be so far reduced in temperature as to descend in snow. The problem, I think, is thus narrowed to the precise issue on which its solution depends.

NOTE.

In moulding ice, it is advisable to first wet the mould with hot water. This facilitates the removal of the compressed substance. The ice-cup, referred to in § 234, may be from 2½ to 3 inches in external diameter, but the thickness of the cup ought not to exceed a quarter of an inch. A conical plug is inserted into my own moulds, the tapping of which soon detaches the ice.

CHAPTER VII.

CONDUCTION A TRANSMISSION OF MOTION — GOOD CONDUCTORS AND BAD CONDUCTORS — CONDUCTIVITY OF THE METALS FOR HEAT: RELATION BETWEEN THE CONDUCTIVITY OF HEAT AND THAT OF ELECTRICITY — INFLUENCE OF TEMPERATURE ON THE CONDUCTION OF ELECTRICITY—INFLUENCE OF MOLECULAR CONSTITUTION ON THE CONDUCTION OF HEAT—RELATION OF SPECIFIC HEAT TO CONDUCTION —PHILOSOPHY OF CLOTHES: RUMFORD'S EXPERIMENTS—INFLUENCE OF MECHANICAL TEXTURE ON CONDUCTION — INCRUSTATIONS OF BOILERS—THE SAFETY LAMP—CONDUCTIVITY OF LIQUIDS AND GASES: EXPERIMENTS OF RUMFORD AND DESPRETZ — COOLING EFFECT OF HYDROGEN GAS — EXPERIMENTS OF MAGNUS ON THE CONDUCTIVITY OF GASES.

(242) I think we are now sufficiently conversant with our subject, to distinguish between the sensible motions produced by heat and heat itself. Heat is not the clash of winds; it is not the quiver of a flame, nor the ebullition of water, nor the rising of a thermometric column, nor the motion which animates steam as it rushes from a boiler, in which it has been compressed. All these are mechanical motions, into which that of heat may be converted; but heat itself is *molecular* motion. The molecules of bodies, when closely grouped, cannot, however, oscillate, without communicating motion from one to the other. To this propagation of the motion of heat from molecule to molecule, we must now devote our attention.

Here is a poker, the temperature of which is scarcely perceptible: I feel it as a hard and heavy body, but it neither warms nor chills me; it has been before the fire, and the motion of its molecules, at the present moment, chances to be the same as that of the molecules of my nerves; there is neither communication nor withdrawal, and hence the temperature of the poker, on the one hand, and my sensations, on the other, remain unchanged. But when the end of the poker is thrust into the fire it is heated; the molecules, in contact with the fire, are thrown into a state of more intense oscillation; the swinging atoms strike their neighbours, these again theirs, and thus, the molecular music rings along the bar. The motion, in this instance, is communicated from atom to atom of the poker, and finally appears at its most distant end. If I now lay hold of the poker, its motion is communicated to my nerves, and produces pain; the bar is what we call hot, and my hand, in popular language, is burned. Convection we have already defined to be the transfer of heat, by sensible masses of matter, from place to place; but this molecular transfer, which consists in each *atom* taking up the motion of its neighbours, and sending it on to others, is called the *conduction* of heat.

(243) Let me exemplify this property of conduction, in a homely way. In this basin, filled with warm water, is placed a cylinder of iron, an inch in diameter, and two

FIG. 56.

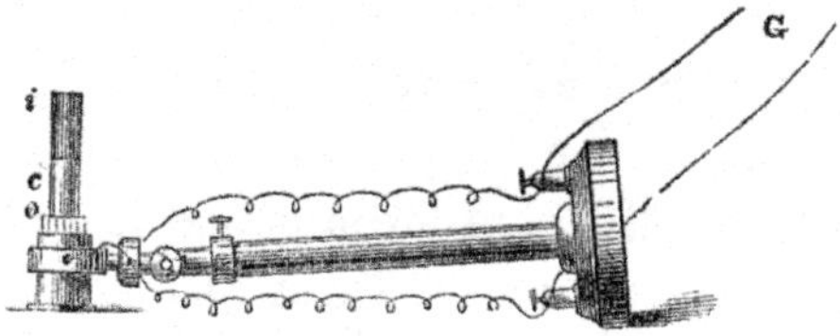

inches in height; this cylinder is to be my source of heat. Laying my thermo-electric pile, *o* (fig. 56), thus flat, with its naked face turned upwards, I place upon that face a cylinder of copper, *c*, which now possesses the temperature of this room. We observe no deflection of the galvanometer. I now place my warm cylinder, *i*, having first dried it, upon the cool cylinder, which is supported by the pile. The upper cylinder is not at more than blood-heat; but you see, almost before this remark is uttered, the needle flies aside, indicating that the heat has reached the face of the pile. Thus, the molecular motion, imparted to the iron cylinder by the warm water, has been communicated to the copper one, through which it has been transmitted, in a few seconds, to the face of the pile.

(244) Different bodies possess different powers of transmitting molecular motion; in other words, of conducting heat. Copper, which we have just used, possesses this power in a very eminent degree. Let us now remove the copper, allow the needle to return to 0°, and then lay upon the face of the pile this cylinder of glass. On the cylinder of glass I place my iron cylinder, which has been re-heated in the warm water. There is, as yet, no motion of the needle, and you would have to wait a long time to see it move. We have already waited thrice the time which the copper required to transmit the heat, and you see the needle continues motionless. Placing cylinders of wood, chalk, stone, and fireclay, in succession, on the pile, and heating their upper ends in the same manner, we find that in the time which we can devote to an experiment, not one of these substances is competent to transmit the heat to the pile. The molecules of these substances are so hampered or entangled, that they are incompetent to pass the motion freely from one to another. These bodies are all *bad conductors* of heat. On the other hand, when cylinders of zinc, iron, lead, bismuth, &c., are placed in succession on the pile, each of them, as you see, has the power of transmitting the motion of heat rapidly through its mass. In comparison with the wood, stone, chalk, glass, and clay, they are all *good conductors* of heat.

(245) As a general rule, not, however, without its exceptions, metals are the best conductors of heat. But

metals differ notably among themselves, as regards their powers of conduction. A comparison of copper and iron will illustrate this point. Behind me are two bars, A B, A C (fig. 57), placed end to end, with balls of wood,

FIG. 57.

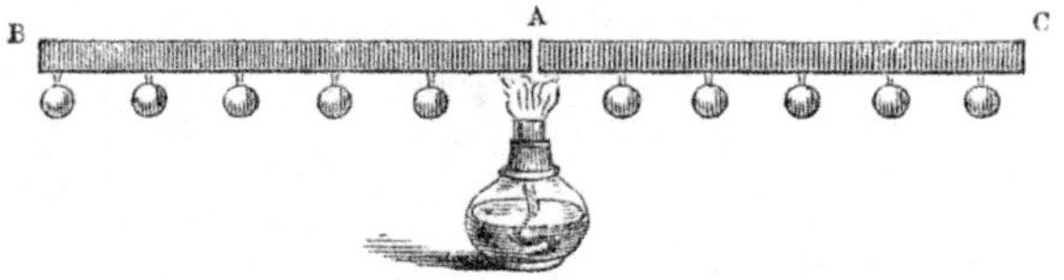

attached by wax at equal distances from the place of junction. Under the junction is placed a spirit-lamp, which heats the ends of the bars: the heat will be propagated right and left through both. The bar A B is iron, the bar A C is copper; the heat travels to a greater distance along the copper, which is the better conductor, and therefore liberates a greater number of its balls.

(246) One of the first attempts to determine, with accuracy, the conductivity of different bodies for heat, was that suggested by Franklin, and carried out by Ingenhausz. He coated a number of bars of various substances with wax, and, immersing the ends of the bars in hot oil, he observed the distance to which the wax was melted, on each of the bars. The good conductors melted the wax to the greatest distance; and the melting distance furnished a measure of the conductivity of the bar.

(247) The second method was that pointed out by Fourier, and followed out experimentally by Despretz. A B (fig. 58) represents a bar of metal, with holes drilled in it, intended to contain small thermometers. At the end of the bar was placed a lamp, as a source of heat; the heat was propagated through the bar, reaching the thermometer *a* first, *b* next, *c* next, and so on. For a certain time, the thermometers continued to rise, but afterwards

the state of the bar became stationary, each thermometer making a constant temperature. The better the conduction, the smaller the difference between any two successive thermometers. The decrement, or *fall* of heat, if I may

FIG. 58.

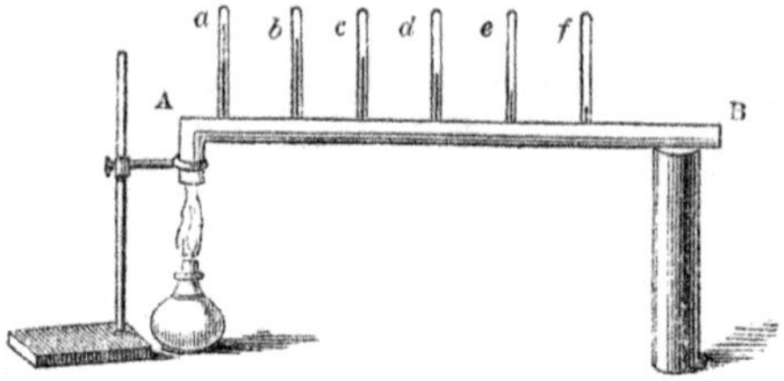

use the term, from the hot end towards the cold, is greater in bad conductors than in good ones, and, from the decrement of temperature shown by the thermometers, we can deduce, and express by a number, the conductivity of the bar. This same method was followed by MM. Wiedemann and Franz, in a very important investigation, but, instead of using thermometers, they employed a suitable modification of the thermo-electric pile. Of the numerous and highly interesting results of this investigation, the following is a résumé :—

Name of Substance	Conductivity For Heat	Conductivity For Electricity
Silver	100	100
Copper	74	73
Gold	53	59
Brass	24	22
Tin	15	23
Iron	12	13
Lead	9	11
Platinum	8	10
German Silver	6	6
Bismuth	2	2

(248) This table shows, that, as regards their conductive powers, metals differ very widely from each other. Calling, for example, the conductive power of silver 100,

that of German silver is only 6. You may illustrate this difference, in a very simple way, by plunging two spoons, one of German silver, and the other of pure silver, into the same vessel of hot water. After a little time, you find the free end of the silver spoon much hotter than that of its neighbour; and if bits of phosphorus be placed on the ends of the spoons, that on the silver will fuse and ignite, in a very short time, while the heat transmitted through the other spoon will never reach an intensity sufficient to ignite the phosphorus.

(249) Nothing is more interesting to the natural philosopher than the tracing out of connections and relations between the various agencies of nature. We know that they are interdependent, we know that they are mutually convertible, but, as yet, we know very little as to the precise form of the conversion. We have every reason to conclude, that heat and electricity are both modes of motion; we know, experimentally, that from electricity we can obtain heat, and from heat, as in the case of our thermo-electric pile, we can obtain electricity. But although we have, or think we have, tolerably clear ideas of the character of the motion of heat, our ideas are very crude as to the precise nature of the change which this motion must undergo, in order to appear as electricity—in fact, we know, as yet, nothing about it.

(250) The above table, however, exhibits one important connection between heat and electricity. Besides the numbers expressing conductivity for heat, MM. Wiedemann and Franz have placed the numbers expressing the conductivity of the same metals for electricity. They run side by side: the good conductor of heat is the good conductor of electricity, and the bad conductor of heat is the bad conductor of electricity.* Thus, we may infer

* Principal Forbes had previously noticed this. See Phil. Mag. 1834, vol. iv. p. 27.

that the same physical quality which interferes with the transmission of heat, interferes, in a proportionate degree, with the transmission of electricity. This common susceptibility of both forces indicates a relation, on which future investigations will no doubt throw light.

(251) It is a proved fact, that the amount of heat developed in a wire, by a current of electricity of a certain strength, is directly proportional to the resistance of the wire.* We may imagine the atoms in this case throwing themselves, like barriers, across the track of the electric current—the current knocking against them, imparting its motion to them, and thus rendering the wire hot. In the case of a good conductor, on the contrary, the current may be pictured as gliding freely among the atoms, without disturbing them in any great degree. Suspended before you are three pieces of platinum wire, each four or five inches long, joined to three pieces of silver wire, of the same length and thickness. I will now send the self-same current, from a battery of twenty of Grove's cells, through this compound wire. You see three spaces white-hot, and dark spaces between them. The white-hot portions of the wire are platinum, and the dark portions are silver. The electric current breaks impetuously upon the molecules of the platinum, while it glides, with little resistance, among the atoms of silver, thus producing, in the two metals, different calorific effects.†

(252) Now it may be shown that the motion of heat interferes with that of electricity. You are acquainted with the platinum lamp, which stands in front of this table. It consists, simply, of a little coil of platinum wire, suitably attached to a brass stand. We can send a current through that coil, and cause it to glow. Into the

* Joule, Phil. Mag. 1841, vol. xix. p. 263.

† May not the condensed ether which surrounds the atoms be the vehicle of electric currents?

circuit are also introduced two additional feet of thin platinum wire, and on establishing the connection, the same current passes through this wire, and through the coil. Both, you see, are raised to redness—both are in a state of intense molecular motion. What I wish now to prove is, that this motion of heat, which the electricity has generated, in these two feet of wire, and in virtue of which the wire glows, offers a hindrance to the passage of the current. The electricity has raised up a foe in its own path. If we cool this wire, we open a wider door for the passage of the electricity. But, if more electricity passes, it will announce itself at the platinum lamp; it will raise that red heat to whiteness, and the change in the intensity of the light will be visible to you all.

FIG. 59.

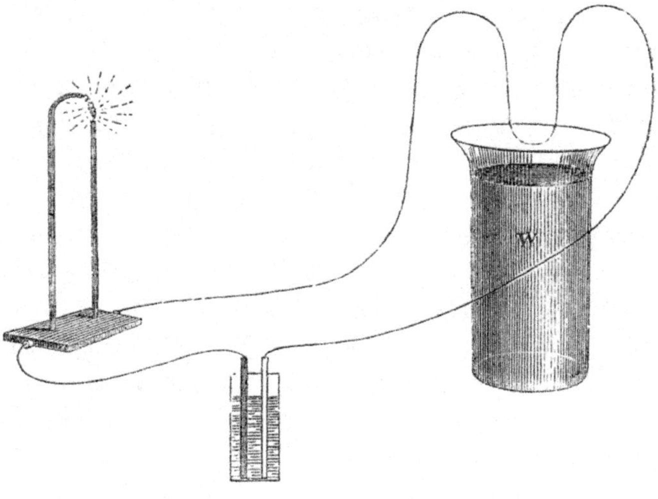

(253) Thus, then, I plunge the red-hot wire into a beaker of water w (fig. 59): the lamp immediately becomes almost too bright to look at. When the wire is raised out of the water, and the heat allowed once more to develope

itself, the current is instantly impeded, and the lamp becomes less bright. I again dip the wire into the cold water, deeper and deeper: observe how the light becomes intensified—deeper still, so as to quench the entire two feet of wire; the augmented current raises the lamp to its maximum brightness, and now it suddenly goes out. The circuit is broken, for the coil has actually been fused by the additional flow of electricity.

(253*a*) And here we may bestow a passing glance at a subject, the complete treatment of which belongs to another department of physics. You know that the electric current which heated the wire in our last experiment is maintained by the chemical action going on in the voltaic battery. In the battery we have, among other things, the combination of zinc with oxygen, a true combustion, though, like that of our own bodies, it is carried on among liquids. Here, as in all other cases, the consumption of a definite amount of zinc generates the same invariable amount of heat. Supposing, then, I connect the two poles of this battery by a stout copper wire, which is an excellent conductor, the current will flow and the zinc will be consumed, and no sensible heat will be developed outside the battery itself. Let the current continue until a pound of zinc has been consumed. A certain measurable amount of heat is generated, and the whole of this heat is confined to the battery. Let us now connect the poles of the battery by a thin platinum wire. It becomes heated and glows before your eyes. Continue the action until a pound of zinc has been consumed. The same amount of heat as before is generated, but it is now distributed in a different manner. Part of it is in the battery, but part of it also is in the connecting wire. Add both these parts together, and you get the same total as before.

(253*b*) Thus, for every unit of heat generated outside

the battery, we have a unit withdrawn from the battery itself; and if, instead of generating this external heat, the electric current be employed to turn a machine, or do any other external work, an amount of heat equivalent to the work performed is withdrawn from the battery. These are not mere theoretic conclusions: they have been established by the excellent experiments of M. Favre, the principle of conservation, as applied to the voltaic battery, being thus vindicated.

(254) Let us now return to the subject of conduction. To all appearance, cold may be conducted, like heat. I warm this copper cylinder a little by holding it, for a moment, in my hand. When placed on the thermo-electric pile, the needle goes up to 90°, declaring heat. On this cylinder, I place a second one, which has been chilled, by sinking it for some time in this mass of ice. We wait a moment, the needle moves: it is now descending to zero, passes it, and goes on to 90°, on the side of cold. Analogy might well lead you to suppose that the cold is conducted downwards, from the top cylinder to the bottom one, as the heat was conducted in our former experiments. No objection need be made to the phrase 'conduction of cold,' if it be used with a clear knowledge of the real physical process involved. The real process is, that the warm intermediate cylinder first delivers up its heat, or motion, to the cold cylinder overhead, and, having thus lost its own heat, it draws upon that of the pile. In our former experiments, we had conduction of motion *to* the pile; in our present one we have conduction of motion *from* the pile. But it is, in both cases, the propagation of motion with which we have to do, the heating and the chilling depending solely upon the *direction* of propagation. I place one of these metal cylinders, which has been purposely cooled, on the face of our pile; a violent deflection follows, declaring the instrument to be chilled.

Are we to suppose cold to be an entity communicated to the pile? No. The pile here is the warm body; its molecular motion is in excess of that possessed by the cylinder; and when both come into contact, the pile seeks to make good the defect. It imparts a quantity of its motion to the cylinder, and, by its own bounty, becomes impoverished: it chills itself, and generates the current.

(255) Substituting for this cold metal cylinder a cylinder of wood, with the same temperature as the metal one, the chill of the wood is very feeble, and the consequent deflection very small. Why does not the cold wood produce an action equal to that of the cold metal? Simply, because the heat, communicated to it by the pile, is accumulated at its under surface; it cannot escape through the bad conducting wood as it escapes through the metal, and thus the quantity of heat withdrawn from the pile by the wood, is less than that withdrawn by the copper. A similar effect is produced when the human nerves are substituted for the pile. When you come into a cold room, and lay your hand upon the fire-irons, the chimney-piece, the chairs, the carpet, in succession, they appear to be of different temperatures: the iron chills you more than the marble, the marble more than the wood, and so on. Your hand is affected exactly as the pile was affected in the last experiment. It is needless to say that the reverse takes place when you enter a hot room; that is to say, a room hotter than your own body. You would certainly suffer, if you lay down upon a plate of metal in a Turkish bath; but you do not suffer when you lie down on a bench of wood. By preserving the body from contact with good conductors, very high temperatures may be endured. Eggs may be boiled, and beefsteaks cooked, by the heat of an apartment, in which the bodies of living men sustain no injury.

(256) The exact philosophy of this last experiment is

worthy of a moment's consideration. With it the names of Blagden and Chantrey are associated, those eminent men having exposed themselves in ovens to temperatures considerably higher than that of boiling water. Let us compare the condition of the two living human beings with that of two marble statues, placed in the same oven. The statues become gradually hotter, until finally they assume the temperature of the air of the oven; the two men, under the same circumstances, do not similarly rise in temperature. If they did, the tissues of the body would be infallibly destroyed, the temperature which they endured being more than sufficient to stew the muscles in their own liquids. But the fact is, that the heat of the blood is scarcely affected by an augmentation of the external heat. This heat, instead of being applied to increase the temperature of the body, is applied to change the aggregation of the body; it prepares the perspiration, forces it through the pores, and, in part, vaporises it. Heat is here converted into potential energy; it is consumed in work. This is the waste-pipe, if I may use the term, through which the excess of heat overflows; and hence it is, that under the most varying conditions of climate, the temperature of the human blood is, practically, constant. The blood of the Laplander is sensibly as warm as that of the Hindoo; while an Englishman, in sailing from the north pole to the south, finds his blood-temperature hardly heightened by his approach to the equator, and hardly diminished by his approach to the antarctic pole.

(257) When the communication of heat is gradual—as it always is, when the body is surrounded by an imperfect conductor—the heat is consumed, in the manner indicated, as fast as it is supplied; but if the supply of heat be so quick (as it would be in the case of contact with a good conductor) that the conversion into this harmless potential

energy cannot be executed with sufficient rapidity, injury to the tissues is the result. Some people have professed to see, in this power of the living body to resist a high temperature, a conservative action, peculiar to the vital force. No doubt, all the actions of the animal organism are connected with what we call its vitality; but the action here referred to is the same in kind as the melting of ice, or the vaporisation of water. It consists, simply, in the diversion of heat from the purposes of temperature to the performance of work.

(258) Thus far, we have compared the conducting power of different bodies together; but the same substance may possess different powers of conduction in different directions. Many crystals are so built, that the motion of heat runs with greater facility along certain lines of atoms than along others. Here, for instance, is a large rock-crystal—

FIG. 60.

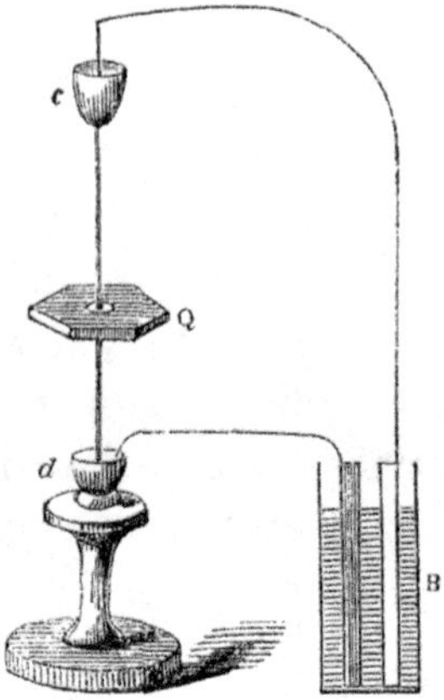

FIG. 61.

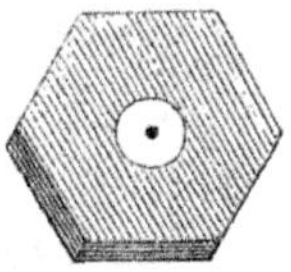

FIG. 62.

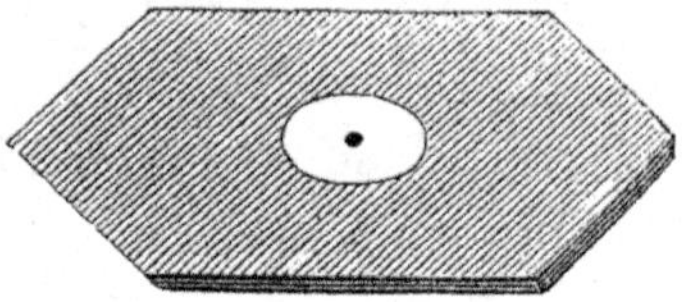

a crystal of quartz—forming a hexagonal pillar, which, if complete, would be terminated by two six-sided pyramids. That heat travels with greater facility along the axis of this crystal, than across it has been proved in a very simple manner by M. de Senarmont. Of these two plates of quartz, one (fig. 61) is cut perpendicularly to the axis of the crystal, and the other (fig. 62) parallel to it. The plates are coated with a layer of white wax, laid on by a camel's-hair pencil. They are pierced at the centre, and into the hole is inserted a small sewing-needle, which can be warmed by an electric current. B (fig. 60) is the battery, whence the current proceeds; *c* is a capsule of wood, through the bottom of which the sewing-needle passes; *d* is a second capsule, into which dips the point of the needle, and Q is the perforated plate of quartz. Each capsule contains a drop of mercury. When the current passes from *c* to *d*, the needle is heated, and the heat is propagated in all directions. The wax melts around the place where the heat is applied; and on this plate, which is cut perpendicularly to the axis of the quartz, the figure of the melted wax is a perfect circle (fig. 61). The heat has travelled with the same rapidity all round, and melted the wax to the same distance in all directions. I make a similar experiment with the other plate: the wax is now melting; but its figure is no longer a circle. The heat travels more speedily along the axis than across it, and hence the wax figure is an ellipse, instead of a circle (fig. 62). When the wax dries, I will project magnified images of these two plates upon the screen, and you will then see the circular figure of the melted wax on the one, and the oval figure on the other. Iceland spar conducts better along the crystallographic axis than at right angles to it, while a crystal of tourmaline conducts best at right angles to its axis. The metal bismuth, with which you are already acquainted, cleaves with great facility in one direction, and, as well shown by MM. Svanberg and Matteucci, it conducts both heat and electricity better along the planes of cleavage than across them.

(259) In wood, we have an eminent example of this

difference of conductivity. Many years ago MM. de la Rive and De Candolle instituted an enquiry into the conductive power of wood,* and, in the case of five specimens examined, established the fact that the velocity of transmission was greater along the fibre than across it. The manner of experiment was that usually adopted in enquiries of this nature, and which was applied to metals by M. Despretz.† A bar of the substance was taken, one end of which was brought into contact with a source of heat, and allowed to remain there until a state of equilibrium was assumed. The temperatures attained by the bar, at various distances from its heated end, were ascertained by means of thermometers fitting into cavities made to receive them; from these data, with the aid of a well-known formula, the conductivity of the wood was determined.

(260) To determine the velocity of calorific transmission, in different directions, through wood, the instrument shown in fig. 63 was devised, some years ago, by myself. Q Q′ R R′ is an oblong piece of mahogany, A is a bar of antimony, B is a bar of bismuth. The united ends of the two bars are kept in close contact by the ivory jaws I I′, and the other ends are let into a second piece of ivory, in which they are firmly fixed. From these ends proceed two pieces of platinum wire to the little ivory cups M M, communicating with a drop of mercury placed in the interior. Two small projections are observed in the figure, jutting from I I′; across, from one projection to the other, a fine membrane is stretched, thus enclosing a little chamber *m*, in front of the wedge-like end of the bismuth and antimony junction; the chamber has an ivory bottom. S is a wooden slider, which can be moved smoothly back and forward along a bevelled groove, by means of the lever L. This

* Mém. de la Soc. de Genève, vol. iv. p. 70.

† Annales de Chim. et de Phys. December 1827.

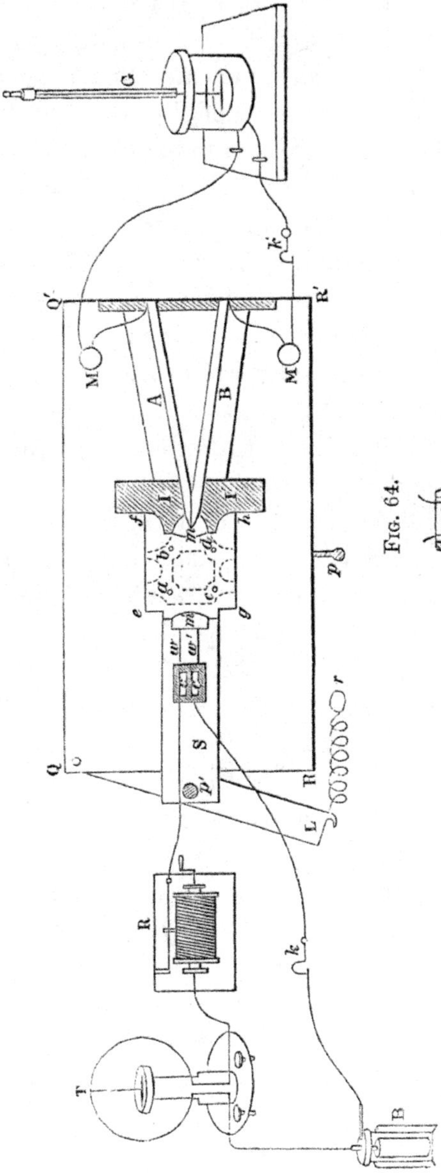

Fig. 63.

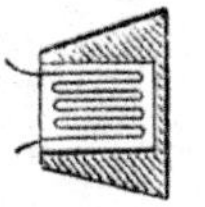

Fig. 64.

lever turns on a pivot at Q, and fits into a horizontal slit in the slider, to which it is attached, by the pin *p′* passing through both; in the lever an oblong aperture is cut, through which *p′* passes, and in which it has a certain amount of lateral play, so as to enable it to push the slider forward in a straight line. Two projections are seen at the end of the slider; across these, from projection to projection, a thin membrane is stretched; a chamber *m′* is thus formed, with three sides and a floor of wood, and bounded in front by the membrane. A thin platinum wire, bent up and down several times, so as to form a kind of grating, was laid against the back of the chamber *m′*, and imbedded in the end of the slider by the stroke of a hammer; the end was then filed down, until about half the wire was removed, and the whole reduced to a uniform flat surface. Against the common surface of the slider and wire, an extremely thin plate of mica was glued, sufficient, simply, to interrupt all contact between the bent wire and a drop of mercury, which the chamber *m′* is destined to contain; the ends *w w′* of the bent wire proceed to two small cisterns *c c′*, hollowed out in a slab of ivory, and filled with mercury. The end of the slider and its bent wire are shown in fig. 64. The rectangular space *e f g h* (fig. 63) is cut quite through the slab of mahogany, and a brass plate is screwed to the latter underneath; from this plate (which is cut away, as shown by the dotted lines in the figure) four conical ivory pillars *a b c d* project upwards; though appearing to be upon the same plane as the upper surfaces of the bismuth and antimony bars, the points of the pillars are in reality 0·3 of an inch below the said surfaces.

(261) The body to be examined is reduced to the shape of a cube, and placed, by means of a pair of pliers, upon the four supports *a b c d*; the slider S is then drawn up against the cube, and the latter becomes firmly clasped

between the projections of the piece of ivory I I′ on the one side, and those of the slider S on the other. The chambers *m* and *m′* being filled with mercury, the membrane in front of each is pressed gently against the cube by the interior fluid mass, and, in this way, a uniform contact, which is absolutely essential, is secured.

The problem which requires solution is the following:—It is required to apply a source of heat, of a strictly measurable character, and always readily attainable, to that face of the cube which is in contact with the membrane *m′* at the end of the slider, and to determine what quantity of this heat crosses the cube to the opposite face, during a minute of time.

(262) To obtain a source of heat, of the nature described, the following method was adopted:—B is a small galvanic battery, from which a current proceeds to the tangent compass T; passes round the ring of the instrument, deflecting in its passage the magnetic needle, which hangs in the centre of the ring. From T the current proceeds to the rheostat R; this instrument consists of a cylinder of serpentine stone, round which a German-silver wire is coiled spirally; by turning the handle of the instrument, any required quantity of this powerfully resisting wire is thrown into the circuit, the current being thus regulated at pleasure. The sole use of these two last instruments, in the present series of experiments, is to keep the current perfectly constant, from day to day. From the rheostat the current proceeds to the cistern *c*, thence through the bent wire, and back to the cistern *c′*, from which it proceeds to the other pole of the battery.

(263) The bent wire, during the passage of the current, becomes gently heated; the heat is transmitted through the mercury in the chamber *m′* to the membrane in front of the chamber; this membrane becomes the proximate source of heat applied to the left-hand face of the cube.

The quantity of heat transmitted from this source, through the mass of the cube, to the opposite face, in any given time, is estimated from the deflection which it is able to produce upon the needle of a galvanometer, connected with the bismuth and antimony pair. G is a galvanometer, used for this purpose; from it proceed wires to the mercury cups M M, which, as before remarked, are connected by platinum wires with A and B.

(264) The action of mercury upon bismuth, as a solvent, is well known; an amalgam is speedily formed when the two metals come into contact. To preserve the thermo-electric couple from this action, their ends are protected by a sheathing of the same membrane as that used in front of the chambers *m m′*.

(265) Previous to the cube's being placed between the two membranes, the latter, by virtue of the fluid masses behind them, bulge out a little, thus forming a pair of soft and slightly convex cushions. When the cube is placed on its supports, and the slider is brought up against it, both cushions are pressed flat, and thus the contact is made perfect. The surface of the cube is larger than the surface of the membrane; * and hence the former is always firmly caught between the opposed rigid projections, the slider being held fast in this position by means of the spring *r*, which is then attached to the pin *p*. The exact manner of experiment is as follows:—Having first seen that the needle of the galvanometer points to zero, when the thermo-circuit is complete, the latter is interrupted by means of the break-circuit key *k′*. At a certain moment, marked by the second hand of a watch, the voltaic circuit is closed by the key *k*, and the current is permitted to circulate for sixty seconds; at the sixtieth second the voltaic circuit is broken, by the left hand at *k*, while, at the same instant, the thermo-electric circuit is closed

* The edge of each cube measured 0·3 inch.

by the right hand at k'. The needle of the galvanometer is instantly deflected, and the limit of the first impulsion is noted. The amount of this impulsion depends, of course, upon the quantity of heat which has reached the bismuth and antimony junction, through the mass of the cube, during the time of action. The limit of the first impulsion being noted, the cube is removed, and the instrument is allowed to cool, until the needle of the galvanometer returns again to zero.

(266) Judging from the description, the mode of experiment may appear complicated, but, in reality, it is not so. A single experimenter has the most complete command over the entire arrangement. The wires from the small galvanic battery (a single cell) remain undisturbed from day to day; all that is to be done is to connect the battery with them, and everything is ready for experiment.

(267) There are in wood three lines, at right angles to each other, which the mere inspection of the substance enables us to fix upon, as the necessary resultants of molecular action: the first line is parallel to the fibre; the second is perpendicular to it, and to the ligneous layers which indicate the annual growth of the tree; while the third is perpendicular to the fibre, and parallel, or rather tangential, to the layers. From each of a number of trees a cube was cut, two of the faces being parallel to the ligneous layers, two perpendicular to them, while the remaining two were perpendicular to the fibre. It was proposed to examine the velocity of calorific transmission through the wood, in these three directions. It may be remarked, that the wood was in all cases well-seasoned and dry.

(268) The cube was first placed upon its four supports a b c d, so that the line of flux from m' to m was parallel to the fibre, and the deflection produced by the heat

transmitted in sixty seconds, was observed. The cube was then placed with its fibre vertical, the line of flux from m' to m being perpendicular to the fibre, and parallel to the ligneous layers; the deflection produced by a minute's action, in this case, was also determined. Finally, the cube was turned 90° round, its fibre being still vertical, so that the line of flux was perpendicular to both fibre and layers, and the consequent deflection was observed. In the comparison of these two latter directions, the chief delicacy of manipulation is necessary. It requires but a rough experiment, to demonstrate the superior velocity of propagation along the fibre, but the velocities in all directions perpendicular to the fibre are so nearly equal, that it is only by great care, and, in the majority of cases, by numerous experiments, that a difference of action can be securely established.

(269) The following table contains some of the results of the enquiry; it will explain itself:—

Description of Wood	Deflections		
	I. Parallel to fibre	II. Perpendicular to fibre and parallel to ligneous layers	III. Perpendicular to fibre and to ligneous layers
	°	°	°
1 American birch . .	35	9·0	11·0
2 Oak	34	9·5	11·0
3 Beech	33	8·8	10·8
4 Coromandel-wood . .	33	9·8	12·3
5 Bird's-eye maple . .	31	11·0	12·0
6 Lance-wood . . .	31	10·6	12·1
7 Box-wood . . .	31	9·9	12·0
8 Teak-wood . . .	31	9·9	12·4
9 Rose-wood . . .	31	10·4	12·6
10 Peruvian-wood . . .	30	10·7	11·7
11 Green-heart . . .	29	11·4	12·6
12 Walnut	28	11·0	13·0
13 Drooping ash . . .	28	11·0	12·0
14 Cocoa-wood . . .	28	11·9	13·6
15 Sandal-wood . . .	28	10·0	11·7
16 Tulip-wood . . .	28	11·0	12·1
17 Camphor-wood . .	28	8·6	10·0
18 Olive-tree . . .	28	10·5	13·2
19 Ash	27	9·5	11·5
20 Black oak . . .	27	8·0	9·4
21 Apple-tree . . .	26	10·0	12·5
22 Iron-wood . . .	26	10·2	12·4
23 Chestnut	26	10·1	11·5
24 Sycamore	26	10·6	12·2
25 Honduras mahogany .	25	9·0	10·0
26 Brazil-wood . . .	25	11·9	13·9
27 Yew	24	11·0	12·0
28 Elm	24	10·0	11·5
29 Plane-tree . . .	24	10·0	12·0
30 Portugal laurel . .	24	10·0	11·5
31 Spanish mahogany . .	23	11·5	12·5
32 Scotch fir . . .	22	10·0	12·0

(270) The above table furnishes us with a corroboration of the result arrived at by De la Rive and De Candolle, regarding the superior conductivity of the wood, in the direction of the fibre. Evidence is also afforded, as to how little mere density affects the velocity of transmission. There appears to be neither law, nor general rule here. American birch, a comparatively light wood, possesses,

undoubtedly, a higher transmissive power than any other in the list. Iron-wood, on the contrary, with a specific gravity of 1·426, stands low. Again, oak and Coromandel-wood—the latter so hard and dense, that it is used for sharp war-instruments by savage tribes—stand near the head of the list, while Scotch fir and other light woods stand low.

(271) If we cast our eyes along the second and third columns of the table, we shall find that, in every instance, the velocity of propagation is greatest in a direction perpendicular to the ligneous layers. The law of molecular action, as regards the transmission of heat through wood, may therefore be expressed as follows:—

At all the points, not situate in the centre of the tree, wood possesses three unequal axes of calorific conduction, which are at right angles to each other. The first and principal axis is parallel to the fibre of the wood; the second and intermediate axis is perpendicular to the fibre, and to the ligneous layers; while the third and least axis is perpendicular to the fibre, and parallel to the layers.

(272) MM. De la Rive and De Candolle have remarked upon the influence which its feeble conducting power in a lateral direction must exert, in preserving within a tree the warmth which it acquires from the soil. In virtue of this property, a tree is able to resist sudden changes of temperature, which would probably be prejudicial to it: it resists alike the sudden abstraction of heat from within, and the sudden accession of it from without. But Nature has gone farther, and clothes the tree with a sheathing of worse-conducting material than the wood itself, even in its worst direction. The following are the deflections, obtained by submitting a number of cubes of bark, of the same size as the cubes of wood, to the same conditions of experiment:—

	Deflection	Corresponding deflection produced by the wood
Beech-tree bark . . .	7°	10·8°
Oak-tree bark	7	11·0
Elm-tree bark	7	11·5
Pine-tree bark	7	12·0

The direction of transmission, in these cases, was from the interior surface of the bark outwards.

(273) The average deflection, produced by a cube of wood, when the flux is lateral, may be taken at

12° ;

a cube of rock-crystal (pure silica), of the same size, produces a deflection of

90°.

(274) There are the strongest experimental grounds for believing that rock-crystal possesses a higher conductive power than some of the metals.

(275) The following numbers express the transmissive power of a few other organic structures :—

Tooth of walrus	16
Tusk of East-Indian elephant . .	17
Whalebone	9
Rhinoceros horn	9
Cow's horn	9

(276) The substances used in the construction of organic tissues are exactly such as are best calculated to resist sudden changes of temperature.

(277) The following results farther illustrate this point. Each of the substances mentioned was reduced to the cubical form, and submitted to an examination, similar in every respect to that of wood and quartz. While, however, a cube of the latter substance produces a deflection of 90°, a cube of

Sealing-wax produces a deflection of .	0°
Sole leather	0
Bees'-wax	0
Glue	0
Gutta-percha	0
India-rubber	0
Filbert-kernel	0
Almond-kernel	0
Boiled ham-muscle	0
Raw veal-muscle	0

(278) The substances here named are animal and vegetable productions; and the experiments demonstrate the extreme imperviousness of every one of them. Starting from the principle, that sudden accessions or deprivations of heat are prejudicial to animal and vegetable health, we see that the materials chosen are precisely those best calculated to avert such changes.

(279) I wish now to direct your attention to what may, at first sight, appear to you a paradoxical experiment. Here is a short prism of bismuth, and here another of iron, of the same size. The ends of both prisms are coated with white wax, and placed, with their coated surfaces upwards, on the lid of this vessel, which contains hot water. The motion of heat will propagate itself through the prisms, and you are to observe the melting of the wax. It is already beginning to yield, but on which? On the bismuth. And now the white has entirely disappeared from the bismuth, the wax overspreading it in a transparent liquid layer, while that on the iron is not yet melted. How is this result to be reconciled with the fact, stated in our table of conductivities, that, the conduction of iron being 12, the conduction of bismuth is only 2? In this experiment, the bismuth seems to be the best conductor. We solve this enigma by turning to our table of specific heats (page 138), where we find that, the specific heat of iron being 0·1138, that of bismuth is only 0·0308; to rise, therefore, a certain

number of degrees in temperature, iron requires more than three times the absolute quantity of heat required by bismuth. Thus, though the iron is really a much better conductor than the bismuth, and is at this moment accepting, in every unit of time, a much greater amount of heat than the bismuth, still, in consequence of the number of its atoms, or the magnitude of its interior work, the augmentation of temperature in its case is slow. Bismuth, on the contrary, can immediately devote a large proportion of the heat imparted to it, to the augmentation of temperature; and thus it apparently outstrips the iron, in the transmission of that motion, to which temperature is due.

(280) You see here, very plainly, the incorrectness of the statements sometimes made in books, and frequently also by candidates in our science examinations, regarding the experiment of Ingenhausz, already referred to. It is usually stated, that the greater the *quickness* with which the wax melts, the better is the conductor. If the bad conductor and the good conductor have the same specific heat, this is true; but in other cases, as proved by our last experiment, it may be entirely incorrect. The proper way of proceeding, as already indicated, is to wait until both the iron and the bismuth have attained a constant temperature—till each of them, in fact, has accepted, and is transmitting, all the motion which it can accept, or transmit, from the source of heat; when this is done, it is found that the quantity, transmitted by the iron, is many times greater than that transmitted by the bismuth. You remember our experiments with the Trevelyan instrument, and know the utility of having a highly expansible body as the bearer of the rocker. Lead is good, because it is thus expansible. But the coefficient of expansion of zinc is slightly higher than that of lead; still zinc does not answer well, as a block. The

reason is, the specific heat of zinc is more than three times that of lead, so that the heat, communicated to the zinc by the contact of the rocker, produces only about one-third the augmentation of temperature, and a correspondingly small amount of local expansion.

(281) These considerations also show that in our experiments on wood the quantity of heat transmitted by our cube in one minute's time, cannot, in strictness, be regarded as the expression of the conductivity of the wood, unless the specific heat of the various woods be the same. On this point, no experiments have been made. But, as regards the influence of molecular structure, the experiments hold good, for here we compare one direction with another, *in the same cube.* With respect to organic structures, I may add that, even allowing them time to accept all the motion, which they are capable of accepting, from a source of heat, their power of transmitting that motion is exceedingly low. They are really bad conductors.

(282) It is the imperfect conductibility of woollen textures, which renders them so eminently fit for clothing. They preserve the body from sudden accessions, and from sudden losses of heat. The same quality of non-conductibility manifests itself, when we wrap flannel round a block of ice. The ice thus preserved is not easily melted. In the case of a human body, on a cold day, the woollen clothing prevents the transmission of motion from within outwards. In the case of the ice, on a warm day, the self-same fabric prevents the transmission of motion from without inwards. Animals which inhabit cold climates are furnished by Nature with their necessary clothing. Birds, especially, need this protection, for they are still more warm-blooded than the mammalia. They are furnished with feathers, and between the feathers the interstices are filled with down, the molecular constitution and

mechanical texture of which render it, perhaps, the worst of all conductors. Here we have another example of that harmonious relation of life to the conditions of life, which is incessantly presented to the student of natural science.

(283) The indefatigable Rumford made an elaborate series of experiments, on the conductivity of the substances used in clothing.* His method was this:—A mercurial thermometer was suspended in the axis of a cylindrical glass tube, ending with a globe, in such a manner that the centre of the bulb of the thermometer occupied the centre of the globe: the space, between the internal surface of the globe and the bulb, was filled with the substance whose conductive power was to be determined; the instrument was then heated in boiling water, and afterwards plunged into a freezing mixture of pounded ice and salt, the times of cooling down 135° Fahr. being noted. They are recorded in the following table:—

Surrounded with	Seconds
Twisted silk	917
Fine lint	1032
Cotton wool	1046
Sheep's wool	1118
Taffety	1169
Raw silk	1264
Beavers' fur	1296
Eider down	1305
Hares' fur	1312
Wood ashes	927
Charcoal	937
Lampblack	1117

(284) Among the substances here examined, hares' fur offered the greatest impediment to the transmission of the heat.

(285) The transmission of heat is powerfully influenced by the mechanical state of the body through which it

* Phil. Trans. 1792, p. 48.

passes. The raw and twisted silk of Rumford's table illustrate this. Pure silica, in the state of hard rock-crystal, is a better conductor than bismuth or lead; but if the crystal be reduced to powder, the propagation of heat through that powder is exceedingly slow. Through transparent rock-salt heat is copiously conducted, through common table-salt very feebly. Here is some asbestos, a substance composed of certain silicates in a fibrous condition; I place it on my hand, and on the asbestos a red-hot iron ball: the ball can be supported without inconvenience. The asbestos intercepts the heat. That this division of the substance should interfere with the transmission might reasonably be inferred; for, heat being motion, anything which disturbs the continuity of the molecular chain, along which the motion is conveyed, must affect the transmission. In the case of the asbestos, the fibres of the silicates are separated from each other by spaces of air; to propagate itself, therefore, the motion has to pass from the solid to the air, a very light body, and again from the air to the solid, a comparatively heavy body; and it is easy to see, that the transmission of motion through this composite texture must be very imperfect. In the case of an animal's fur, this is more especially the case; for here, not only do spaces of air intervene between the hairs, but the hairs themselves, unlike the fibres of the asbestos, are very bad conductors. Lava has been known to flow over a layer of ashes, underneath which was a bed of ice, and the non-conductivity of the ashes has saved the ice from fusion. Red-hot cannon-balls may be wheeled to the gun's mouth in wooden barrows partially filled with sand. Ice is packed in sawdust, to prevent it from melting; powdered charcoal is also an eminently bad conductor. But there are cases where sawdust, chaff, or charcoal, could not be used with safety, on account of their combustible nature. In such cases, powdered gypsum may be used with advantage; in the solid crystalline state, it is an incomparably worse conductor than silica, and it may be safely inferred, that, in the powdered state, its imperviousness far transcends that of sand, each grain of which is a good conductor. A jacket of gypsum powder, round a steam boiler, would materially lessen its loss of heat.

(286) Water usually holds certain minerals in solution. In percolating through the earth, it dissolves more or less of the substances with which it comes into contact. For example, in chalk districts, the water always contains a quantity of carbonate of lime; such water is called *hard water*. Sulphate of lime is also a common ingredient of water. In evaporating, the water only is driven off, the mineral is left behind, often in quantities too great to be held in solution by the water. Many springs are strongly impregnated with carbonate of lime, and the consequence is, that when the waters of such springs reach the surface, and are exposed to the air, where they can partially evaporate, the mineral is precipitated, and forms incrustations on the surfaces of plants and stones, over which the water trickles. In boiling water, the same occurs; the minerals are precipitated, and there is scarcely a kettle in London, which is not internally coated with a mineral incrustation. This is an extremely serious difficulty, as regards steam boilers; the crust is a bad conductor, and it may become so thick, as materially to intercept the passage of heat to the water. Before you is an example of this mischief. This is a portion of a boiler belonging to a steamer, which was all but lost through the exhaustion of her coals: to bring this vessel into port, her spars, and every other piece of available wood, were burnt. On examination, this formidable incrustation was found within the boiler: it is mainly carbonate of lime, which by its non-conducting power rendered a prodigal

expenditure of fuel necessary, to generate the required quantity of steam. Doubtless, the slowness of many kettles in boiling would be found due to a similar cause.

(287) One or two instances of the action of good conductors, in preventing the local accumulation of heat, will not be out of place here. These two spheres are of the same size, and are both covered closely with white paper. One of them is copper, the other is wood. I place a spirit lamp underneath each of them. The motion of heat is, of course, communicating itself to each ball, but in one, it is quickly conducted away from the place of contact with the flame, through the entire mass of the ball; in the other, this quick conduction does not take place, the motion therefore accumulates at the point where the flame plays upon the ball; and here you have the result. On turning up the wooden ball, the white paper is seen to be charred; the other ball, so far from being charred, is *wet*, at its under surface, by the condensation of the aqueous vapour generated by the lamp. Here is a cylinder covered closely with paper; I hold its centre, thus, over the lamp, turning it so that the flame shall play all round the cylinder: you see a well-defined black mark, on one side of which the paper is charred, on the other side not. The cylinder is half brass and half wood, and this black mark shows their line of junction; where the paper covers the wood, it is charred; where it covers the brass, it is not sensibly affected.

(288) If the entire moving force of a common rifle bullet were communicated to a heavy cannon-ball, it would produce in the latter a very small amount of motion. Supposing the rifle bullet to weigh two ounces, and to have a velocity of 1,600 feet a second, the moving force of this bullet, communicated to a 100 lb. cannon-ball, would impart to the latter a velocity of only 32 feet a second. Thus with regard to a flame; its molecular motion is

very intense, but its weight is extremely small, and if communicated to a heavy body, the intensity of the motion must fall. Here, for example, is a sheet of wire gauze, with meshes wide enough to allow air to pass freely through them; and here is a jet of gas, burning brilliantly. I bring down the wire gauze upon the flame; you would imagine, that the flame could readily pass through the meshes of the gauze: but no, not a flicker gets through (fig. 65). The combustion is

FIG. 65.

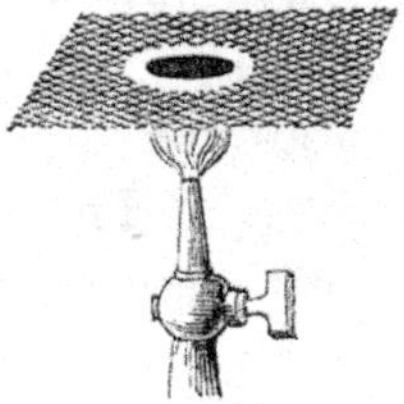

FIG. 66.

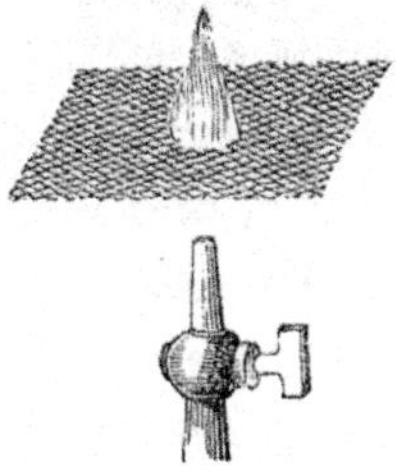

entirely confined to the space under the gauze. I extinguish the flame, and allow the unignited gas to stream from the burner. I place the wire gauze, thus, above the burner: the gas is now freely passing through the meshes. On igniting the gas above, you have the flame, but it does not propagate itself downwards to the burner (fig. 66). You see a dark space of four inches, between the burner and the gauze, a space filled with gas in a condition eminently favourable to ignition, but still it does not ignite. Thus, you see, this metallic gauze, which allows the gas to pass freely through, intercepts the flame. And why? A certain heat is necessary, to cause the gas to ignite; but by placing the wire gauze over the flame, or the flame over the wire gauze, you transfer the motion of that light and quivering thing to the comparatively heavy metal. The intensity of the molecular motion is greatly lowered: so much lowered, indeed, that it is incompetent to propagate the combustion to the opposite side of the gauze.

(289) We are all, unhappily, too well acquainted with the terrible accidents that occur, through explosions in coal mines. You know that the cause of these explosions is the presence of a certain gas—a compound of carbon and hydrogen—generated in the coal strata. When this gas is mixed with a sufficient quantity of air, it explodes on ignition, the carbon of the gas uniting with the oxygen of the air, to produce carbonic acid; the hydrogen of the gas uniting with the oxygen of the air, to produce water. By the flame of the explosion, the miners are burnt; but even should this not destroy life, they are often suffocated afterwards, by the carbonic acid produced. The original gas is the miner's 'fire-damp,' the carbonic acid is his 'choke-damp.' Sir Humphry Davy, after having assured himself of the action of wire gauze, just exhibited before you, applied it to the construction of a lamp, which should enable the miner to carry his light into an explosive atmosphere. Previous to the introduction of the *safety lamp*, the miner had to content himself with the light from sparks produced by the collision of flint and steel, for these sparks were found incompetent to ignite the fire-damp.

(290) Davy surrounded a common oil lamp by a cylinder of wire gauze (fig. 67). So long as this lamp is fed by pure air, the flame burns with the ordinary brightness of an oil-flame; but when the miner comes into an atmosphere containing 'fire-damp,' his flame enlarges, and becomes less luminous; instead of being fed by the pure oxygen of the air, it is now, in part, surrounded by inflammable gas. This enlargement of the flame he ought to take as a warning to retire. Still, though a continuous explosive atmosphere may extend from the air outside, through the meshes of the gauze, to the flame within, igni-

tion is not propagated across the gauze. The lamp may be filled with an almost lightless flame; still, explosion does not occur. A defect in the gauze, the destruction of the wire at any point by oxidation, hastened by the flame playing against it, would cause explosion. The motion of the lamp through the air might also force, mechanically, the flame through the meshes. In short, a certain amount of intelligence and caution is necessary in using the lamp. This intelligence, unhappily, is not always possessed, nor this caution always exercised, by the miner; and the consequence is, that even with the safety-lamp, explosions still occur. Before permitting a man or boy to enter a mine, would it not be well to place these results, by experiment, visibly before him? Mere advice will not enforce caution; but let the miner have the physical image of what he is to expect, clearly and vividly before his mind, and he will find it a restraining and a monitory influence, long after the effect of cautioning *words* has passed away.

Fig. 67.

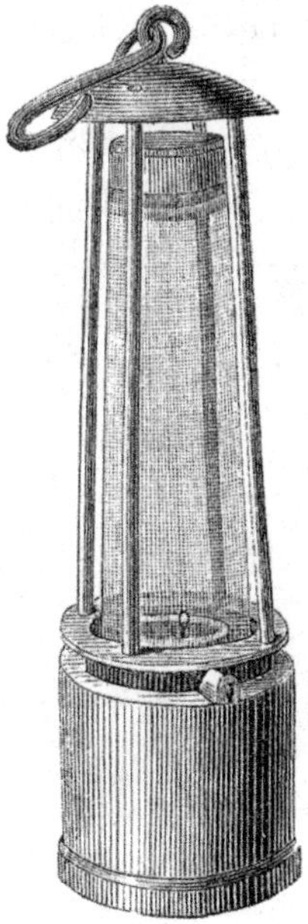

(291) A word or two, now, on the conductivity of liquids and gases. Rumford made numerous experiments on this subject, showing at once clearness of conception, and skill of execution. He supposed liquids to be non-conductors, clearly distinguishing the 'transport' of heat, by convection, from true conduction; and in order to prevent convection in his liquids, he heated them *at the top*. In this way, he found the heat of a warm iron cylinder incompetent to pass downwards, through 0·2 of an inch of olive oil; he also boiled water in a glass tube, over ice, without melting the latter substance. The later experiments of

M. Despretz apparently show that liquids possess true, though extremely feeble powers of conduction. Rumford also denied the conductivity of gases, though he was well acquainted with their convection.* The subject of gaseous conduction has been recently taken up by Professor Magnus, of Berlin, and this distinguished philosopher considers his experiments prove that hydrogen gas conducts heat like a metal.

(292) The cooling action of air by convection may be thus illustrated. On sending a voltaic current through this coil of platinum wire, it glows bright red. I now stretch out the coil, so as to form a straight wire; the glow instantly sinks —you can hardly see it. This effect is due to the freer access of the cold air to the stretched wire. Here, again, is a receiver R (fig. 68) which can be exhausted at pleasure; attached to the bottom is a vertical metal rod, *m n*, and through the top another rod, *a b*, passes, which can be moved up and down through an air-tight collar, so as to bring the ends of the two rods within any required distance of each other. At present, the rods are united by two inches of platinum wire, *b m*, which may be heated to any required degree of intensity by a voltaic current. On establishing connection with this small battery, the wire is barely luminous enough to be seen; in fact, the current from a single cell only is now sent through it. It is surrounded by air, which is carrying off a portion of its heat. When the receiver is exhausted, the wire glows more brightly than before. I allow air to re-enter—

FIG. 68.

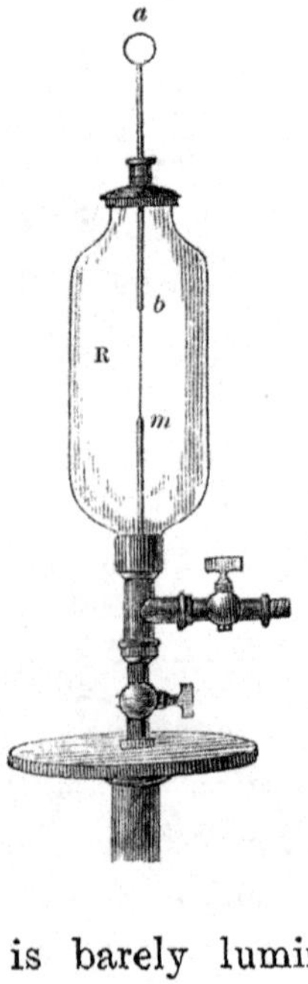

* Phil. Trans. 1792; Essays, vol. ii. p. 56.

the wire, for a time, is quite quenched, in fact, rendered perfectly black; but after the air has ceased to enter, its first feeble glow is restored. The current of air here passing over the wire, and destroying its glow, acts like the current established by the wire itself, by heating the air in contact with it. The cooling of the wire, in both cases, is due to convection, not to true conduction.

(293) The same effect is obtained, in a greatly increased degree, if hydrogen be used instead of air. We owe this interesting observation to Mr. Grove, and it formed the starting-point of M. Magnus's investigation. The receiver is now exhausted, the wire being almost white-hot. Air cannot do more than reduce that whiteness to bright redness; but observe what hydrogen can do. On the entrance of this gas, the wire is totally quenched, and even after the receiver has been filled with the gas, and the inward current has ceased, the glow of the wire is not restored. The electric current, now passing through the wire, is from two cells; I try three cells, the wire glows feebly; five cause it to glow more brightly, but even with five, it is but a bright red. Were no hydrogen there, the current now passing through the wire would infallibly fuse it. Let us see whether this is not the case. On exhausting the receiver the first few strokes of the pump produce a scarcely sensible effect; but the effect of rarefaction soon begins to be visible. The wire whitens, and appears to thicken. To those at a distance it is now as thick as a goose-quill; and now it glows, upon the point of fusion; I continue to work the pump,—the light suddenly vanishes: the wire is fused.

(294) This extraordinary cooling power of hydrogen has been usually ascribed to the mobility of its particles, which enables currents to establish themselves in this gas, with greater facility than in any other. But Professor Magnus conceives the chilling of the wire to be an effect of con-

duction. To impede, if not to prevent, the formation of currents, he passes his platinum wire along the axis of a narrow glass tube, filled with hydrogen. Although, in this case, the wire is surrounded by a mere film of the gas, and the presence of currents, in the ordinary sense, is scarcely to be assumed, the film shows itself just as competent to quench the incandescence as when the wire is caused to pass through a large vessel containing the gas. Professor Magnus also heated the closed top of a vessel, and found that the heat was conveyed more quickly from it to a thermometer, placed at some distance below the source of heat, when the vessel was filled with hydrogen, than when it was filled with air. He found this to be the case even when the vessel was loosely filled with cotton wool or eider down. Here, he contends, currents could not be formed; the heat must be conveyed to the thermometer by the true process of conduction, and not by convection.

(295) Beautiful and ingenious as these experiments are, I do not think they establish the conductivity of hydrogen. Let us suppose the wire, in Professor Magnus's first experiment, to be stretched along the axis of a wide cylinder containing hydrogen, we should have convection, in the ordinary sense, on heating the wire. Where does the heat thus dispersed ultimately go? It is manifestly given up to the sides of the cylinder, and if we narrow our cylinder, we simply hasten the transfer. The process of narrowing may continue, till a narrow tube is the result,—the convection between centre and sides will continue, and produce the same cooling effect as before. The heat of the gas being instantly lowered, by communication to the heavy tube, it is prepared to re-abstract the heat from the wire. With regard, also, to the vessel heated at the top, it would require a surface mathematically horizontal, and a perfectly uniform application of heat to that surface—it would, moreover, be necessary to cut the heat sharply off

from the sides of the vessel—to prevent convection. Even in the interstices of the eider down and of the cotton wool, the convective mobility of hydrogen will make itself felt, and, taking everything into account, I think the experimental question of gaseous conduction is still an open one.*

* In my opinion, the question of liquid conduction also demands further investigation. This opinion is founded on numerous experiments which I have myself made in connection with this question.

CHAPTER VIII.

COOLING A LOSS OF MOTION: TO WHAT IS THIS MOTION IMPARTED?—EXPERIMENTS ON SOUND BEARING ON THIS QUESTION—EXPERIMENTS ON LIGHT BEARING ON THIS QUESTION—THE THEORIES OF EMISSION AND UNDULATION—LENGTH OF WAVES AND NUMBER OF IMPULSES OF LIGHT—PHYSICAL CAUSE OF COLOUR—INVISIBLE RAYS OF THE SPECTRUM—THE CALORIFIC RAYS BEYOND THE RED—THE CHEMICAL RAYS BEYOND THE BLUE—DEFINITION OF RADIANT HEAT—REFLECTION OF RADIANT HEAT FROM PLANE AND CURVED SURFACES: LAWS THE SAME AS THOSE OF LIGHT—CONJUGATE MIRRORS.

APPENDIX:—SINGING FLAMES.

(296) WE have this day reached the boundary of one of the two great divisions of our subject. Hitherto we have dealt with heat, while associated with solid, liquid, or gaseous bodies. We have found it competent to produce changes of volume in all these bodies. We have also observed it reducing solids to liquids, and liquids to vapours; we have seen it transmitted through solids, by the process of conduction, and distributing itself through liquids and gases by the process of convection. We have now to follow it into conditions of existence, different from any which we have heretofore examined.

(297) This heated copper ball hangs in the air; you see it glow, the glow sinks, the ball becomes obscure; in popular language, the ball cools. Bearing in mind what has been said on the nature of heat, we must regard this cooling as a loss of molecular motion. But motion cannot

be lost: it must be imparted to something; to what then is the molecular motion of this ball transferred? You would, perhaps, answer, to the air; and this is partly true: over the ball air is passing, and rising in a heated column, quite visible against the screen, if we allow the electric beam to pass through the warmed air. But not the whole, not even the chief part, of the molecular motion of the ball is dissipated in this way. If the ball were placed *in vacuo*, it would still cool. Rumford, of whom we have heard so much, contrived to hang a small thermometer, *by a single fibre of silk*, in the middle of a glass globe, exhausted by means of mercury, and he found that the calorific rays passed to and fro across the vacuum; thus proving the transmission of the heat to be independent of air. Davy, with the apparatus now before you, showed that the heat rays from the electric light pass freely through an air-pump vacuum; and we can repeat his experiment substantially for ourselves. It is only necessary to take the receiver already employed (fig. 68), and, removing the remains of the platinum wire, then destroyed, attach to each end of the two rods, *m n* and *a b*, a bit of retort carbon. I now exhaust the receiver, bring the coal-points together, and send a current from point to point. The moment the points are drawn a little apart, the electric light shines forth: and here is the thermo-electric pile ready to receive a portion of the rays. The galvanometer needle at once flies aside, and this has been accomplished by rays which have crossed the vacuum.

(298) But if not to air, to what is the motion of our cooling ball communicated? We must reach by easy stages the answer to this question. Men had taken a very considerable step in science, when they first obtained a clear conception of the way in which sound is transmitted through air, and a very important experiment was made by Hauksbee before the Royal Society in 1705, when he

showed that sound could *not* propagate itself through a vacuum. Now I wish, in the first instance, to make manifest to you this conveyance of the vibrations of sound by the air. This bell is turned upside-down, and supported by a stand. When a fiddle-bow is drawn across the edge of the bell you hear its tone; the bell is now vibrating, and when sand is thrown upon its flattish bottom it arranges itself there, so as to form a definite figure; or if it were filled with water, we should have the surface fretted with beautiful crispations. These crispations would show that the bell in emitting this note, divides itself into four swinging parts, separated from each other by lines of no swinging. Here is a sheet of tracing paper, drawn tightly over a hoop, so as to form a kind of fragile drum. When held over the vibrating bell, but not so as to touch the latter, you hear the shivering of the membrane. It is a little too slack; I tighten it by warming it before the fire, and repeat the experiment. You no longer hear a shivering, but a loud musical tone, superadded to that of the bell. When the membrane is raised and lowered, or moved to and fro, you hear the rising and the sinking of the tone. Here is a smaller drum, which I pass round the bell, holding the membrane vertical; it actually bursts into a roar, when brought within half an inch of the bell. The motion of the bell, communicated to the air, has been transmitted to the membrane, and the latter is thus converted into a sonorous body.

(299) The two plates of brass, A B (fig. 69), are united together by a metal rod. The plates have been darkened by bronzing, and on both of them is strewn a quantity of fine white sand. I now take the connecting brass rod by its centre, between the finger and thumb of my left hand, and holding it upright, draw, with my right, a piece of flannel, over which a little powdered resin has been shaken, along the rod. You hear the sound; but observe the

behaviour of the sand: a single stroke has caused it to jump into a series of concentric rings, which must be quite visible to you all. Operating more gently; you hear the clear, weak musical sound, and see the sand shivering, and creeping, by degrees, to the lines which it formerly occupied. The curves now there are as sharply drawn upon the surface of the lower disc, as if they had been arranged with a camel's-hair pencil. On the upper disc, you see a series of concentric circles of the same kind. The vibrations here imparted to the rod have communicated themselves to both the discs, and divided each of them into a series of vibrating segments, separated from each other by lines of no vibration, on which lines the sand finds peace.

FIG. 69.

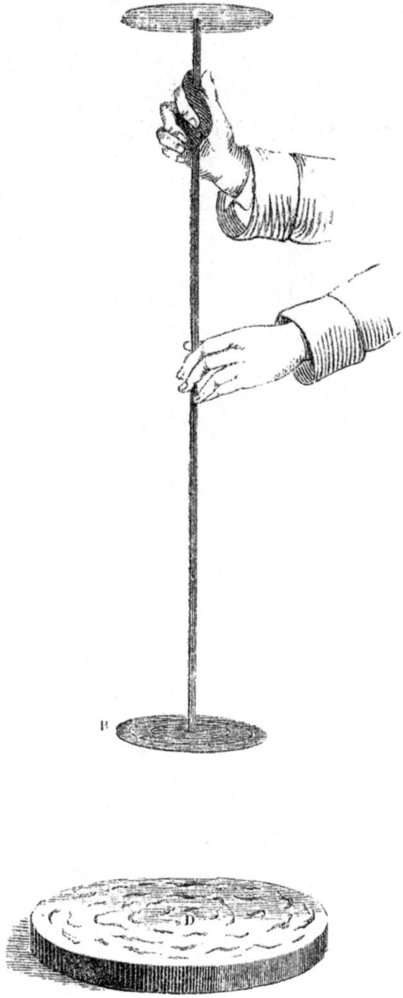

(300) Now let me show you the transmission of these vibrations, from the lower disc through the air. On the floor is a paper drum, D, with dark-coloured sand strewn uniformly over it; I might stand on the table,—or indeed as high

as the ceiling, and produce the effect which you are now to witness. Pointing the rod which unites the plates, in the direction of the paper drum, I draw the resined rubber vigorously over the rod: a single stroke has caused the sand to spring into a reticulated pattern. A precisely similar effect is produced, by sound, on the drum of the ear; the tympanic membrane is caused to shiver, in the same manner as that drum-head of paper, and its motion, conveyed to the auditory nerves, and transmitted thence to the brain, awakes in us the sensation of sound.

FIG. 70.

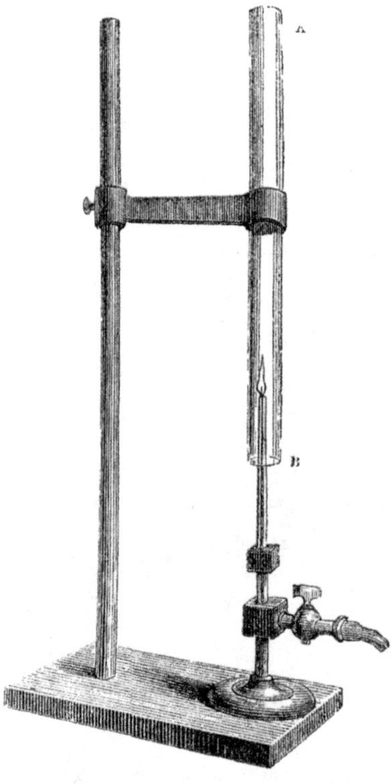

(301) Here is a still more striking example of the conveyance of the motion of sound through air. By permitting a jet of gas to issue through a small orifice, a slender flame is obtained, and by turning the cock, the flame is reduced to a height of about half an inch. The flame is then introduced into this glass tube, A B (fig. 70), which is twelve inches long. Give me your permission to address that flame. If I be skilful enough to pitch my voice to a certain note, the flame will respond by suddenly starting into a melodious song, and it will continue singing, as long as the gas continues to burn. The burner is now arranged within the tube, which covers it to a depth of a couple of inches. If the tube were lower, the flame would sing of its own accord, as in the well-known case of the

hydrogen harmonica; but, with the present arrangement, it cannot sing until ordered to do so. I emit a sound, which you will pardon, if it be not musical. The flame does not respond; it has not been spoken to in the proper language. But a note of somewhat higher pitch causes the flame to stretch, and every individual in this large audience now hears its song. I stop the sound, and stand at a greater distance from the flame: now that the proper pitch has been ascertained, the experiment is sure to succeed, and from a distance of twenty or thirty feet, the flame is caused to sing. I turn my back upon it, and strike the note as before: when called to, it answers, and, with a little practice, one is able to command a flame to sing and to stop, while it strictly obeys the injunction. Here, then, we have a striking example of the conveyance of the vibrations of the organ of voice through air, and of their communication to a body eminently sensitive to their action.*

(302) Why are these experiments on sound performed? Simply for the purpose of giving you clear conceptions regarding what takes place in the case of heat; to lead you from the tangible to the intangible; from the region of sense into that of theory.

(303) After philosophers had become aware of the manner in which sound was produced and transmitted, analogy led some of them to suppose that light might be produced and transmitted in a somewhat similar manner. And perhaps, in the whole history of science, there was never a question more hotly contested than this one. Sir Isaac Newton supposed light to consist of minute particles,

* Though not belonging to our present subject, so many persons have evinced an interest in this experiment, that I have been induced to reprint two short papers in the Appendix to this Chapter, in which the experiment is more fully described. Flames and jets of smoke are acoustic tests of astonishing sensibility. Water jets may also be rendered exceedingly sensitive. For a full account of their action, see *Tyndall on Sound*, Lecture VI. Longmans.

darted out from luminous bodies: this was the celebrated Emission Theory. Huyghens, the contemporary of Newton, found great difficulty in conceiving this cannonade of particles; or in realising that they could shoot with inconceivable velocity through space, and yet not disturb each other. This celebrated man entertained the view that light was produced by vibrations, similar to those of sound. Euler supported Huyghens, and one of his arguments, though not truly physical, is so quaint and curious, that I will repeat it here. He considers our various senses, and the manner in which they are affected by external objects. 'With regard to smell,' he says, 'we know that it is produced by material particles, which issue from a volatile body. In the case of hearing, nothing is detached from the sounding body, and in the case of feeling we must touch the body itself. The distance at which our senses perceive bodies is, in the case of touch, no distance; in the case of smell, a small distance; in the case of hearing, a considerable distance; but, in the case of sight, greatest of all. It is therefore more probable that the same mode of propagation subsists for sound and light, than that odours and light should be propagated in the same manner;—that luminous bodies should behave, not as volatile substances, but as sounding ones.'

(304) The authority of Newton bore these men down, and not until a man of genius within these walls took up the subject, had the Theory of Undulation any chance of coping with the rival Theory of Emission. To Dr. Thomas Young, formerly Professor of Natural Philosophy in this Institution, belongs the immortal honour of stemming this tide of authority, and of establishing, on a safe basis, the Theory of Undulation. Great things have been done in this edifice; but scarcely a greater thing than this. And Young was led to his conclusion regarding light, by a series of investigations on sound. He, like ourselves at the present

moment, rose from the known to the unknown, from the tangible to the intangible. This subject has been illustrated and enriched by the labours of genius ever since the time of Young; but one name only will I here associate with his,—a name which, in connection with this question, can never be forgotten: that is, the name of Augustin Fresnel.

(305) According to the theory now universally received, light consists of a vibratory motion of the particles of the luminous body; but how is this motion transmitted to our organs of sight? Sound has the air as its medium, and a close examination of the phenomena of light, by the most refined and demonstrative experiments, has led philosophers to the conclusion, that space is occupied by a substance almost infinitely elastic, through which the pulses of light make their way. Here your conceptions must be perfectly clear. The intellect knows no difference between great and small: it is just as easy, as an intellectual act, to picture a vibrating atom as to picture a vibrating cannon-ball; and there is no more difficulty in conceiving this *ether*, as it is called, which fills space, than imagining all space filled with jelly. You must, then, imagine the atoms of luminous bodies vibrating, and their vibrations you must figure as communicated to the ether in which they swing, being propagated through it in waves; these waves enter the pupil, cross the ball, and impinge upon the retina, at the back of the eye. The act, remember, is as real, and as truly mechanical, as the stroke of sea waves upon the shore. The motion of the ether is communicated to the retina, transmitted thence along the optic nerve of the brain, and there announces itself to consciousness, as light.

(306) On the screen in front of you I project an image of the incandescent coal points, which produce the electric light. The points are first brought together, and then separated. Observe the effect. You have first the place

of contact rendered luminous, then you see the glow conducted downwards, to a certain distance along the stem of coal. This, as you know, is in reality the conduction of motion. When the circuit is interrupted, the points continue to glow for a short time. Their light is now subsiding, and now they are quite dark; but have they ceased to radiate? By no means. At the present moment, there is a copious emission from these points, which, though incompetent to affect sensibly the nerves of vision, is quite competent to affect other nerves of the human system. To the eye of the philosopher, who looks at such matters without reference to sensation, these obscure radiations are precisely the same in kind, as those which produce the impression of light. You must, therefore, figure the particles of the heated body as in a state of motion; you must figure that motion as communicated to the surrounding ether, and transmitted through it with a velocity, which we have the strongest reason for believing to be the same as that of light. Thus, when you turn towards a fire on a cold day, and expose your chilled hands to its influence, the warmth which you feel is due to the impact of these ethereal billows upon your skin; they throw the nerves into motion, and the consciousness, corresponding to this motion, is what we popularly call warmth. Our task, during the lectures which remain to us, is to examine heat thus propagated through the ether. In this form it is called *Radiant Heat.*

(307) For the investigation of this subject, we possess our invaluable thermo-electric pile, the face of which is now coated with lampblack, a powerful absorber of radiant heat. I hold the instrument before my cheek; it is a radiating body, and the pile drinks in the rays. They generate electricity, and the needle of the galvanometer moves up to 90°. Withdrawing the pile from the source of heat, and allowing the needle to come to rest, I now place this

slab of ice in front of the pile. You have a deflection in the opposite direction, as if rays of cold were striking on the instrument. But in this case the pile is the hot body; it radiates its heat against the ice; the face of the pile is thus chilled, and the needle moves up to 90° on the side of cold. Our pile is, therefore, not only available for the examination of heat communicated to it by direct contact, but also for the examination of radiant heat. Let us apply it at once to a most important investigation, and examine, by means of it, the distribution of thermal power in the electric spectrum.

(308) Let me, in the first place, show you this spectrum. It is formed by sending a slice of pure white light from the

FIG. 71.

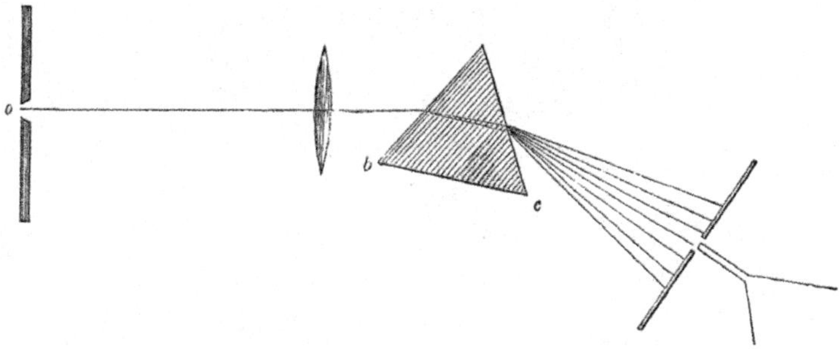

orifice, *o* (fig. 71), through a double convex lens, and through a prism, *a b c*, built up of plane glass sides, and filled with the liquid bisulphide of carbon. This liquid gives a richer display of colour than glass does, and this is one reason for its employment in preference to glass. The white beam is now reduced to its component colours—red, orange, yellow, green, and blue; the long blue space being usually subdivided into blue, indigo, and violet. I will now cause a thermo-electric pile of particular construction to pass gradually through all these colours in succession, so as to test their heating powers, and you shall observe the consequent action of the galvanometer.

(309) The experiment is made with this beautiful piece of apparatus (fig. 72), designed by Melloni, and executed, with his accustomed skill, by M. Ruhmkorff.* You observe here a polished brass plate, A B, attached to a stem; this stem is mounted on a horizontal bar, which, by means of a screw, may have motion imparted to it. By turning this ivory handle in one direction, the plate of brass is caused to approach; by turning it in the other, it is caused to recede, the motion being so fine and gradual, that we can, with ease and certainty, push the screen through a space less than $\frac{1}{2000}$th of an inch. You observe a narrow vertical slit in the middle of the plate A B, and something dark behind it. That dark line is the blackened face of a thermo-electric pile, P, the elements of which are ranged in a single row, and not in a square, as in our other instrument. We will allow distinct slices of the spectrum to fall on that slit; each will impart whatever heat it possesses to the pile, and the quantity of heat will be marked by the needle of our galvanometer.

FIG. 72.

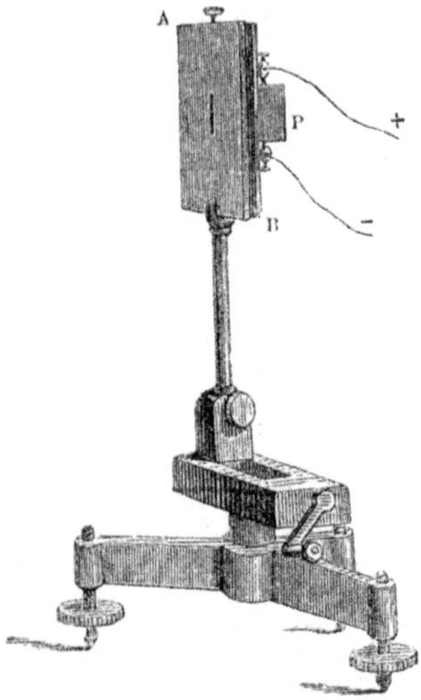

(310) At present, a small but brilliant spectrum falls upon the plate, A B, but the pile is quite out of the spectrum. On turning the handle, the slit gradually approaches the violet; the light now falls upon it, but the needle does not move sensibly. In the indigo,

* Lent to me by my friend M. Gassiot.

the needle is still quiescent; the blue also shows no action. Nor does the green: the yellow falls upon the slit; the motion of the needle is now perhaps for the first time visible to you; but the deflection is small, though the pile is exposed *to the most luminous part of the spectrum.** I pass on to the orange, which is less luminous than the yellow, but you observe, though the light diminishes, the heat increases; the needle moves still farther; while in the red, which is still less luminous than the orange, we have the greatest thermal power of the visible spectrum.

(311) The appearance, however, of this burning red might lead you to suppose it natural for such a colour to be better than any of the others. But now pay attention. I cause the pile to pass entirely out of the spectrum, quite beyond the extreme red. The needle goes promptly up to the stops. So that we have here a heat-spectrum which we cannot see, and whose thermal power is far greater than that of any part of the visible spectrum. In fact, the electric light, with which we deal, emits an infinity of rays converged by our lens, refracted by our prism, forming the prolongation of our spectrum, but utterly incompetent to excite the optic nerve. It is the same with the sun. Our orb is rich in these obscure rays; and though they are, for the most part, cut off by our atmosphere, multitudes of them still reach us. To the celebrated Sir William Herschel we are indebted for this discovery.

(312) This is sufficient for our present purpose. I propose, in a future lecture, to sift the composite emission of our electric lamp, detaching the visible from the invisible rays, and illustrating the discoveries which have been recently made in connection with the subject of obscure radiation.

* I am here dealing with a large lecture-room galvanometer.

(313) The visible spectrum, then, simply marks an interval of radiant action, in which the rays are so related to our organisation, as to excite the impression of light. Beyond this interval, *in both directions*, radiant power is exerted—obscure rays fall—those falling beyond the red being powerful to produce heat, while those falling beyond the violet are powerful to promote chemical action. These latter rays can actually be rendered visible; more strictly speaking, the undulations or waves which are now striking here beyond the violet against the screen, though they are incompetent to excite vision, may be caused to impinge upon another body, to impart their motion to it, and actually to convert the dark space beyond the violet into a brilliantly illuminated one. The means of making this experiment are at hand. The lower half of this sheet of white paper has been washed with a solution of sulphate of quinine, the upper half being left in its natural state. Holding the sheet so that the straight line dividing its prepared from its unprepared half shall be horizontal, it will cut the spectrum into two equal parts; the upper half will remain unaltered, and you will be able to compare with it the under half, on which you will find the spectrum elongated. You see the effect. We have here a splendid fluorescent band, several inches in width, where a moment ago there was nothing but darkness. When the prepared paper is removed, the light disappears; but on reintroducing it, the light flashes out again, showing you, in the most emphatic manner, that the visible limits of the ordinary spectrum by no means mark the limits of radiant action. I dip my brush in this solution of sulphate of quinine, and dab it against the paper; wherever the solution falls, light flashes forth. The existence of these extra violet rays has been long known; it was known to Thomas Young, who actually experimented on them; but to the excellent researches of Professor Stokes we are

indebted for our present complete knowledge of this subject. He rendered the rays thus visible.

(314) How then are we to picture to ourselves the rays, visible and invisible, which fill this large space upon the screen? Why are some of them visible and others not? Why are the visible ones distinguished by various colours? Is there anything that we can lay hold of, in the undulations of the ether, to which, as a physical cause, we must assign the colour? Observe first, that the entire beam of white light is drawn aside or refracted by the prism, but the violet is pulled aside more than the indigo, the indigo more than the blue, the blue more than the green, the green more than the yellow, the yellow more than the orange, and the orange more than the red. The colours are differently refrangible, and upon this depends the possibility of their separation. To every particular degree of refraction belongs a definite colour, and no other. But why should light of one degree of refrangibility produce the sensation of red, and light of another degree the sensation of green? This leads us to consider more closely the cause of these sensations.

(315) A reference to the phenomena of sound will materially help us here. Figure clearly to your minds a harp-string vibrating to and fro; it advances and causes the particles of air in front of it to crowd together, thus producing a *condensation* of the air. It retreats, and the air particles behind it separate more widely, thus producing a *rarefaction* of the air. The string again advances and produces a condensation as before, it again retreats and produces a rarefaction. In this way, the air, through which the sound of the string is propagated, is moulded into a regular sequence of condensations and rarefactions, which travel with a velocity of about 1,100 feet a second.

(316) The condensation and rarefaction constitute what is called a sonorous pulse or *wave*. The length of the

wave is measured from the centre of one condensation to the centre of the next one. Now, the quicker a string vibrates, the more quickly will these pulses follow each other, and the shorter, at the same time, will be the length of each individual wave. Upon these differences the *pitch* of a note in music depends. If a violin player wishes to produce a higher note, he shortens the string by pressing his finger on it; thereby augmenting the rapidity of vibration. If his point of pressure exactly *halves* the length of his string, he obtains the *octave* of the note which the string emits, when vibrating as a whole. 'Boys are chosen as choristers to produce the shrill notes, men to produce the bass notes; the reason being, that the boy's organ vibrates more rapidly than the man's;' so, the hum of a gnat is shriller than that of a beetle, because the smaller insect can send a greater number of impulses per second to the ear.

(317) We have now cleared our way towards the full comprehension of the physical cause of colour. This spectrum is to the eye what the musical scale is to the ear; its different colours represent notes of different pitch. The vibrations which produce the impression of red are slower, and the ethereal waves which they generate are longer, than those which produce the impression of violet, while the other colours are excited by waves of some intermediate length. The length of the waves both of sound and light, and the number of shocks which they respectively impart to the ear and eye, have been strictly determined. Let us here go through a simple calculation. Light travels through space at a velocity of 192,000 miles a second. Reducing this to inches, we find the number to be 12,165,120,000. Now it is found that 39,000 waves of red light, placed end to end, would make up an inch; multiplying the number of inches in 192,000 miles by 39,000, we obtain the number of waves of red light

embraced in a distance of 192,000 miles: this number is 474,439,680,000,000. *All these waves enter the eye in a single second.* To produce the impression of red in the brain, the retina must be hit at this almost incredible rate. To produce the impression of violet, a still greater number of impulses is necessary. It would take 57,500 waves of violet to fill an inch, and the number of shocks required to produce the impression of this colour amounts to six hundred and ninety-nine millions of millions per second. The other colours of the spectrum, as already stated, rise gradually in pitch from red to violet.

(318) But beyond the violet we have rays of too high a pitch to be visible, and beyond the red we have rays of too low a pitch to be visible. The phenomena of light are in this case also paralleled by those of sound. If it did not involve a contradiction, we might say that there are musical sounds of too high a pitch to be heard, and also sounds of too low a pitch to be heard. Speaking strictly, there are waves transmitted through the air from vibrating bodies, which, though they strike upon the ear in regular recurrence, are incompetent to excite the sensation of a musical note. Probably, sounds are heard by insects, which entirely escape our perceptions; and, indeed, as regards human beings, the self-same note may be of piercing shrillness to one person, while it is absolutely unheard by another. Both as regards light and sound, our organs of sight and hearing embrace a certain practical range, beyond which, on both sides, though the objective cause exists, our nerves cease to be influenced by it.

(319) When, therefore, I place this red-hot copper ball before you, and watch the waning of its light, you will have a perfectly clear conception of what is occurring here. The atoms of the ball oscillate, but they oscillate in a resisting medium, on which their motion is expended, and which transmits it on all sides with inconceivable

velocity. The oscillations competent to produce light are now exhausted; the ball is quite dark, still its atoms oscillate, and still their oscillations are taken up and transmitted on all sides by the ether. The ball cools as it thus loses its molecular motion, but no cooling to which it can be practically subjected can entirely deprive it of its motion. That is to say, all bodies, whatever may be their temperature, are radiating heat. From the body of every individual here present, waves are speeding away, some of which strike upon this cooling ball, and restore a portion of its lost motion. But the motion thus received by the ball is far less than that which it communicates, and the difference between them expresses the ball's loss of motion. As long as this state of things continues, the ball will continue to show an ever-lowering temperature: its temperature will sink until the quantity it emits is equal to the quantity which it receives, and at this point its temperature becomes constant. Thus, though you are conscious of no reception of heat, when you stand before a body of your own temperature, an interchange of rays is passing between you. Every superficial atom of each mass is sending forth its waves, which cross those that move in the opposite direction, every wave asserting its own individuality, amid the entanglement of its fellows. When the sum of motion received is greater than that given out, warming is the consequence; when the sum of motion given out is greater than that received, chilling takes place. This is Prevost's Theory of Exchanges, expressed in the language of the Wave Theory.

(320) Let us occupy ourselves now, for a short time, in illustrating experimentally the analogy between light and radiant heat, as regards reflection. You observed when the thermo-electric pile was placed in front of my cheek there was attached to it an open cone which was not employed in our former experiments. This cone is sil-

vered inside, and it is intended to augment the action of feeble radiations, by converging them upon the face of the thermo-electric pile. It does this by reflection. Instead of shooting wide of the pile, as many of them would do if the reflector were removed, the rays meet the silvered surface, and glance from it against the pile. The augmentation of the effect is thus shown. I place the pile at this end of the table without its reflector, and at a distance of four or five feet this copper ball, hot—but not red-hot; you observe scarcely any motion of the needle of the galvanometer. Disturbing nothing, I now attach the reflector to the pile; the needle instantly goes up to 90°, declaring the augmented action.

(321) The law of this reflection is precisely the same as that of light. Let me devote a few minutes to the illustration of this subject. Observe this apparently solid luminous cylinder, issuing horizontally from our electric lamp, and marking its track thus vividly upon the dust of our darkened room. Permitting the beam to fall upon this mirror, it is reflected, and now strikes the ceiling. The horizontal beam here is the *incident* beam, the vertical one is the *reflected* beam, and the law of light, as many of you know, is, that the angle of incidence is equal to the angle of reflection. The incident and reflected beams now enclose a right angle, and when this is the case we may be sure that both beams form, with a perpendicular to the surface of the mirror, an angle of 45°.

(322) I place the electric lamp at this corner, E, of the table (fig. 73); behind the table is fixed a looking-glass, L, and on the table, you observe, is drawn a large arc, *a b*. Attached to the mirror is a long straight lath, *m n*, so that the mirror, resting upon rollers, can be turned by the lath, which is to serve as an index. Those in front may see that the lath itself and its reflection in the mirror now form a straight line, which proves that the lath is perpendicular

to the mirror. Right and left of the central line, *m n*, the arc is divided into ten equal parts; commencing at the end E with 0°, the arc is graduated up to 20°. I now turn the lath index, so that it shall be in the line of the beam emitted by the lamp. The beam falls upon the mirror, striking it as a perpendicular, and you see that it

FIG. 73.

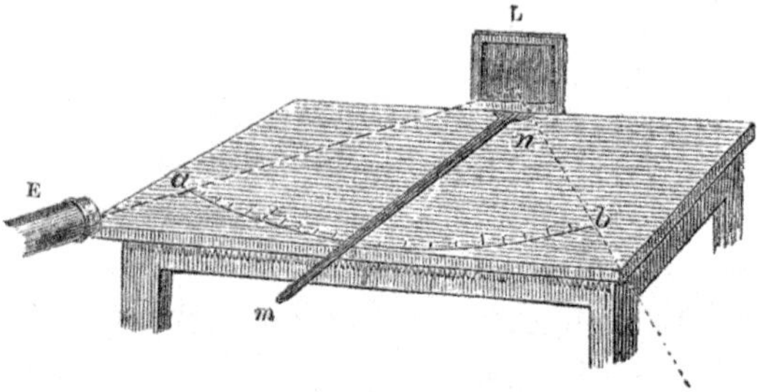

is reflected back along the line of incidence. I now move the index to 1; the reflected beam, as you observe, draws itself along the table, cutting the figure 2. I move the index to 2, the beam is now at 4; I move the index to 3, the beam is now at 6; I move it to 5, the beam is now at 10; I move it to 10, the beam is now at 20. Standing midway between the incident and reflected beams, and stretching out my arms, my finger tips touch each of them. One lies as much to the left of the perpendicular as the other does to the right. The angle of incidence is equal to the angle of reflection. But we have also demonstrated that the beam moves twice as fast as the index; and this is usually expressed in the statement, that the angular velocity of a reflected beam is twice that of the mirror which reflects it.

(323) You have already seen that these incandescent coal-points emit an abundance of obscure rays—rays of pure heat, which have no illuminating power. We have now to learn that those rays of heat emitted by the lamp, have obeyed precisely the same laws as the rays of light.

Here is a piece of black glass, so black that when you look through it at the electric light, or even at the noonday sun, you see nothing. You observe the disappearance of the beam, when the glass is placed in front of the lamp. It cuts off every ray of light; but, strange as it may appear to you, it is, to some extent, transparent to the obscure rays of the lamp. I extinguish the light by interrupting the current, and lay my thermo-electric pile on the table at the number 20, where the luminous beam fell a moment ago. The pile is connected with the galvanometer, and the needle of the instrument is now at zero. On igniting the lamp, no light makes its appearance, but the needle of the galvanometer has already swung to 90°, through the action of the non-luminous rays upon the pile. When the pile is moved right or left from its present position, the needle immediately sinks; the calorific rays have pursued the precise track of the luminous rays; and for them, also, the angle of incidence is equal to the angle of reflection. Repeating the experiments already executed with light,—bringing the index in succession to 1, 2, 3, 5, &c., it may be proved that, in the case of radiant heat also, the angular velocity of the reflected beam is twice that of the mirror.

(324) The heat of a fire obeys the same law. This sheet of tin is a homely reflector, but it will answer my purpose. At this end of the table is placed the thermo-electric pile, and at the other end the tin reflector. The needle of the galvanometer is now at zero. I turn the reflector, so as to cause the heat striking it to rebound towards the pile: it now meets the instrument, and the needle at once declares its arrival. Observe the positions of the fire, of the reflector, and of the pile; you see that they are such as make the angle of incidence equal to that of reflection.

(325) But in these experiments the heat is, or has

been, associated with light. Let me now show, that the law holds good for rays emanating from a truly obscure body. Here is a copper ball, heated to dull redness.

FIG. 74.

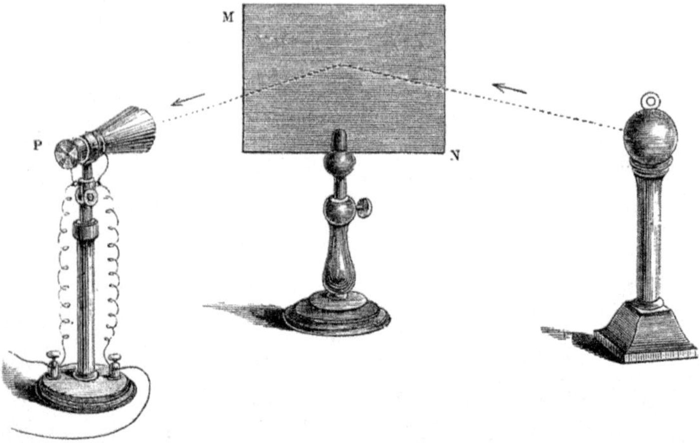

Plunging it into water for a moment, its light totally disappears, but it is still warm. It is still giving out radiant heat. I set it on a candlestick C (fig. 74), as a support, and now I place my pile, P, turning its conical reflector away from C, so that no *direct* ray from the ball in position can reach the pile. The needle remains at zero. I now introduce the tin reflector, M N, so that a line drawn to it from the ball shall make the same angle with a perpendicular to the reflector, as a line drawn from the pile. The axis of the conical reflector lies in this latter line. True to the law, the heat-rays emanating from the ball rebound from it, and strike the pile, and you observe the consequent prompt motion of the needle.

(326) Like the rays of light, the rays of heat emanating from our ball proceed in straight lines through space, diminishing in intensity, exactly as light diminishes. Thus, this ball, which when close to the pile causes the needle of the galvanometer to fly up to 90°, at a distance of 4 feet

6 inches shows scarcely a sensible action. Its rays are squandered on all sides, and comparatively few of them reach the pile. But I now introduce between the pile and the ball this tin tube, A B (fig. 75), four feet long. It is

FIG. 75.

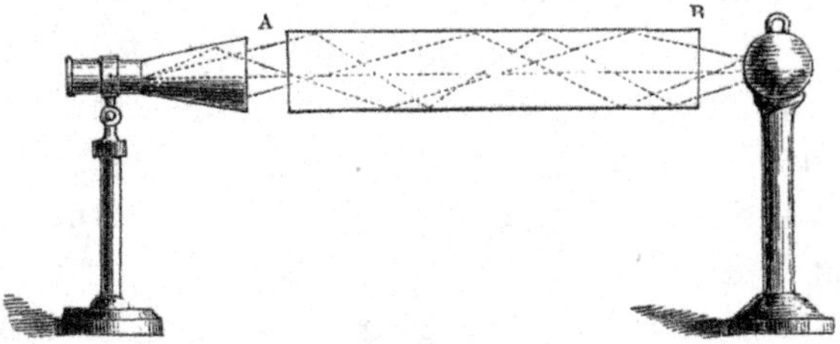

polished within, and therefore capable of reflection. The calorific rays strike the interior surface obliquely, are reflected from side to side of the tube, and thus enabled to reach the pile. You see the result: the needle, which a moment ago showed no sensible action, moves promptly to its stops.

(327) We have now dwelt sufficiently long on the reflection of radiant heat by *plane surfaces*; let us turn for

FIG. 76.

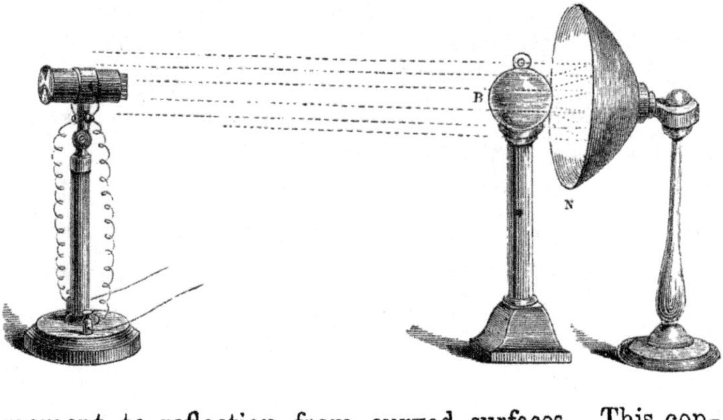

a moment to reflection from curved surfaces. This concave mirror, M N (fig. 76), is formed of copper, but is coated with silver. The warm copper ball, B, is placed at a distance of eighteen inches from the pile, whose conical reflector is now removed; you observe scarcely any motion of the needle. If the reflector, M N, were placed properly behind a candle, its rays would be collected, and sent back in a cylinder of light. The mirror thus collects and reflects the calorific rays emitted by the ball B; you cannot, of course, see the track of these obscure rays, as you can that of the luminous ones; but the galvanometer reveals the action, the needle of the instrument going promptly up to 90°.

(328) A pair of much larger mirrors is now before you, one of them being placed flat upon the table. The curvature of this mirror is so regulated, that if a light be placed at its focus, the rays which fall divergent upon the curved surface are reflected upward from it parallel. Let us make the experiment: In the focus I place our electric coal-points, bring them into contact, and then draw them a little apart; the electric light flashes against the mirror, a splendid vertical cylinder, marked by the shining dust of the room, being cast upwards by the reflector. If we reversed the experiment, and allowed a parallel beam to fall upon the mirror, the rays of that beam, after reflection, would be collected in the focus. We can actually make this experiment by introducing a second mirror, which indeed is already here suspended from the ceiling. Drawing it up to a height of 20 or 25 feet, the vertical beam, which previously fell upon the ceiling, is now received by the reflector above. In the focus of the upper mirror is hung a bit of oiled paper, to enable you to see the collection of the rays at the focus. You observe how intensely that piece of paper is now illuminated, not by the direct light from below, but by the reflected light converged upon it by the mirror above.

(329) Some of you have probably witnessed the extra-

ordinary action of light upon a mixture of hydrogen and chlorine, and I will now exhibit this action in a novel way. This transparent collodion balloon is filled with the mixed gases; I lower the upper reflector, and suspend the balloon from a hook attached to it, so that the little globe shall swing in the focus. We will now draw the mirror quite up to the ceiling (fig. 77). Placing, as before, the coal-points in the focus of the lower mirror, the moment they are drawn apart, the gases explode. And remember, this is the action of the *light*; you know that collodion is an inflammable substance, and hence might suppose it to be the *heat* of the coal-points which ignites it, communicating its combustion to the gases. But you see the flakes of the balloon

Fig. 77.

descending on the table, proving that the luminous rays went harmlessly through it, caused the gases to explode, the hydrochloric acid, formed by their combustion, having actually preserved the inflammable envelope.

(330) I again lower the mirror, and hang in its focus a second balloon, containing a mixture of oxygen and hydrogen, on which light has no sensible effect; I raise the mirror, and in the focus of the lower one place this red-hot copper ball. The calorific rays are now reflected and converged, as the luminous ones were reflected and converged in the last experiment; but they act upon the *envelope*, which has been purposely blackened to enable it to intercept the heat rays; the action is not so sudden as in the last case, but *there* is the explosion, and you now see no trace of the balloon; the inflammable substance is entirely dissipated.

(331) Let us lower the upper mirror once more, and suspend in its focus a flask of hot water. The thermo-electric pile is now placed at the focus of the lower mirror; its face being turned upwards, and exposed to the direct radiation of the warm flask, there is no sensible action produced by the direct rays. But when the face of the pile is turned downwards, if light and heat behave alike, the rays from the flask which strike the reflector will be collected at its focus. You see that this is the case; the needle, which was not sensibly affected by the direct rays, goes up to its stops. The direction of that deflection is to be noted; the red end of the needle moves towards you.

(332) In the place of the flask of hot water, I now suspend a second one containing a freezing mixture. Placing, as in the former case, the pile in the focus of the lower mirror; when turned directly towards the upper flask, there is no action. Turned downwards, the needle moves: observe the direction of the motion—the red end comes towards me.

(333) Does it not appear as if this body in the upper focus were now emitting rays of cold, which are converged by the lower mirror, like the rays of heat in our former experiment? The facts are exactly complementary, and it would seem that we have precisely the same right to infer from the experiments the existence and convergence of cold rays, as we have to infer the existence and convergence of heat rays. Many of you, no doubt, have already perceived the real state of the case. The pile is a warm body, but in the last experiment, the heat which it lost by radiation was more than made good by that received from the hot flask above. *Now* the case is reversed; the quantity which the pile radiates is in excess of the quantity which it receives, and hence the pile is chilled;—the exchanges are against it, its loss of heat is only partially compensated—and the deflection due to cold is the necessary consequence.

APPENDIX TO CHAPTER VIII.

ON THE SOUNDS PRODUCED BY THE COMBUSTION OF GASES IN TUBES.*

In the first volume of Nicholson's Journal, published in 1802, the sounds produced by the combustion of hydrogen in tubes are referred to as having been 'made in Italy;' Dr. Higgins, in the same place, shows that he had discovered them in the year 1777, while observing the water formed in a glass vessel, by the slow combustion of a slender stream of hydrogen. Chladni, in his 'Akustik,' published in 1802, page 74, speaks of their being mentioned, and incorrectly explained, by De Luc in his 'New Ideas on Meteorology:' I do not know the date of the volume. Chladni himself showed that the tones produced were the same as those of an open pipe, of the same length as the tube which encompassed the flame. He also succeeded in obtaining a tone and its octave from the same tube, and in one case obtained the fifth of the octave. In a paper published in the 'Journal de Physique' in 1802, G. De la Rive endeavoured to account for the sounds by referring them to the alternate contraction and expansion of aqueous vapour; basing his opinion upon a series of experiments of great beauty and ingenuity, made with the bulbs of thermometers. In 1818 Mr. Faraday took up the subject,† and showed that the tones were produced when the glass tube was enveloped by an atmosphere higher in temperature than 212° Fahr. That they were not due to aqueous vapour was farther shown by the fact that they could be produced by the combustion

* From the Philosophical Magazine for July 1857. By John Tyndall, F.R.S.

† Journal of Science and the Arts, vol. v. p. 274.

of carbonic oxide. He referred the sounds to successive explosions, produced by the periodic combination of the atmospheric oxygen with the issuing jet of hydrogen gas.

I am not aware that the dependence of the pitch of the note on the size of the flame has yet been noticed. To this point I will, in the first place, briefly direct attention.

A tube 25 inches long was placed over an ignited jet of hydrogen: the sound produced was the fundamental note of the tube.

A tube 12½ inches long was brought over the same flame, but no sound was obtained.

The flame was lowered, in order to make it as small as possible, and the tube last mentioned was again brought over it; it gave a clear melodious note, the octave of that obtained with the 25-inch tube.

The 25-inch tube was now brought over the same flame; it no longer gave its fundamental note, but exactly the same as that obtained from the tube of half its length.

Thus we see, that although the speed with which the explosions succeed each other depends upon the length of the tube, the flame has also a voice in the matter: that to produce a musical sound, its size must be such as to enable it to explode in unison, either with the fundamental pulses of the tube, or with the pulses of its harmonic divisions.

With a tube 6 feet 9 inches long, by varying the size of the flame, and adjusting the depth to which it reached within the tube, I have obtained a series of notes in the ratio of the numbers 1, 2, 3, 4, 5.

These experiments explain the capricious nature of the sounds sometimes obtained by lecturers upon this subject. It is, however, always possible to render the sounds clear and sweet, by suitably adjusting the size of the flame to the length of the tube.*

Since the experiments of Mr. Faraday, nothing, that I am aware of, has been added to this subject, until quite recently. In a recent number of Poggendorff's 'Annalen,' an interesting

* With a tube 14½ inches in length and an exceedingly minute jet of gas, I obtained, without altering the quantity of gas, a note and its octave: the flame possessed the power of changing its own dimensions to suit both notes.

experiment is described by M. von Schaffgotsch, and made the subject of some remarks by Prof. Poggendorff himself. A musical note was obtained with a jet of ordinary coal-gas, and it was found that when the voice was pitched to the same note, the flame assumed a lively motion, which could be augmented until the flame was actually extinguished. M. von Schaffgotsch does not describe the conditions necessary to the success of his experiment; and it was while endeavouring to find out these conditions that I alighted upon the facts which form the principal subject of this brief notice. I may remark, that M. von Schaffgotsch's result may be produced, with certainty, if the gas be caused to issue, under sufficient pressure, through a very small orifice.

In the first experiments I made use of a tapering brass burner, $10\frac{1}{2}$ inches long, having a superior orifice about $\frac{1}{20}$th of an inch in diameter. The shaking of the singing flame within the glass tube, when the voice was properly pitched, was so manifest as to be seen by several hundred people at once.

I placed a syren within a few feet of the singing flame, and gradually heightened the note produced by the instrument. As the sounds of the flame and syren approached perfect unison, the flame shook, jumping up and down within the tube. The interval between the jumps became greater, until the unison was perfect, when the motion ceased for an instant; the syren still increasing in pitch, the motion of the flame again appeared, the jumping became quicker and quicker, until finally it ceased to be discernible.

This experiment showed that the jumping of the flame, observed by M. von Schaffgotsch, is the optical expression of the *beats* which occur at each side of the perfect unison: the beats could be heard in exact accordance with the shortening and lengthening of the flame. Beyond the region of these beats, in both directions, the sound of the syren produced no visible motion of the flame. What is true of the syren is true of the voice.

While repeating and varying these experiments, I once had a silent flame within a tube, and on pitching my voice to the note of the tube, the flame, to my great surprise, instantly started into song. Placing the finger on the end of the tube, and silencing the melody, on repeating the experiment the same result was obtained.

I placed the syren near the flame, as before. The latter was burning tranquilly within its tube. Ascending gradually from the lowest notes of the instrument, at the moment when the sound of the syren reached the pitch of the tube which surrounded the gas flame, the latter suddenly stretched itself and commenced its song, which continued indefinitely after the syren had ceased to sound.

With the burner which I have described, and a glass tube 12 inches long, and from $\frac{1}{2}$ to $\frac{3}{4}$ of an inch in internal diameter, this result can be obtained with ease and certainty. If the voice be thrown a little higher or lower than the note due to the tube, no visible effect is produced upon the flame: the pitch of the voice must lie within the region of the audible beats.

By varying the length of the tube, we vary the note produced, and the voice must be modified accordingly.

That the shaking of the flame, to which I have already referred, proceeds in exact accordance with the beats, is beautifully shown by a tuning-fork which gives the same note as the flame. Loading the fork, so as to throw it slightly out of unison with the flame, when the former is sounded and brought near the flame, the jumpings are seen at exactly the same intervals as those in which the beats are heard. When the tuning-fork is brought over a resonant jar or bottle, the beats may be heard and the jumpings seen, by a thousand people at once. By changing the load upon the tuning-fork, or by slightly altering the size of the flame, the quickness with which the beats succeed each other may be changed, but in all cases, the jumpings address the eye at the same moment that the beats address the ear.

With the tuning-fork I have obtained the same results, as with the voice and syren. Holding a fork over a tube which responds to it, and which contains within it a silent flame of gas, the latter immediately starts into song. I have obtained this result with a series of tubes varying from $10\frac{1}{2}$ to 29 inches in length. The following experiment could be made:—A series of tubes, capable of producing the notes of the gamut, might be placed over suitable jets of gas; all being silent, let the gamut be run over by a musician, with an instrument sufficiently powerful, placed at a distance of twenty or thirty yards. At the sound of each particular note, the gas-jet contained in the corresponding tube would instantly start into song.

I must remark, however, that, with the jet which I have used, the experiment is most easily made with a tube about 11 or 12 inches long: with longer tubes, it is more difficult to prevent the flame from singing spontaneously, that is, without external excitation.

The principal point to be attended to is this. With a tube, say of 12 inches in length, the flame requires to occupy a certain position in the tube, in order that it shall sing with a maximum intensity. Let the tube be raised, so that the flame may penetrate it to a less extent; the energy of the sound will be thereby diminished, and a point (A) will at length be attained, where it will cease altogether. Above this point for a certain distance, the flame may be caused to burn tranquilly and silently for any length of time, but when excited by the voice it will sing.

When the flame is too near the point (A), on being excited by the voice or by a tuning-fork, it will respond for a short time, and then cease. A little above the point where this cessation occurs, the flame burns tranquilly, if unexcited; but if once caused to sing, it will continue to do so. With such a flame, which is not too sensitive to external impressions, I have been able *to reverse the effect hitherto described*, and to stop the song at pleasure by the sound of my voice, or by a tuning-fork, without quenching the flame itself. Such a flame, I find, may be made to obey the word of command, and to sing or cease to sing, as the experimenter pleases.

The mere clapping of the hands, producing an explosion, shouting at an incorrect pitch, shaking of the tube surrounding the flame, are, when the arrangements are properly made, ineffectual. Each of these modes of disturbance doubtless affects the flame, but the impulses do not accumulate, as in the case where the note of the tube itself is struck. It appears as if the flame were *deaf* to a single impulse, as the tympanum would probably be, and, like the latter, needs the accumulation of impulses to give it sufficient motion. A difference of half a tone, between two tuning-forks, is sufficient to cause one of these to set the flame singing, while the other is powerless to produce this effect.

I have said that the voice must be pitched to the note of the tube which surrounds the flame; it would be more correct to say, the note produced by the flame when singing. In all cases, this note is sensibly higher than that due to the open tube which surrounds the flame; this ought to be the case, because of the high temperature of the vibrating column. An open tube, for example, which, when a tuning-fork is held over its end, gives a maximum reinforcement, produces, when surrounding a singing flame, a note higher than that of the fork. To obtain the latter note, the tube must be sensibly longer.

What is the constitution of the flame of gas, while it produces these musical sounds? This is the next question to which I will briefly call attention. Looked at with the naked eye, the sounding flame appears constant; but is the constancy real? Supposing each pulse to be accompanied by a physical change of the flame, such a change would not be perceptible to the naked eye, on account of the velocity with which the pulses succeed each other. The light of the flame would appear continuous, on the same principle that the troubled portion of a descending liquid jet appears continuous, although, by proper means, this portion of a jet can be shown to be composed of isolated drops. If we cause the image of the flame to pass speedily over different portions of the retina, the changes accompanying the periodic impulses will manifest themselves, in the character of the image thus traced.

I took a glass tube, 3 feet 2 inches long, and about an inch and a half in internal diameter, and placing it over a very small flame of olefiant gas (common gas will also answer), obtained the fundamental note of the tube: on moving the head to and fro, the image of the sounding flame was separated into a series of distinct images; the distance between the images depended upon the velocity with which the head was moved. This experiment is suited to a darkened lecture-room. It was still easier to obtain the separation of the images in this way, when a tube 6 feet 9 inches in length, and a larger flame, were made use of.

The same result is obtained, when an opera-glass is moved to and fro before the eye.

But the most convenient mode of observing the flame is with a mirror; and it can be seen either directly in the mirror, or by projection upon a screen.

A lens of 33 centimetres focus was placed in front of a flame of common gas, upwards of an inch long, and a paper screen was hung at about 6 or 8 feet distance behind the flame. In front of the lens, a small looking-glass was held, which received the light that had passed through the lens, and reflected it back upon the screen placed behind the latter. By adjusting the position of the lens, a well-defined inverted image of the flame was obtained upon the screen. On moving the mirror, the image was displaced, and owing to the retention of the impression by the retina, when the movement was sufficiently speedy, the image described a continuous luminous track. Holding the mirror motionless, the 6-foot 9-inch tube was placed over the flame: the latter changed its shape, the moment it commenced sounding, remaining, however, well defined upon the screen. On now moving the mirror, a totally different effect was produced: instead of a continuous track of light, a series of distinct images of the sounding flame was observed. The distance of these images apart varied with the motion of the mirror; and, of course, could be made, by suitably turning the reflector, to form a ring of images. The experiment is beautiful, and in a dark room may be made visible to a large audience.

The experiment was also varied in the following manner:—A triangular prism of wood had its sides coated with rectangular pieces of looking-glass: it was suspended by a thread, with its axis vertical; torsion was imparted to the thread, and the prism, acted upon by this torsion, caused to rotate. It was so placed, that ts three faces received in succession the beam of light, sent from the flame through the lens in front of it, and threw the images upon the screen. On commencing its motion, the images were but slightly separated, but became more and more so, as the motion approached its maximum. This once past, the images drew closer together again, until they ended in a kind of luminous ripple. Allowing the acquired torsion to react, the same series of effects could be produced, the motion being in an opposite direction. In these experiments, that half of the tube which was turned towards the screen was coated with lamp-black, so as to cut off the direct light of the jet from the screen.*

* Since these experiments were made, Mr. Wheatstone has drawn my attention to the following passage, which proves that he had already made use of the rotating mirror in examining a singing flame: 'A flame of hydrogen gas burning in the open air presents a continuous circle in the mirror; but while producing a sound within a glass tube, regular intermissions of intensity are observed, which present a chain-like appearance, and indicate alternate contractions and dilatations of the flame corresponding with the sonorous vibrations of the column of air.' (Phil. Trans. 1834, p. 586.)

But what is the state of the flame, in the interval between two images? The flame of common gas, or olefiant gas, owes its brightness to the solid particles of carbon discharged into it. If we blow against a luminous gas flame, a sound is heard, a small explosion in fact, and by such a puff the light may be caused to disappear. During a windy night, the exposed gas-jets in the shops are often deprived of their light, and burn blue. In like manner, the common blowpipe jet deprives burning coal-gas of its brilliant light. I hence concluded, that the explosions, the repetition of which produces the musical sound, rendered, at the moment they occurred, the combustion so perfect as to extinguish the solid carbon particles; but I imagined that the images on the screen would, on closer examination, be found united by spaces of blue, which, owing to their dimness, were not seen by the method of projection. This in many instances was found to be the case.

I was not, however, prepared for the following result:—A flame of olefiant gas, rendered almost as small as it could be, was procured. The 3-foot 2-inch tube was placed over it; the flame, on singing, became elongated, and lost some of its light, still it was bright at its top; looked at in the moving mirror, a beaded line of great beauty was observed; in front of each bead was a little luminous star, after it, and continuous with it, a spot of rich blue light, which terminated, and left, as far as I could judge, a perfectly dark space, between it and the next following luminous star. I shall examine this farther when time permits me, but as far as I can at present judge, the flame was actually extinguished and relighted, in accordance with the sonorous pulsations.

When a silent flame, capable, however, of being excited by the voice in the manner already described, is placed within a tube, and the continuous line of light produced by it in the moving mirror is observed, I know no experiment more pretty than the resolution of this line into a string of richly luminous pearls, at the instant the voice is pitched to the proper note. This may be done at a considerable distance from the jet, and with the back turned towards it.

The change produced in the line of beads, when a tuning-fork, capable of giving beats with the flame, is brought over the tube, or over a resonant jar near it, is also extremely interesting to observe. I will not at present enter into a more minute description of these results. Sufficient, I trust, has been said, to induce experimenters to reproduce the effects for themselves; the sight of them will give more pleasure than any description of mine could possibly do.

TRANSLATION OF A PAPER ON ACOUSTIC EXPERIMENTS.*

A glass tube open at both ends, when simply blown upon by the mouth, gives its fundamental tone, i.e. the deepest tone belonging to it, as an open organ-pipe, feebly but distinctly. On placing the open hand upon one of the openings, and rapidly withdrawing it, the tube yields two notes, one after the other; first the fundamental note of the closed pipe, and then the note of the open pipe, already mentioned, which is *an octave higher*. By the application of heat, these fundamental tones, of which only the higher one will be taken into consideration here, are raised, as is well known; this is observed immediately on blowing upon a tube heated externally, or by a gas-flame burning in its interior. For example, a tube 242 millims. in length, and 20 millims. in diameter, heated throughout its whole length, when blown upon even before it reaches a red heat, gives a tone raised a major third, namely, the second G sharp in the treble clef, instead of the corresponding E. If a gas-flame 14 millims. in length, and 1 millim. in breadth at the bottom, is burning in the tube, the tone rises to the second treble F sharp. The same gas-flame raises the tone of a tube 273 millims. in length, and 21 millims. in width, from the second treble D to the corresponding E. These two tubes, which for brevity will hereafter be referred to as the E tube and the D tube, served for all the following experiments, the object of which was to show a well-known and by no means surprising fact, in a striking manner, namely, that the column of air in a tube is set in vibration when its fundamental tone, or one nearly allied—for example, an octave—is sounded outside the tube. The existence of the aërial vibrations was rendered perceptible by a column of smoke, a current of gas, and a gas flame.

* By Count Schaffgotsch: Phil. Mag., December 1857.

1. A glimmering smoky taper was placed close under the E tube, held perpendicularly, and the smoke passed through the tube, in the form of a uniform thread. At a distance of 1·5 metre from the tube, the first treble E was sung. The smoke curled, and it appeared as if a part of it would be forced out at the upper, and the other part at the lower opening of the tube.

2. Two gas-burners, 1 millim. in the aperture, were applied near each other to the same conducting tube. Common gas flowed from both of them; one projected from below into the D tube, for about one-fifth of its length; the gas flame of the other was 3 millims. in height. At a distance of 1·5 metre therefrom, the first treble D was sung; the flame increased several times in breadth and height, and, consequently, in size generally; a larger quantity of gas, therefore, flowed out of the outer burner, which can only be explained by a diminution of the stream of gas in the inner burner, that is, in the one surrounded by the glass tube.

3. A burner, with an aperture of 1 millim. projecting from below into the D tube, about 80 millims., yielded a gas flame 14 millims. in length. At 5·6 metres therefrom, the first treble E was sung: the flame was instantaneously extinguished. The same thing took place at 7 metres, when the flame was only 10 millims. in height, and the first treble D sharp was sung.

4. The last-mentioned flame is also extinguished by the note G sharp, sounded close to it. Noises, such as the clapping of hands, pushing a chair, or shutting a book, do not produce this effect.

5. A burner with an aperture of 0·5 millim., projecting from below 60 millims. into the D tube, yielded a globular gas flame 3 to 3·5 millims. in diameter. By gradually closing a stopcock, the passage of gas was more and more limited. The flame suddenly became much longer, but at the same time narrower, and nearly cylindrical, acquiring a bluish colour throughout, and from the tube a piercing second treble D was sounded; this is the phenomenon of the so-called chemical harmonica, which has been known for eighty years. When the stopcock is still farther closed, the tone becomes stronger, the flame longer, narrower, and nearly spindle-shaped; at last it disappears.

An effect, exactly similar to that caused by cutting off the gas, is produced upon the small gas flame by a D, or the first treble

D, sung or sounded from instruments; and, in this case, it is to be observed that the flame generally becomes the more sensitive, the smaller it is, and the farther the burner projects into the glass tube.

6. The flame in the D tube was 2 or 3 millims. in length; at a distance of 16·3 metres (more than 51 feet) from it, the first treble D was sounded. The flame immediately acquired the unusual form, and the second treble D sounded, and continued to sound, from the tube.

7. While the second treble D of the preceding experiment was sounding, the first treble D was sounded loudly, close to the tube, when the flame became excessively elongated, and then disappeared.

8. The flame being only 1·5 millim. in length, the first treble D was sounded. The flame gave out the second treble D (and perhaps sometimes also a higher D) only for a moment, and disappeared. The flame is also affected by various D's of an adjustable labial pipe, by the contra D, D, D, the first treble D, and the second treble D of a harmonium, but by no single C sharp or D sharp of this powerful instrument. It is also affected by the third treble D of a clarionet, although only when quite close. The sung note also acts, when produced by inspiration (in this case the second treble), or when the mouth is turned from the flame.

9. In immediate proximity, the note G sung is effective.

Some influence is exerted by noises, but not by all, and often not by the strongest and nearest, evidently because the exciting tone is not contained in them.

10. The flame burning quietly in the interior of the D tube was about 2·5 millims. in length. In the next room, the door of which was open, the four legs of a chair were stamped simultaneously upon the wooden floor. The phenomenon of the chemical harmonica immediately occurred. A very small flame is, of course, extinguished, after sounding for an instant, by the noise of a chair. A tambourine, when struck, acts sometimes, but not in general.

11. The flame burning, in the excited singing condition, in the interior of the D tube, the latter was slowly raised as high as possible, without causing the return of the flame to the ordinary

condition. The note, the first treble D, was sung strongly and *broken off suddenly*, at a distance of 1·5 metre. The harmonic tone ceased, and the flame fell into a state of repose, without being extinguished.

12. The same result was produced by acting upon the draught of air in the tube by a fanning motion of the open hand, close above the upper aperture of the tube.

13. In the D tube there were two burners close together; one of them, 0·5 millim. in aperture, opened five millims. below the other, the diameter of which was 1 millim. or more. Currents of gas, independent of each other, flowed out of both; that flowing from the narrower burner being very feeble, and burning, when ignited, with a flame about 1·5 millim. in length, nearly invisible in the day; the first treble D was sung at a distance of three metres. The strong current of gas was immediately inflamed, because the little flame situated below it, becoming elongated, flared up into it. By a stronger action of the tone, the small flame itself is extinguished, so that an actual transfer of the flame from one burner to the other takes place. Soon afterwards, the feeble current of gas is usually again inflamed by the large flame, and if the latter be again extinguished alone, everything is ready for a repetition of the experiment.

14. The same result is furnished by stamping with the chair, &c. It is evident that, in this way, gas-flames of any desired size, and any mechanical action, may be produced by musical tones and noises, if a wire stretched by a weight be passed through the glass tube, in such a way that the flaring gas-flame must burn upon it.

15. If the flame of the chemical harmonica be looked at steadfastly, and if, at the same time, the head be moved rapidly to the right and left alternately, an uninterrupted streak of light is not seen, such as is given by every other luminous body, but a series of closely approximated flames, and often dentated and undulated figures, especially when tubes of a metre, and flames of a centimetre, in length are employed.

This experiment also succeeds very easily without moving the eyes, when the flame is looked at through an opera-glass, the object-glass of which is moved rapidly to and fro, or in a circle; and also when the picture of the flame is observed in a hand-mirror shaken about. It is, however, only a variation of the experiment long since described and explained by Wheatstone, for which a mirror turned by watchwork was employed.

[It is perhaps but right that I should draw attention to the relation of the foregoing paper to one that I have published on the same subject. On May 6, and the days immediately following, the principal facts described in my paper were discovered; but on April 30, the foregoing results had been communicated by Prof. Poggendorff to the Academy of Sciences in Berlin. Through the kindness of M. von Schaffgotsch himself, I received his paper at Chamouni, many weeks after the publication of my own, and until then I was not aware of his having continued his experiments upon the subject.

We thus worked independently of each other; but, as far as the described phenomena are common to both, all the merit of priority rests with Count Schaffgotsch.—J. T.]

CHAPTER IX.

LAW OF DIMINUTION WITH THE DISTANCE—THE WAVES OF SOUND LONGITUDINAL; THOSE OF LIGHT TRANSVERSAL—WHEN THEY OSCILLATE, THE MOLECULES OF DIFFERENT BODIES COMMUNICATE DIFFERENT AMOUNTS OF MOTION TO THE ETHER—RADIATION THE COMMUNICATION OF MOTION TO THE ETHER; ABSORPTION THE ACCEPTANCE OF MOTION FROM THE ETHER—THOSE SURFACES WHICH RADIATE WELL ABSORB WELL—A CLOSE WOOLLEN COVERING FACILITATES COOLING—PRESERVATIVE INFLUENCE OF GOLD-LEAF—THE ATOMS OF BODIES INTERCEPT CERTAIN WAVES, AND ALLOW OTHERS TO PASS—TRANSPARENCY AND DIATHERMANCY—DIATHERMIC BODIES BAD RADIATORS—DEFINITION OF THE TERM 'QUALITY' AS APPLIED TO RADIANT HEAT—THE RAYS WHICH PASS WITHOUT ABSORPTION DO NOT HEAT THE MEDIUM—PROPORTION OF LUMINOUS AND OBSCURE RAYS IN VARIOUS FLAMES.

(334) THE intensity of radiant heat diminishes with the distance, in the same manner as that of light. What is the law of diminution for light? Each side of this square sheet of paper measures two feet; folded thus, it forms a smaller square, each side of which is a foot in length. The electric lamp now stands at a distance of sixteen feet from the screen; and at a distance of eight feet, that is, exactly midway between the screen and the lamp, I hold this square of paper. The lamp is naked, unsurrounded by its camera, and the rays, uninfluenced by any lens, are emitted in straight lines on all sides. You see the shadow of the square of paper on the screen; let us measure the boundary of that shadow, and then unfold the sheet of paper, so as to obtain the original large square. You see, by the creases, that it is exactly

four times the area of the smaller one. This large sheet, when placed against the screen, exactly covers the space formerly occupied by the shadow of the small square.

(335) On the small square, therefore, when it stood midway between the lamp and screen, a quantity of light fell which, when the small square is removed, is diffused over four times the area upon the screen. But if the same quantity of light is diffused over four times the area, it must be diluted to one-fourth of its original intensity. Hence, by doubling the distance from the source of light, we diminish the intensity to one-fourth. By a precisely similar mode of experiment, we could prove, that by trebling the distance, we diminish the intensity to one-ninth; and by quadrupling the distance we reduce the intensity to one-sixteenth: in short, we thus demonstrate the law that the intensity of light diminishes, as the square of the distance increases. This is the celebrated law of Inverse Squares, as applied to light.

(336) But it has been stated that heat diminishes, according to the same law. Observe the experiment which I am now about to perform before you. Here is a narrow tin vessel, M N (fig. 78), presenting a side, coated with lampblack, a square yard in area. The vessel is filled with hot water, which converts this large surface into a source of radiant heat. I now place the conical reflector on the thermo-electric pile, P, but instead of permitting it to remain a reflector, I push into the hollow cone this lining of black paper, which fits exactly, and which, instead of reflecting any heat that may fall obliquely on it, effectually cuts off the oblique radiation. The pile is now connected with the galvanometer, and its reflector is close to the radiating surface, the face of the pile itself being about six inches distant from the surface.

(337) The needle of the galvanometer moves; it now

points steadily to 60°, and there it will remain as long as the temperature of the radiating surface remains sensibly constant. I will now gradually withdraw the pile from the surface, and ask you to observe the effect upon the galvanometer. You will naturally expect that as the pile is withdrawn, the intensity of the heat will diminish, and that the deflection of the galvanometer will fall, in a corresponding degree. The pile is now at double the

FIG. 78.

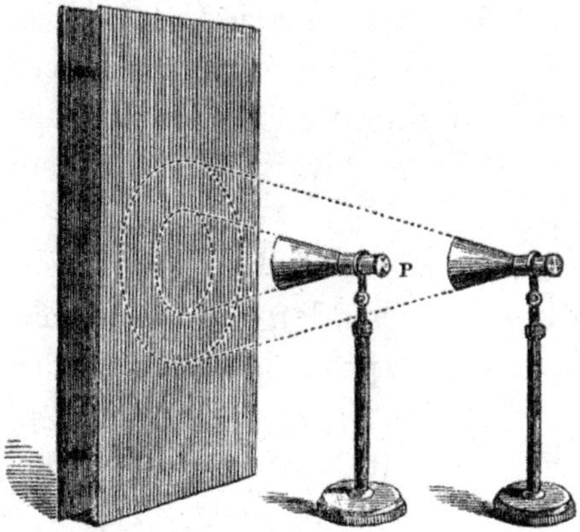

distance, but the needle does not move; at treble the distance, the needle is still stationary; we may successively quadruple, quintuple—go to ten times the distance, but the needle is rigid in its adherence to the deflection of 60°. There is, to all appearance, no diminution at all of intensity with the increase of distance.

(338) From this experiment, which might at first sight appear fatal to the law of inverse squares, as applied to heat, Melloni, in the most ingenious manner, proved the law. I will here follow his reasoning. Imagine the hollow cone in front of the pile prolonged; it would cut the radiating surface in a circle, and this circle is the only portion of that surface whose rays can reach the pile.

All the other rays are cut off by the non-reflecting lining of the cone. When the pile is moved to double the distance, the section of the cone prolonged encloses a circle exactly four times the area of the former one; at treble the distance, the radiating surface is augmented nine times; at ten times the distance, the radiating surface is augmented 100 times. Now, the constancy of the deflection proves that the augmentation of the surface must be exactly neutralised by the diminution of the intensity. But the radiating surface augments as the square of the distance, hence the intensity of the heat must diminish as the square of the distance; and thus the experiment, which might at first sight appear fatal to the law, demonstrates that law in the most simple and conclusive manner.

(338*a*) I have spoken of the dilution suffered by light when it is diffused over a large surface. This, however, is but a vague way of expressing the real fact. The diminution of intensity in the case both of light and radiant heat is, in reality, a diminution of motion. Every ether particle, as a wave passes it, makes a complete oscillation to and fro. At the two limits of its excursion it is brought momentarily to rest, midway between those limits its velocity is a maximum. Now *the intensity of the light is proportional to the square of this maximum velocity.* The range of the vibration of an ether particle is technically called its *amplitude*; and the intensity of the light is also proportional to the square of the amplitude. It can be proved that both the maximum velocity and the amplitude vary inversely as the distance from the radiant point; hence the intensity of the light and heat emitted by that point must vary inversely as the square of the distance. The problem is one of pure mechanics.

(339) Let us now revert for a moment to our fundamental conceptions regarding radiant heat. Its origin is

an oscillatory motion of the ultimate particles of matter —a motion taken up by the ether, and propagated through it in waves. The particles of ether in these waves do not oscillate in the same manner as the particles of air, in the case of sound.* The air particles move to and fro, in the direction in which the sound travels; the ether particles move to and fro, *across* the line in which the light travels. The undulations of the air are longitudinal, those of the ether transversal. The ether waves resemble more the ripples of water than they do the aërial pulses which produce sound; that this is the case has been inferred from optical phenomena. But it is manifest that the disturbance produced in the ether must depend upon the character of the oscillating molecule; one atom may be more unwieldy than another, and a single atom could not be expected to produce so great a disturbance as a group of atoms oscillating as a system. Thus, when different bodies are heated, we may fairly expect that their atoms will not all create the same amount of disturbance in the ether. It is probable that some will communicate a greater amount of motion than others: in other words, that some will radiate more copiously than others; for radiation, strictly defined, *is the communication of motion from the particles of a heated body, to the ether in which these particles are immersed, and through which the motion is propagated.*

(340) Let us now test this idea by experiment. This cubical vessel C (fig. 79) is called a 'Leslie's cube,' because vessels of this shape were used by Sir John Leslie, in his beautiful researches on radiant heat. The cube is of pewter, but one of its sides is coated with a layer of gold, another with a layer of silver, a third with a layer of copper, while the fourth is coated with a varnish of isinglass. Let

* The intensity in both cases varies in accordance with the same law. See 'Tyndall on Sound,' p. 11. Longmans.

us fill the cube with hot water, and keeping it at a constant distance from the thermo-electric pile, P, allow its four faces to radiate, in succession, against the pile. The hot gold surface, you see, produces scarcely any deflection; the hot silver is equally inoperative; the same is the case with the copper; but when this varnished surface is turned towards the pile, the gush of heat becomes suddenly so great that the needle moves up to its stops. Hence we

FIG. 79.

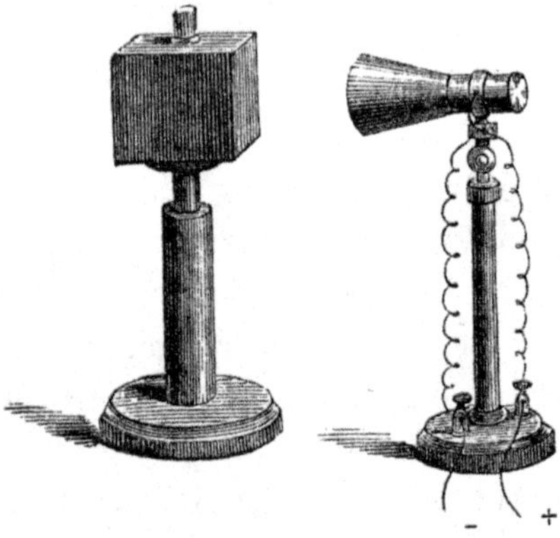

infer, that through some physical cause or other, the molecules of the varnish, when agitated by the hot water within the cube, communicate more motion to the ether than do the atoms of the metals; in other words, the varnish is a better radiator than the metals are. A similar result is obtained when this silver teapot is compared with this earthenware one; both being filled with boiling water, the silver produces but little effect, while the radiation from the earthenware is so copious, as to drive the needle up to 90°. Thus, also, if this pewter pot be compared with this glass beaker, when both are filled with hot water, the radiation from the glass proves to be much more powerful than that from the pewter.

(341) You have often heard of the effect of colours on

radiation, and heard a good deal, as shall afterwards be shown, that is unwarranted by experiment. One of the sides of this cube is coated with whiting, another with carmine, a third with lampblack, while the fourth is left uncoated. On presenting the black surface to the pile, the cube being filled with boiling water, the needle moves up, and now points steadily to 65°. The cube rests upon a little turn-table, and by turning the support, the white face is presented to the pile: the needle remains stationary, proving the radiation from the white surface to be just as copious as that from the black. When the red surface is turned towards the pile, there is no change in the position of the needle. I now turn the uncoated side; the needle instantly falls, proving the inferiority of the metallic surface as a radiator. Precisely the same experiments may be repeated with this cube, the sides of which are covered with velvet; one face with black, another with white, and a third with red. The three velvet surfaces radiate alike, while the naked surface radiates less than any of them. These experiments show that the radiation from the clothes which cover the human body is independent of their colour; that of an animal's fur being equally incompetent to influence the radiation. These are the conclusions arrived at by Melloni *for obscure heat.* We shall subsequently push the investigation of this subject much beyond the point at which Melloni left it.

(342) Now if the coated surface in the foregoing experiments communicates more motion to the ether than the uncoated one, it necessarily follows, that the coated vessel will cool more quickly than the uncoated one. Here are two cubes, one of which is covered with lampblack, while the other is bright. Three-quarters of an hour ago boiling water was poured into these vessels, a thermometer being placed in each of them. Both thermometers then showed the same temperature, but now one of them is two

degrees below the other, the vessel which has cooled most rapidly being the coated one. Here are two vessels, one of which is bright, and the other closely coated with flannel. Half an hour ago two thermometers, plunged in these vessels, showed the same temperature, but the covered vessel has now a temperature two or three degrees lower than the naked one. It is not unusual to preserve the heat of teapots by a woollen covering, but the 'cosy' must fit very loosely. A closely fitting cosy, which has the heat of the teapot freely imparted to it by contact, would, as we have seen, promote the loss which it is intended to diminish, and thus do more harm than good.

(343) One of the most interesting points connected with this subject is the reciprocity which exists between the power of a body to communicate motion to the ether, or to radiate; and its power to accept motion from the ether, or to absorb. As regards radiation, we have already compared lampblack and whiting with metallic surfaces: we will now compare the same substances, with reference to their powers of absorption. Of these two sheets of tin, M N, O P (fig. 80), one O P, is coated with whiting, and the other M N, left uncoated; I place them thus parallel to each other, and at a distance of about two feet asunder. To the edge of each sheet is soldered a screw, and from one sheet to the other is stretched a copper wire, *a, b*. At the back of each sheet is soldered one end of a little bar of bismuth, to the other end, *e*, of which a wire is attached, and terminating by a binding screw. With these two binding screws are connected the two ends of the wire, coming from the galvanometer beyond G, and you observe that we have now an unbroken circuit, in which the galvanometer is included. You know already what the bismuth bars are intended for. When the warm finger is placed on this left-hand one, a current is immediately developed, which passes from the bismuth to

Fig. 80.

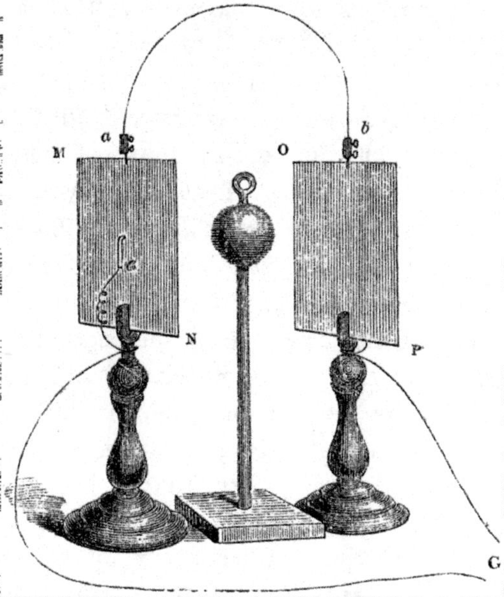

equally warmed, and one of them would neutralise the other. But if one surface be a more powerful absorber than the other, a deflection of the galvanometer needle will be the consequence, and the direction of the deflection will tell us which is the best absorber. The ball is now upon the stand, and the prompt and energetic deflection of the needle informs us that the coated surface is the most powerful absorber. In the same way I compare lampblack and varnish with tin, and find the two former to be by far the best absorbers.

(345) The thinnest metallic coating furnishes a powerful defence against the absorption of radiant heat. The back of this sheet of 'gold paper,'—the gold being merely copper reduced to great tenuity—is coated with the red iodide of mercury. This iodide, as many of you know, has its red colour discharged by heat, the powder becoming a pale yellow. I lay the paper flat on a board, with the coloured surface downwards: on its upper metallic surface are pasted pieces of paper so as to form a complicated pattern. I now pass a red-hot spatula several times over the sheet; the spatula radiates strongly against the sheet, but its rays are absorbed in very different degrees. The metallic surface absorbs but little; the paper surfaces absorb greedily; and, on turning up the sheet, you see that the iodide underneath the metallic portion is perfectly unchanged, while under every bit of paper the colour is discharged. An exact copy of the figures pasted on the opposite surface of the sheet is thus formed. For another example of the same kind, I am indebted to Mr. Hill. A fire sent its rays against this painted piece of wood (fig. 81), on which the number 338 was printed in gold-leaf letters; the paint is blistered and charred all round the letters, but underneath the letters the wood and paint are quite unaffected. This thin film of gold has been quite sufficient to prevent the absorp-

tion, to which the destruction of the surrounding surface is due.

(346) The luminiferous ether fills stellar space; it makes the universe a whole, and renders possible the intercommunication of light and energy between star and star. But the subtle substance penetrates farther; it surrounds the very atoms of solid and liquid sub-

FIG. 81.

stances. Transparent bodies are such, because the ether and the atoms of such bodies are so related to each other, that the waves which excite light can pass through them without transferring their motion to the atoms. In coloured bodies, certain waves are absorbed; but those which give the body its colour pass without absorption. Through this solution of sulphate of copper, for example, the blue waves speed unimpeded, while the red waves are destroyed. A brilliant spectrum is now formed upon the screen; when the beam is sent through this solution, the red end of the spectrum is cut away. This piece of red glass, on the contrary, owes its colour to the fact that its substance can be traversed freely by the longer undulations of red, while the shorter waves are absorbed. Placed in the path of the light, it leaves merely a vivid red band upon the screen. This blue liquid, then, cuts off the rays transmitted by the red glass; and the red glass cuts off those transmitted by the liquid; by the union of both we ought to have perfect opacity, and so we have. When both are placed in the path of the beam, the entire spectrum disappears; the union of the two partially transparent bodies producing an opacity, equal to that of pitch or metal.

(347) A solution of the permanganate of potash placed in the path of the beam permits the two ends of the spectrum to pass freely through; you have the red and the blue, but between both a space of intense blackness. The yellow of the spectrum is pitilessly destroyed by this liquid; among its atoms these yellow rays cannot pass, while the red and the blue get through the inter-atomic spaces, without sensible hindrance. And hence the gorgeous colour of this liquid. Turning the lamp round, and projecting a disc of light two feet in diameter upon the screen, I introduce this liquid. Can anything be more splendid than the colour of that disc? I turn the lamp obliquely, and introduce a prism; the violet component of that beautiful colour has slidden away from the red. You see two definite discs of these two colours which overlap in the centre, and exhibit there the tint of the composite light which passes through the liquid.

(348) Thus, as regards the waves of light, bodies exercise, as it were, an elective power, singling out certain waves for destruction, and permitting others to pass. Transparency to waves of one length does not at all imply transparency to waves of another length, and from this we might reasonably infer, that transparency to light does not necessarily imply transparency to radiant heat. This conclusion is entirely verified by experiment. This tin screen, M N (fig. 82), is pierced by an aperture, behind which is soldered a small stand S. I place this copper ball, B, heated to dull redness, on a proper stand, at one side of the screen. At the other side is placed the thermo-electric pile, P; the rays from the ball now pass

through the aperture in the screen and fall upon the pile —the needle moves, and finally comes to rest with a steady

FIG. 82.

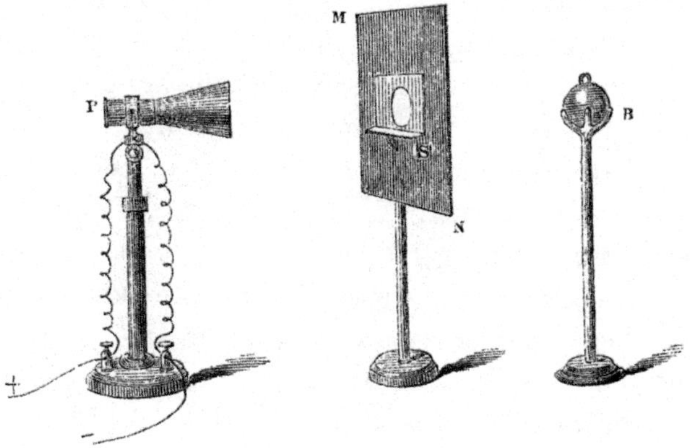

deflection of 80°. I place this glass cell, a quarter of an inch wide, filled with distilled water, on the stand S, so that all rays reaching the pile must pass through the water. What takes place? The needle steadily sinks to zero; scarcely a ray from the ball can cross the water; to the undulations issuing from the ball the water is practically opaque, though so extremely transparent to the rays of light. Before removing the cell of water, I place behind it a similar cell, containing transparent bisulphide of carbon; so that now, when the water-cell is removed, the aperture is still barred by the new liquid. What occurs? The needle promptly moves upwards, and describes a large arc; so that the selfsame rays which found the water impenetrable, find easy access through the bisulphide of carbon. In the same way, when alcohol is compared with chloride of phosphorus, we find the former almost opaque to the rays emitted by our warm ball, while the latter permits them to pass freely.

(349) So, also, as regards solid bodies. This plate of

very pure glass is now placed on the stand, between the pile and this cube of hot water. No movement of the needle is perceptible. I now displace the plate of glass by a plate of rock-salt of ten times the thickness; you see how promptly the needle moves, until arrested by its stops. To these rays, then, rock-salt is eminently transparent, while glass is practically opaque to them.

(350) For these, and numberless similar results, we are indebted to Melloni, who may be almost regarded as the creator of this branch of our subject. To express the power of transmitting radiant heat, he proposed the word *diathermancy*. Diathermancy bears the same relation to radiant heat that transparency does to light. Instead of giving you, at this state of our enquiries, determinations of my own of the diathermancy of solids and liquids, I will make a selection from the tables of the eminent Italian philosopher just referred to. In these determinations, Melloni uses four different sources of heat: the flame of a Locatelli lamp; a spiral of platinum wire, kept incandescent by the flame of an alcohol lamp; a plate of copper heated to 400° Cent., and a plate of copper heated to 100° Cent., the last-mentioned source being the surface of a copper tube, containing boiling water. The experiments were made in the following manner:—First, the radiation of the source, that is to say the galvanometric deflection produced by it, was determined, when nothing but air intervened between the source of heat and the pile. This deflection expressed the total radiation. Then the substance whose diathermancy was to be examined was introduced, and the consequent deflection noted; this deflection expressed the quantity of heat transmitted by the substance. Calling the total radiation 100, the proportionate quantities transmitted by twenty-five different substances are given in the annexed table.

Names of Substances—reduced to a common thickness of $\frac{1}{10}$th of an inch (2·6 millims.)	Transmissions : percentage of the total radiation			
	Locatelli lamp	Incandescent platinum	Copper at 400° C.	Copper at 100° C.
1 Rock-salt	92·3	92·3	92·3	92·3
2 Sicilian sulphur . . .	74	77	60	54
3 Fluor spar	72	69	42	33
4 Beryl	54	23	13	0
5 Iceland spar . . .	39	28	6	0
6 Glass	39	24	6	0
7 Rock-crystal (clear) . .	38	28	6	3
8 Smoky quartz . . .	37	28	6	3
9 Chromate of potash . .	34	28	15	0
10 White topaz . . .	33	24	4	0
11 Carbonate of lead . .	32	23	4	0
12 Sulphate of baryta . .	24	18	3	0
13 Felspar	23	19	6	0
14 Amethyst (violet) . .	21	9	2	0
15 Artificial amber . . .	21	5	0	0
16 Borate of soda . . .	18	12	8	0
17 Tourmaline (deep green) .	18	16	3	0
18 Common gum . . .	18	3	0	0
19 Selenite	14	5	0	0
20 Citric acid	11	2	0	0
21 Tartrate of potash . .	11	3	0	0
22 Natural amber . . .	11	5	0	0
23 Alum	9	2	0	0
24 Sugar-candy . . .	8	1	0	0
25 Ice	6	0·5	0	0

(351) This table shows, in the first place, what very different transmissive powers different solid bodies possess. It shows us also that, with a single exception, the diathermancy of the bodies mentioned varies with the quality of the heat. Rock-salt, only, is equally transparent to heat from the four sources. It must here be borne in mind that the luminous rays are also calorific rays; that the selfsame ray, falling upon the nerve of vision, produces the impression of light; while, impinging upon other nerves of the body, it produces the impression of heat. The luminous calorific rays have, however, a shorter wave-length than the obscure calorific rays; and knowing,

as we do, how differently waves of different lengths are absorbed by bodies, we are in a measure prepared for the results of the foregoing table. Thus, while glass of the thickness specified permits 39 per cent. of the rays of Locatelli's lamp, and 24 per cent. of the rays from the incandescent platinum to pass, it transmits only 6 per cent. of the rays from a source of 400° C., while it is absolutely opaque to all rays emitted from a source of 100° C. We also see that limpid ice, so highly transparent to light, transmits only 6 per cent. of the rays of the lamp, and 0·5 per cent. of the rays of the incandescent platinum, while it cuts off all rays issuing from the other two sources. We have here an intimation, that by far the greater portion of the rays emitted by the lamp of Locatelli must be obscure. Luminous rays pass through ice, of the thickness here given, without sensible absorption, and the fact that 94 per cent. of the rays issuing from Locatelli's flame are destroyed by the ice, proves that this proportion of these rays has no light-giving power. As regards the influence of transparency, clear and smoky quartz are very instructive. Here are the two substances, one perfectly pellucid, the other a dark brown; still, for the luminous rays only do these two specimens show a difference of transmission. The clear quartz transmits 38 per cent., and the smoky quartz 37 per cent. of the rays from the lamp, while, for the other three sources, the transmissions of both substances are identical.

(352) Melloni supposed rock-salt to be perfectly transparent to all kinds of calorific rays, the 7·7 per cent. less than a hundred which the foregoing table exhibits, being due, not to absorption but to reflection at the two surfaces of the plate of salt. But the accurate experiments of MM. de la Provostaye and Desains prove that this substance is permeable in different degrees to heat of dif-

ferent kinds; while Mr. Balfour Stewart has established the important fact, that rock-salt is particularly opaque to rays issuing from a heated piece of the same substance. We shall return to this important subject.

(353) In the following table, which is also borrowed from Melloni, the calorific transmissions of nineteen different

	Names of Liquids; thickness, 0·36 in.	Transmission: percentage of total radiation
1	Bisulphide of carbon	63
2	Bichloride of sulphur	63
3	Protochloride of phosphorus	62
4	Essence of turpentine	31
5	Olive oil	30
6	Naphtha	28
7	Essence of lavender	26
8	Sulphuric ether	21
9	Sulphuric acid	17
10	Hydrate of ammonia	15
11	Nitric acid	15
12	Absolute alcohol	15
13	Hydrate of potash	13
14	Acetic acid	12
15	Pyroligneous acid	12
16	Concentrated solution of sugar	12
17	Solution of rock-salt	12
18	White of egg	11
19	Distilled water	11

liquids are given. The source of heat was an Argand lamp, furnished with a glass chimney, and the liquids were enclosed in a cell with glass sides, the thickness of the liquid layer being 9·21 millimetres, or 0·36 of an inch. Liquids are here shown to be as diverse in their powers of transmission as solids; and it is also worthy of remark, that water maintains its position as regards opacity, notwithstanding the change in its state of aggregation.

(354) The reciprocity, which we have already demonstrated between radiation and absorption, in the case of metals, varnishes, &c., may now be extended to the bodies contained in Melloni's tables. One or two illustrations,

borrowed from an extremely suggestive memoir by Mr. Balfour Stewart, will be sufficient. In this copper vessel water is kept in a state of gentle ebullition. On the flat copper lid of the vessel are laid plates of glass and of rock-salt, until they assume the temperature of the lid. When the plate of heated rock-salt is fixed upon a stand, in front of the thermo-electric pile, the deflection produced is so small as to be scarcely sensible. Removing the rock-salt, I put in its place a plate of heated glass; the needle moves through a large arc, thus conclusively showing that the glass, which is the more powerful absorber of obscure heat, is also the more powerful radiator. Alum, unfortunately, melts at a temperature lower than that here made use of; but though its temperature is not so high as that of the glass, you can see that it transcends the glass as a radiator; the action on the galvanometer is still more energetic than in the last experiment.

(355) Absorption takes place *within* the absorbing body; a certain thickness being requisite to effect the absorption. This is true of both light and radiant heat. A very thin stratum of pale ale is almost as colourless as a stratum of water, the absorption being too inconsiderable to produce the decided tint which larger masses of the ale exhibit. When distilled water is poured into a drinking glass, it exhibits no trace of colour; but an experiment is

FIG. 83.

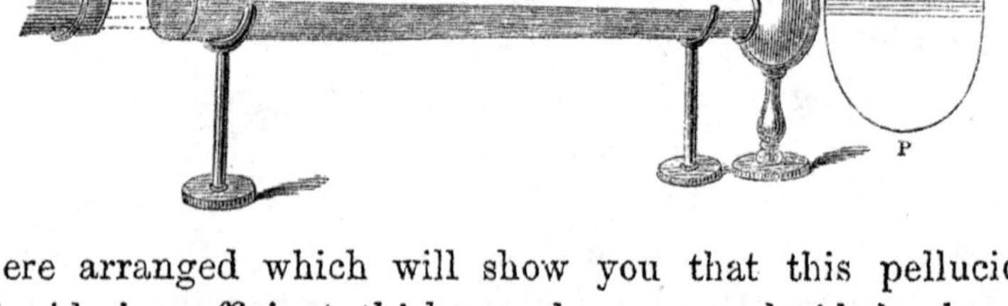

here arranged which will show you that this pellucid liquid, in sufficient thickness, has a very decided colour. This tube A B (fig. 83), fifteen feet long, is placed horizontally, its ends being stopped by pieces of plate glass. At one end of the tube stands an electric lamp, L, from which a cylinder of light will be sent through the tube. It is now half filled with water, the upper surface of which cuts the tube in two equal parts horizontally. Thus, half of the beam is sent through air, and half through water, and with this lens, C, a magnified image of the adjacent end of the tube is projected on the screen. You now see the image, O P, composed of two semicircles, one of which is formed by the light which has passed through the water, the other the light which has passed through the air. Placed thus, side by side, you can compare them, and you notice that while the air semicircle is a pure white, the water semicircle is a bright and delicate blue-green. Thus, by augmenting the thickness through which the light has to pass, we deepen the colour; this proves that the destruction of the light rays takes place *within* the absorbing body, and that it is not an effect of surface merely.

(356) Melloni shows the same to be true of radiant heat. In his experiments, already recorded at page 281, the thickness of the plates used was 2·6 millimetres, but by rendering the plate thinner, we enable a greater quantity of heat to get through it, and by rendering a very opaque substance sufficiently thin, we may almost reach the transmission of rock-salt. The following table shows the influence of thickness on the transmissive power of a plate of glass.

Thickness of Plates in millimetres	Transmissions by Glass of different thicknesses : percentage of the total Radiation			
	Locatelli Lamp	Incandescent Platinum	Copper at 400° C.	Copper at 100° C.
2·6	39	24	6	0
0·5	54	37	12	1
0·07	77	57	34	12

(357) Thus, we see that by diminishing the thickness of the plate from 2·6 to 0·07 millimetres, the quantity of heat transmitted rises, in the case of the lamp of Locatelli, from 39 to 77 per cent.; in the case of the incandescent platinum, from 24 to 57 per cent.; in the case of copper at 400° C., from 6 to 34 per cent.; and in the case of copper at 100° C., from absolute opacity to a transmission of 12 per cent.

(358) The influence of the thickness of a plate of selenite on the quantity of heat which it transmits, is exhibited in the following table:—

Thickness of Plates in millimetres	Transmissions by Selenite of different thicknesses: percentage of the total Radiation			
	Locatelli Lamp	Incandescent Platinum	Copper at 400° C.	Copper at 100° C.
2·6	14	5	0	0
0·4	38	18	7	0
0·01	64	51	32	21

These experiments prove conclusively that the absorption of heat takes place within the body, and is not a surface action.

(359) The decomposition of the solar beam produces the solar spectrum; luminous in the centre, calorific at one end, and chemical at the other. The sun is, therefore, a source of heterogeneous rays, and there can scarcely be a doubt that most other sources of heat, luminous and obscure, partake of this heterogeneity. In general, when such mixed rays enter a diathermic substance, some are intercepted, others permitted to pass. Supposing, then, that we take a sheaf of calorific rays, which have already passed through a diathermic plate, and permit them to fall upon a second plate of the same material, the transparency of this second plate to the heat incident upon it, must be greater than

the transparency of the first plate to the heat incident on *it*. The first plate, if sufficiently thick, has already extinguished, in great part, the rays which the substance is capable of absorbing; and the residual rays, as a matter of course, pass freely through a second plate of the same substance. The original beam is *sifted* by the first plate, and the purified beam possesses, for the same substance, a higher penetrative power than the original beam.

(360) This power of penetration has usually been taken as a test of the *quality* of heat; the heat of the purified beam is said to be different in quality from that of the unpurified beam. It is not, however, that any individual ray or wave has changed its character, but that from the beam, as a whole, certain components have been withdrawn; and that this withdrawal has altered the proportion of the incident heat transmitted by a second substance. This is the true meaning of the term 'quality,' as applied to radiant heat. In the path of the beam from a lamp, let plates of rock-salt, alum, bichromate of potash, and selenite be successively placed, each plate 2·6 millimetres in thickness: let the heat emergent from these respective plates fall upon a second series of the same thickness; out of every hundred units of this heat, the following proportions are transmitted:—

Rock-salt . . .	92·3
Alum	90
Chromate of potash . .	71
Selenite	91

(361) Referring to the table, p. 281, we find, that of the whole heat emitted by the Locatelli lamp, only 34 per cent. is transmitted by the chromate of potash; here we find the percentage 71. Of the entire radiation, selenite transmits only 14 per cent., but of the beam which has been purified by a plate of its own substance, it transmits 91 per cent. The same remark applies to the alum,

which transmits only 9 per cent. of the unpurified beam, and 90 per cent. of the purified beam. In rock-salt, on the contrary, the transmissions of the sifted and unsifted beam are the same, because the substance is sensibly equally transparent to rays of all the qualities here employed. In these cases, I have supposed the beam emergent from rock-salt to pass through rock-salt; the beam emergent from alum to pass through alum, and so of the others; but, as might be expected, the sifting of the beam by any substance, will alter the proportion in which it will be transmitted by almost any other second substance.

(362) I will conclude these observations with an experiment, which will show you the influence of sifting, in a very striking manner. Here is a sensitive differential air-thermometer with a clean glass bulb. The slightest touch of my hand causes a depression of the thermometric column. Let us now converge the powerful beam of our electric lamp on the bulb of that thermometer. The focus falls directly on the bulb, and the air within it is traversed by a beam of intense power; but not the slightest depression of the thermometric column is discernible. When this experiment was first shown to a person here present, he almost doubted the evidence of his senses; but the explanation is simple. The beam, before it reaches the bulb, is already sifted by the glass lens used to concentrate it; and having passed through 12 or 14 feet of air, it contains no constituent, which can be sensibly absorbed by the air within the bulb. Hence, the hot beam passes through both air and glass, without warming either. It is competent, however, to warm the thermo-electric pile, whose exposure to it, for a single instant, suffices to drive the needle violently aside. I now coat with lampblack the portion of the glass bulb struck by the beam; you see the effect: the heat is now absorbed, the air expands, and the thermometric column is forcibly depressed.

(363) We use glass fire-screens, which allow the pleasant light of the fire to pass, while they cut off the heat; the reason is, that by far the greater part of the heat emitted by a fire is obscure, and to this the glass is opaque. But in no case is there any loss. The heat absorbed by the glass warms it; the motion of the ethereal waves is here transferred to the molecules of the solid body.

FIG. 84.

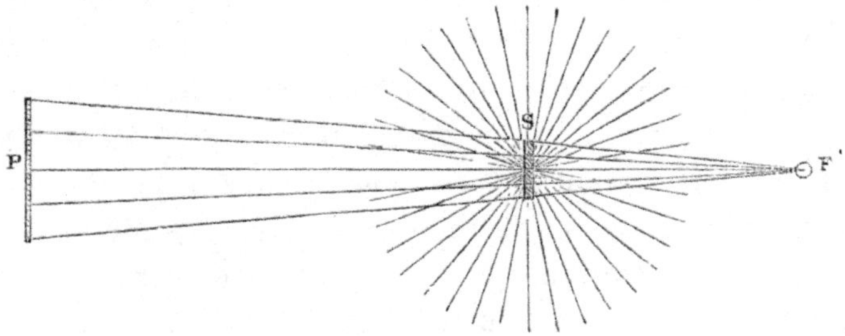

But you may be inclined to urge, that, under these circumstances, the glass itself ought to become a source of heat, and that, therefore, we ought to derive no benefit from the absorption. The fact is so, but the conclusion is unwarranted. The philosophy of the screen is this:—Let F (fig. 84) be a point of a fire, from which the rays proceed in straight lines, towards a person at P. Before the screen is introduced, each ray pursues its course direct to P; but now let a screen be placed at S. The screen intercepts the heat, and becomes warmed; but instead of sending on the rays in their original direction only, it, as a warm body, emits them *in all directions*. Hence, it cannot transmit to the person at P all the heat intercepted. A portion of the heat is restored, but by far the greater part is diverted from P, and distributed in other directions.

(364) Where the waves pursue their way unabsorbed, no motion of heat is imparted, as we have seen in the

case of the air thermometer. A joint of meat might be roasted before a fire, the air around the joint being cold as ice. The air on high mountains may be intensely cold, while a burning sun is over head; the solar rays which, striking on the human skin, are almost intolerable, are incompetent to heat the air sensibly, and we have only to withdraw into perfect shade, to feel the chill of the atmosphere. I never, on any occasion, suffered so much from solar heat, as in descending from the 'Corridor' to the Grand Plateau of Mont Blanc, on August 13, 1857; though my companion and myself were at the time hip deep in snow, the sun blazed against us with unendurable power. Immersion in the shadow of the Dôme du Goûté at once changed my feelings; for here the air was at a freezing temperature. It was not, however, sensibly colder than the air through which the sunbeams passed; and we suffered, not from the contact of hot air, but from radiant heat, which had reached us through an icy cold medium.

(365) The beams of the sun penetrate glass without sensibly heating it; the reason is, that having passed through our atmosphere, the heat has been in a great measure deprived of those constituents liable to be absorbed by glass. An experiment was made on a former occasion, which you will now completely understand. A beam was sent from the electric lamp through a plate of ice, without melting it. The beam had been previously sifted by sending it through a vessel of water, in which the heat, capable of being absorbed by the ice, was lodged, and lodged so copiously, that the water was raised almost to the boiling point during the experiment. It is here worthy of remark, that the liquid water and the solid ice appear to be pervious and impervious to the same rays; the one may be used as a *sieve* for the other: a result which indicates that the quality of the absorption is not influenced, in this case, by the difference of aggregation. It is easy

to prove, that the beam which has traversed ice without melting it, is really a calorific beam. Allowing it to fall upon our thermo-electric pile, it causes the needle to move with energy to its stops.

(366) When the calorific waves are intercepted, they, as a general rule, raise the temperature of the body by which they are absorbed; but when the absorbing body is ice, at a temperature of 32° Fahr., it is impossible to raise its temperature. How, then, does the heat absorbed by the ice employ itself? It produces internal liquefaction, it takes down the crystalline atoms, and thus forms those lovely liquid flowers shown to you on a former occasion.

(367) We have seen that transparency is not at all a test of diathermancy; that a body, highly transparent to the luminous undulations, may be highly opaque to the non-luminous ones. A body may, as we have seen, be absolutely opaque to light, and still, in a considerable degree, transparent to heat. Here is another example of the same kind. The convergent beam of the electric lamp now marks its course through the dust of the room: you see the point of convergence of the beam, at a distance of fifteen feet from the lamp. Let us mark that point accurately. This plate of rock-salt is coated so thickly with soot that the light, not only of every gas lamp in this room, but the electric light itself, is cut off by it. When this plate of smoked salt is placed in the path of the beam, the light is intercepted, but the mark enables me to find the place where the focus fell. I place the pile at this focus: you see no beam falling on it, but the violent action of the needle instantly reveals to the mind's eye a focus of heat, at the point from which the light has been withdrawn.

(368) You might, perhaps, be disposed to think that the heat falling on the pile has been first absorbed by the soot, and then radiated from it, as from an independent source.

Melloni has removed every objection of this kind; but none of his experiments, I think, are more conclusive, as a refutation of the objection, than that now performed before you. For if the smoked salt were the source, the rays could not converge here to a focus, the salt being *at this side* of the converging lens. You also see, that when the pile is displaced a little, laterally, but still turned towards the smoked salt, the needle sinks to zero. The heat, moreover, falling on the pile, is, as shown by Melloni, practically independent of the position of the plate of rock-salt; you may cut off the beam, at a distance of fifteen feet from the pile, or a distance of one foot: the result is sensibly the same, which could not be the case, if the smoked salt itself were the source of heat.

(369) When the experiment is repeated with this black glass, the result, as you see, is the same. Now, the glass *reflects* a considerable portion of the light and heat, from the lamp; when held a little obliquely to the beam you can see the reflected portion. While the glass is in this position, I will coat it with an opaque layer of lamp-black, thereby causing it to absorb, not only all the luminous rays which are now entering it, but also the portion which it formerly reflected. What is the result? Though the glass plate has become the seat of augmented absorption, it has ceased to affect the pile, and the needle descends to zero, thus furnishing additional proof that the heat which, in the first place, acted upon the pile, came direct from the lamp, and traversed the black glass, as light traverses a transparent substance.

(370) Rock-salt transmits all rays, luminous and obscure; alum, of the thickness already given, transmits, according to Melloni, only the luminous rays;* hence, the difference between alum and rock-salt will give the value of the

* We shall subsequently learn that this is an error.

obscure radiation. Tested in this way, Melloni finds the following proportions of luminous to obscure rays, for the three sources mentioned:—

Source	Luminous	Obscure
Flame of oil	10	90
Incandescent platinum . .	2	98
Flame of alcohol . . .	1	99

Thus, of the heat radiated from the flame of oil, 90 per cent. is due to obscure rays; of the heat radiated from incandescent platinum, 98 per cent. is due to obscure rays; while of the heat radiated from the flame of alcohol, fully 99 per cent. is due to obscure radiation.

CHAPTER X.

ABSORPTION OF HEAT BY GASEOUS MATTER—APPARATUS EMPLOYED—EARLY DIFFICULTIES—DIATHERMANCY OF AIR AND OF THE TRANSPARENT ELEMENTARY GASES—ATHERMANCY OF OLEFIANT GAS AND OF OTHER COMPOUND GASES—ABSORPTION OF RADIANT HEAT BY VAPOURS—RADIATION OF HEAT BY GASES—RECIPROCITY OF RADIATION AND ABSORPTION—INFLUENCE OF MOLECULAR CONSTITUTION ON THE PASSAGE OF RADIANT HEAT—TRANSMISSION OF HEAT THROUGH OPAQUE BODIES—HEAT-SPECTRUM DETACHED FROM LIGHT-SPECTRUM BY AN OPAQUE PRISM.

(371) WE have now examined the diathermancy, or transparency to heat, of solid and liquid bodies; and learned, that closely as the atoms of such bodies are packed together, the interstitial spaces between the atoms afford free play and passage to the ethereal undulations, which are in many cases transmitted without sensible hindrance among the atoms. In other cases, however, we found that the molecules stopped the waves of heat which impinged upon them; but that in so doing, they themselves became centres of motion. Thus we learned, that while perfectly diathermic bodies allowed the heat undulations to pass through them, without suffering any change of temperature, those bodies which stopped the calorific flux became heated by the absorption. Through ice itself we sent a powerful calorific beam; but because the beam was of such a quality as not to be intercepted by the ice, it passed through this highly sensitive substance without melting it. We have now to deal with *gaseous* bodies; and here the interatomic spaces are so vastly augmented, the molecules are so completely released from

all mutual entanglement, that we should be almost justified in concluding that gases and vapours furnish a perfectly open door for the passage of the calorific waves. This, indeed, until quite recently, was the universal belief; and the conclusion was verified by such experiments as had been made on atmospheric air, which was found to give no evidence of absorption.

(372) But each succeeding year augments our experimental power; the invention of improved methods enabling us to renew our enquiries with increased chances of success. Let us, then, test once more the diathermancy of atmospheric air. We may make a preliminary essay in the following way: We have here a hollow tin cylinder, A B (fig. 85), 4 feet long, and nearly 3 inches in diameter, through which we may send our calorific beam. We must, however, be able to compare the passage of the heat through the air with its passage through a vacuum, and hence we must have some means of stopping the ends of our cylinder, so as to be able to exhaust it. Here we encounter our first experimental difficulty. As a general rule, obscure heat is more greedily absorbed than luminous heat, and as our object is to make the absorption of a highly diathermic body sensible, we are most likely to effect this object by employing the radiation from an obscure source.

(373) Our tube, therefore, must be stopped by a substance which permits of the free passage of such heat. Shall we use glass for the purpose? An inspection of the table at page 281 shows us, that for such heat plates of glass would be perfectly opaque; we might as well stop our tubes with plates of metal. Observe here how one investigator's results are turned to account by another;—how science grows by the continual degradation of ends to means. Had not Melloni discovered the diathermic properties of rock-salt, we should now be utterly at a loss.

For a time, however, the difficulty of obtaining plates of salt sufficiently large and pure to stop the ends of my tube was extreme. But a scientific worker, if his wants are made known, does not long lack help; and, thanks to such friendly aid, I have here plates of this precious substance,

FIG. 85.

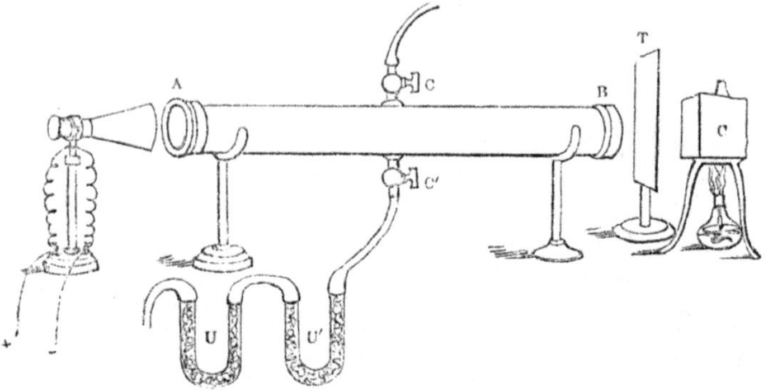

which, by means of the caps A and B, can be screwed air-tight on to the ends of my cylinder.* You observe two stopcocks attached to the cylinder. One of them, C, is

* At a time when I was greatly in need of a supply of rock-salt, I stated my wants in the 'Philosophical Magazine,' and met with an immediate response from Sir John Herschel. He sent me a block of salt, accompanied by a note, from which, as it refers to the purpose for which the salt was originally designed, I will make an extract. I am also greatly indebted to Dr. Szabo, the Hungarian Commissioner to the International Exhibition of 1862, by whom I have been raised to comparative opulence, as regards the possession of rock-salt. To the Messrs. Fletcher of Northwich, and to Mr. Corbett of Bromsgrove, my best thanks are also due for their ready kindness.

To these acknowledgments I have now to add my respectful thanks to the government of Würtemburg, for the splendid block of salt placed in their department in the late Paris Exhibition.

Here follows the extract from Sir J. Herschel's note:—'After the publication of my paper in the Phil. Trans. 1840, I was very desirous to disengage myself from the influence of glass prisms and lenses, and ascertain, if possible, whether in reality my insulated heat spots β γ δ ϵ in the spectrum were of solar or terrestrial origin. Rock-salt was the obvious resource, and after many and fruitless endeavours to obtain sufficiently large and pure specimens, the late Dr. Somerville was so good as to send me (as I understood from a friend in Cheshire) the very fine block which I now forward.

connected with an air-pump, by which the tube can be exhausted; while through the other one, C′, air, or any other gas, can be allowed to enter the tube.

(374) At one end of the cylinder is placed a Leslie's cube C, containing boiling water, and coated with lamp-black, to augment its power of radiation. At the other end of the cylinder stands our thermo-electric pile, from which wires lead to the galvanometer. Between the end B of the cylinder and the cube C, is introduced a tin screen, T, which, when withdrawn, will allow the calorific rays to pass from C through the tube to the pile. We first exhaust the cylinder, then draw the screen a little aside, and now the rays are traversing a vacuum and falling upon the pile. The tin screen, you observe, is only partially withdrawn, and the steady deflection, produced by the heat at present transmitted, is 30 degrees.

(375) Let us now admit dry air; we can do so by means of the cock C′, from which a piece of flexible tubing leads to the bent tubes U, U′, the use of which shall be now explained. The tube U is filled with fragments of pumice stone, moistened with a solution of caustic potash; it is employed to withdraw whatever carbonic acid may be contained in the air. The tube U′ is filled with fragments of pumice stone, moistened with sulphuric acid; it is intended to absorb

It is, however, much cracked, but I have no doubt pieces large enough for lenses and prisms (especially if cemented together) might be got from it.

'But I was not prepared for the working of it—evidently a very delicate and difficult process (I proposed to *dissolve* off the corners, &c., and, as it were, *lick* it into shape), and though I have never quite lost sight of the matter, I have not yet been able to do anything with it; meanwhile, I put it by. On looking at it a year or two after, I was dismayed to find it had lost much by deliquescence. Accordingly, I potted it up in salt in an earthen dish, with iron rim, and placed it on an upper shelf in a room with an Arnott stove, where it has remained ever since.

'If you should find it of any use, I would ask you, if possible, to repeat my experiment as described, and settle that point, which has always struck me as a very important one.'

the aqueous vapour of the air. Thus, the air reaches the cylinder deprived both of its aqueous vapour and its carbonic acid. It is now entering,—the mercury gauge of the pump is descending, and, as it enters, you will observe the needle. If the entrance of the air diminish the radiation through the cylinder—if air be a substance which is competent to intercept the waves of ether in any sensible degree—this will be declared by the diminished deflection of the galvanometer. The tube is now full, but you see no change in the position of the needle, nor could you see any change, even if you were close to the instrument. The air thus examined seems as transparent to radiant heat as the vacuum itself.

(376) By changing the screen we can alter the amount of heat falling upon the pile; thus, by gradually withdrawing it, the needle can be caused to stand at 40°, 50°, 60°, 70°, and 80°, in succession; and while it occupies each position, the experiment just performed before you can be repeated. In no instance could you recognise the slightest movement of the needle. The same is the case if the screen be pushed forward, so as to reduce the deflection to 20 or 10 degrees.

(377) The experiment just made is a question addressed to Nature, and her silence might be construed into a negative reply. But a natural philosopher must not lightly accept a negative, and I am not sure that we have put our question in the best possible language. Let us analyse what we have done, and first consider the case of our smallest deflection of 10 degrees. Supposing that the air is *not* perfectly diathermic; that it really intercepts a small portion—say the thousandth part of the heat passing through the tube—that out of every thousand rays it intercepted one; should we be able to detect this action? Such absorption, if it took place, would lower the deflection the thousandth part of ten degrees, or the hundredth

part of one degree, a diminution which it would be impossible for you to see, even if you were close to the galvanometer.* In the case here supposed, *the total quantity of heat falling upon the pile is so inconsiderable, that a small fraction of it, even if absorbed, might well escape detection.*

(378) But we have not confined ourselves to a small quantity of heat; the result was the same when the deflection was 80°, as when it was 10°. Here I must ask you to sharpen your attention and accompany me, for a time, over rather difficult ground. I want now to make clearly intelligible to you an important peculiarity of the galvanometer.

(379) The needle being at zero, let us suppose a quantity of heat to fall upon the pile, sufficient to produce a deflection of one degree. Suppose the quantity of heat to be afterwards augmented, so as to produce deflections of two degrees, three degrees, four degrees, five degrees; then the quantities of heat which produce these deflections stand to each other in the ratios of 1 : 2 : 3 : 4 : 5: the quantity of heat which produces a deflection of 5° being exactly five times that which produces a deflection of 1°. But this proportionality exists only so long as the deflections do not exceed a certain magnitude. For, as the needle is drawn more and more aside from zero, the current acts upon it at an ever augmenting disadvantage. The case is illustrated by a sailor working a capstan; he always applies his strength at right angles to the lever, for, if he applied it obliquely, only a portion of that strength would be effective in turning the capstan round. And in the case of our electric current, when the needle is very oblique to the current's direction, only a portion of its force is effective in moving the needle. Thus it

* It will be borne in mind that I am here speaking of *galvanometric*, not of *thermometric* degrees.

happens, that though the quantity of heat may be, and, in our case, *is*, accurately expressed by the strength of the current which it excites, still the larger deflections, inasmuch as they do not give us the action of the whole current, but only of a part of it, cannot be a true measure of the amount of heat falling upon the pile.

(380) The galvanometer now before you is so constructed, that the angles of deflection, up to 30° or thereabouts, are proportional to the quantities of heat; the quantity necessary to move the needle from 29° to 30°, is sensibly the same as that required to remove it from 0° to 1°. But beyond 30° the proportionality ceases. The quantity of heat required to move the needle from 40° to 41° is three times that necessary to move it from 0° to 1°; to deflect it from 50° to 51° requires five times the heat necessary to move it from 0° to 1°; to deflect it from 60° to 61° requires about seven times the heat necessary to move it from 0° to 1°; to deflect it from 70° to 71° requires eleven times, while to move it from 80° to 81° requires more than fifty times the heat necessary to move it from 0° to 1°. Thus, the higher we go, the greater is the quantity of heat represented by a degree of deflection; the reason being, that the force which then moves the needle is only a fraction of the force of the current really circulating in the wire, and hence represents only a fraction of the heat falling upon the pile.

(381) By a process, to be afterwards described,* the higher degrees of the galvanometer can be expressed in terms of the lower ones. We thus learn, that while deflections of 10°, 20°, 30°, respectively express quantities of heat represented by the numbers 10, 20, 30, a deflection of 40° represents a quantity of heat expressed by the number 47; a deflection of 50° expresses a quan-

* See Appendix to this Chapter.

tity of heat expressed by the number 80; while the deflections 60°, 70°, 80°, express quantities of heat which increase in a much more rapid ratio than the deflections themselves.

(382) What is the upshot of this analysis? It will lead us to a better method of questioning Nature. It suggests the reflection that, when we make our angles *small*, the quantity of heat falling on the pile is so inconsiderable, that even if a fraction of it were absorbed, it might escape detection; while, if we make our deflections large, by employing a powerful flux of heat, the needle is in a position from which it would require a considerable addition or abstraction of heat to move it. The 1,000th part of the whole radiation, in the one case, would be too small, absolutely, to be measured: the 1,000th part in the other case might be very considerable, without, however, being considerable enough to affect the needle in any sensible degree. When, for example, the deflection is over 80°, an augmentation or diminution of heat, equivalent to 15 or 20 of the lower degrees of the galvanometer, would be scarcely sensible.

(383) We are now face to face with our problem: it is this, to work with a flux of heat so large that a small fractional part of it will not be infinitesimal, and still to keep our needle in its most sensitive position. If we can accomplish this, we shall augment indefinitely our experimental power. If a fraction of the heat, however small, be intercepted by the gas, *we can augment the absolute value of that fraction by augmenting the total of which it is a fraction.*

(384) The problem, happily, admits of an effective practical solution. You know that when we allow heat to fall upon the opposite faces of the thermo-electric pile, the currents generated neutralise each other more or less; and, if the quantities of heat falling upon the two faces be per-

fectly equal, the neutralisation is complete. Our galvanometer needle is now deflected to 80° by the flux of heat passing through the tube; I uncover the second face of the pile, which is also furnished with a conical reflector, and place a second tube of boiling water in front of it; the needle, as you see, descends instantly.

(385) By means of a proper adjusting screen the quantity of heat falling upon the posterior face of the pile can be so regulated that it shall exactly neutralise the heat incident upon its other face: this is now effected; and the needle points to zero.

(386) Here, then, we have two powerful and perfectly equal fluxes of heat, falling upon the opposite faces of the pile, one of which passes through our exhausted cylinder. If air be allowed to enter the cylinder, and if this air exert any appreciable action upon the rays of heat, the equality now existing will be destroyed; a portion of the heat passing through the tube being intercepted by the air, the second source of heat will triumph; the needle, now in its most sensitive position, will be deflected; and from the magnitude of the deflection we can accurately calculate the absorption.

(387) I have thus sketched, in rough outline, the apparatus by which our researches on the relation of radiant heat to gaseous matter must be conducted. The necessary tests, are, however, at the same time so powerful and so delicate, that a rough apparatus like that just described would not answer our purpose. But you will now experience no difficulty in comprehending the construction and application of the more perfect apparatus, with which the experiments on gaseous absorption and radiation have been actually made.

(388) Between S and S′ (Plate I. at the end of the book) stretches the *experimental cylinder*, a hollow tube of brass, polished within; at S and S′ are the plates of rock-salt

which close the cylinder air-tight; the length from S to S′, in the experiments to be first recorded, is 4 feet. The source of heat, C, is a cube of cast copper, filled with water, which is kept continually boiling by the lamp L. Attached to the cube C by brazing is the short cylinder F, of the same diameter as the experimental cylinder, and capable of being connected air-tight with the latter at S. Thus, between the source C and the end S′ of the experimental tube, we have *the front chamber* F, from which the air can be removed, so that the rays from the source will enter the cylinder S S′ unsifted by air. To prevent the heat of the source C from passing by conduction to the plate at S, the chamber F is caused to pass through the vessel V, in which a stream of cold water continually circulates, entering through the pipe *i i*, which dips to the bottom of the vessel, and escaping through the waste-pipe *e e*. The experimental tube and the front chamber are connected, independently, with the air-pump A A, so that either of them may be exhausted or filled, without interfering with the other. I may remark that, in later arrangements, the experimental cylinder was supported apart from the pump, being connected with the latter by a flexible tube. The tremulous motion of the pump, which occurred when the connection was rigid, was thus completely avoided. P is the thermo-electric pile, placed on its stand at the end of the experimental cylinder, and furnished with its two conical reflectors. C′ is the *compensating cube*, used to neutralise the radiation from C; H is the *adjusting screen*, which is capable of an exceedingly fine motion to and fro. N N is a delicate galvanometer connected with the pile P, by the wires *w w′*. The graduated tube O O (to the right of the plate), and the appendage M K (attached to the centre of the experimental tube) shall be referred to more particularly by-and-by.

(389) It would hardly sustain your interest, were I to

state the difficulties which at first beset the investigation conducted with this apparatus, or the numberless precautions, which the exact balancing of the two powerful sources of heat here resorted to, rendered necessary. I believe the experiments, made with atmospheric air alone, might be numbered by tens of thousands. Sometimes for a week, or even for a fortnight, coincident and satisfactory results would be obtained; the strict conditions of accurate experimenting would appear to be found, when an additional day's experience would destroy the superstructure of hope, and necessitate a recommencement, under changed conditions, of the whole enquiry. It is this which daunts the experimenter; it is this preliminary fight with the entanglements of a subject, so dark, so doubtful, so uncheering, —without any knowledge whether the conflict is to lead to anything worth possessing,—which renders discovery difficult and rare. But the experimenter, especially the *young* experimenter, ought to know that, as regards his own moral manhood, he cannot but win, if he only contend aright. Even with a negative result, the consciousness that he has gone fairly to the bottom of his subject, as far as his means allowed—the feeling that he has not shunned labour, though that labour may have resulted in laying bare the nakedness of his case—reacts upon his own mind, and gives it firmness for future work.

(390) But to return;—I first neglected atmospheric vapour and carbonic acid altogether; concluding, as others afterwards did, that the quantities of these substances being so small, their effect upon radiant heat must be quite inappreciable; after a time, however, this assumption was found to be leading me quite astray. Chloride of calcium was first used as a drying agent, but I had to abandon it. Pumice stone, moistened with sulphuric acid, was next used, but it also proved unsuitable. I finally resorted to pure glass broken into small fragments, wetted with sulphuric

acid, and inserted by means of a funnel into a U tube. This arrangement was found to be the best, but even here the greatest care was needed. It was necessary to cover each column with a layer of dry glass fragments, for the smallest particle of dust from the cork, or a quantity of sealing wax not more than the twentieth part of a pin's head in size, was quite sufficient, if it reached the acid, to vitiate the results. The drying-tubes, moreover, had to be frequently changed, as the organic matter of the atmosphere, infinitesimal though it was, after a time introduced disturbance.

(391) To remove the carbonic acid, pure carrara marble was broken into fragments, wetted with caustic potash, and introduced into a U tube. These, then, are the agents now employed for drying the gas and removing the carbonic acid; but previous to their final adoption, the arrangement shown in Plate I. was made use of. The glass tubes marked Y Y, each three feet long, were filled with chloride of calcium, after them were placed two U tubes, R Z, filled with pumice stone and sulphuric acid. Hence, the air, in the first place, had to pass over 18 feet of chloride of calcium, and afterwards through the sulphuric acid tubes, before entering the experimental tube S S′. A gasholder, G G′, was employed for other gases than atmospheric air. In the investigation on which I am at present engaged, this arrangement, as already stated, is abandoned, a simpler one being found more effectual.

(392) Both the front chamber F, and the experimental tube S S′ being exhausted, the rays pass from the source C through the front chamber; across the plate of rock-salt at S, through the experimental tube, across the plate at S′, afterwards impinging upon the anterior surface of the pile P. This radiation is neutralised by that from the compensating cube C′. The needle, you will observe,

is at zero. We will commence our experiments by applying this severe test to dry air. It is now entering the experimental cylinder; but at your distance you see no motion of the needle, and thus our more powerful mode of experiment fails to detect any absorption on the part of the air. Its atoms, apparently, are incompetent to stop a single calorific wave; *it is a practical vacuum, as regards the rays of heat.* Oxygen, hydrogen, and nitrogen, when carefully purified, exhibit the action of atmospheric air; they are sensibly neutral.

(393) This is the deportment which prior to the researches now to be described was ascribed to gases generally. Let us see whether rightly or not. This gasholder contains olefiant gas,—common coal gas would also answer my purpose. The perfect transparency of this gas to light is demonstrated by discharging it into the air; you see nothing, the gas is not to be distinguished from the air. The experimental tube is now exhausted, and the needle points to zero. Observe the effect when the olefiant gas is permitted to enter. The needle moves in a moment; the transparent gas intercepts the heat, like an opaque body—the final and permanent deflection, when the tube is full of gas, amounting to 70°.

(394) Let us now interpose a metal screen between the pile P and the end S′ of the experimental tube, thus entirely cutting off the radiation through the tube. The face of the pile turned towards the metal screen wastes its heat speedily by radiation; it is now at the temperature of this room, and the radiation from the compensating cube alone acts on the pile, producing a deflection of 75°. But at the commencement of the experiment the radiations from both cubes were equal; hence, the deflection 75° corresponds to the *total radiation* through the experimental tube, when the latter is exhausted.

(395) Taking as unit the quantity of heat necessary to

move the needle from 0° to 1°, the number of units expressed by a deflection of 75° is

276.

The number of units expressed by a deflection of 70° is

211.

Out of a total, therefore, of 276, olefiant gas has intercepted 211; that is, about seven-ninths of the whole, or about 80 per cent.

(396) Does it not seem to you as if an opaque layer had been suddenly precipitated on our plates of salt, when the gas entered? The substance, however, deposits no such layer. When a current of the dried gas is discharged against a polished plate of salt, you do not perceive the slightest dimness. The rock-salt plates, moreover, though necessary for exact measurements, are not necessary to show the destructive power of this gas. Here is an open tin cylinder, interposed between the pile and our radiating source; when olefiant gas is forced gently into the cylinder from this gasholder, you see the needle fly up to its stops. Observe the smallness of the quantity of gas now to be employed. First cleansing the open tube, by forcing a current of air through it, and bringing the needle to zero; I turn this cock on and off, as speedily as possible. A mere bubble of the gas enters the tube in this brief interval; still you see that its presence causes the needle to swing to 70°. Let us now abolish the open tube, and leave nothing but the free air between the pile and source; from the gasometer I discharge olefiant gas into this open space. You see nothing in the air; but the swing of the needle through an arc of 60° declares the presence of this invisible barrier to the calorific rays.

(397) Thus, it is shown that the ethereal undulations which glide among the atoms of oxygen, nitrogen, and

hydrogen, without hindrance, are powerfully intercepted by the molecules of olefiant gas. We shall find other transparent gases, also, almost immeasurably superior to air. We can limit at pleasure the number of the gaseous atoms, and thus vary the amount of destruction of the ethereal waves. Attached to the air-pump is a barometric tube, by means of which measured portions of the gas can be admitted. The experimental cylinder is now exhausted: turning this cock slowly on, and observing the mercury gauge, olefiant gas enters, till the mercurial column has been depressed an inch. I observe the galvanometer, and read the deflection. Determining thus the absorption produced by one inch, another inch is added, and the absorption effected by two inches of the gas is determined. Proceeding thus, we obtain for tensions from 1 to 10 inches the following absorptions:—

Olefiant Gas.

Pressures in inches	Absorption
1	90
2	123
3	142
4	157
5	168
6	177
7	182
8	186
9	190
10	193

(398) The unit here used is the amount of heat absorbed, when a *whole atmosphere* of dried air is allowed to enter the tube. The table, for example, shows that one thirtieth of an atmosphere of olefiant gas exercises ninety times the absorption of a whole atmosphere of air. The deflection produced by the tubeful of dry air is here taken to be one degree: it is probably less even than this infinitesimal amount.

(399) The table also informs us that each additional inch of olefiant gas produces less effect than the preceding one. A single inch, at the commencement, intercepts 90 rays, but a second inch absorbs only 33, while the addition of an inch, when nine inches are already in the tube, effects the destruction of only 3 rays. This is what might reasonably be expected. The number of rays emitted is finite, and the discharge of the first inch of olefiant gas amongst them has so thinned their ranks, that the execution produced by the second inch is naturally less than that of the first. This execution must diminish, as the number of rays capable of being destroyed by the gas becomes less; until, finally, all absorbable rays being removed, the residual heat passes through the gas unimpeded.*

(400) But supposing the quantity of gas first introduced to be so inconsiderable, that the heat intercepted by it is a vanishing quantity, compared with the total amount, we might then reasonably expect that, for some time at least, the quantity of heat intercepted would be proportional to the quantity of gas present. That a double quantity of gas would produce a double effect, a treble quantity a treble effect; or, in general terms, that the absorption would, for a time, be found proportional to the density.

(401) To test this idea, we will make use of a portion of the apparatus omitted in the general description. O O (Plate I.) is a graduated glass tube, the end of which dips into the basin of water B. The tube is closed above by

* A wave of ether starting from a radiant point in all directions, in a uniform medium, constitutes a spherical shell, which expands with the velocity of light or of radiant heat. A *ray* of light, or a *ray* of heat, is a line perpendicular to the wave, and, in the case here supposed, the rays would be the radii of the spherical shell. The word 'ray,' however, is used in the text, to avoid circumlocution, as equivalent to the term *unit of heat*. Thus, calling the amount of heat intercepted by a whole atmosphere of air 1, the amount intercepted by $\frac{1}{30}$th of an atmosphere of olefiant gas is 90.

means of the stopcock *r*; *d d* is a tube containing fragments of chloride of calcium which dries the gas. The tube o o is first filled with water up to the cock *r*, and the water is afterwards carefully displaced by olefiant gas, introduced in bubbles from below. The gas is admitted into the experimental cylinder by the cock *r*, and as it enters, the water rises in o o, each of the divisions of which represents a volume of $\frac{1}{50}$th of a cubic inch. Successive measures of this capacity are permitted to enter the tube, and the absorption in each particular case is determined.

(402) In the following table, the first column contains the quantity of gas admitted into the tube; the second contains the corresponding absorption; the third column contains the absorption, calculated on the supposition that it is proportional to the density.

Olefiant Gas.

Unit-measure, $\frac{1}{50}$th of a cubic inch.

Measures of Gas	Absorption Observed	Absorption Calculated
1	2·2	2·2
2	4·5	4·4
3	6·6	6·6
4	8·8	8·8
5	11·0	11·0
6	12·0	13·2
7	14·8	15·4
8	16·8	17·6
9	19·8	19·8
10	22·0	22·0
11	24·0	24·2
12	25·4	26·4
13	29·0	28·6
14	30·2	29·8
15	33·5	33·0

(403) This table proves the correctness of the surmise, that when very small quantities of the gas are employed, the absorption is sensibly proportional to the density. But consider for a moment the tenuity of the gas with which

we have here operated. The volume of our experimental tube is 220 cubic inches; imagine $\frac{1}{50}$th of a cubic inch of gas diffused in this space, and you have the atmosphere through which the calorific rays passed in our first experiment. This atmosphere possesses a pressure not exceeding $\frac{1}{11000}$th of that of ordinary air. It would depress the mercurial column connected with the air-pump not more than $\frac{1}{367}$th of an English inch. Its action, however, upon the calorific rays is perfectly measurable, being more than twice that of a whole atmosphere of dry air.

(404) But the absorptive energy of olefiant gas, extraordinary as it is shown to be by the foregoing experiments, is exceeded by that of various vapours, the action of which on radiant heat is now to be illustrated. This glass flask, G (fig. 86), is provided with a brass cap, into which a stopcock can be screwed air-tight. A small quantity of sulphuric ether is poured into the flask, and by means of an air-pump, the air which fills the flask above the liquid is completely removed. I attach the flask to the experimental tube, which is now exhausted—the needle pointing to zero—and permit the vapour from the flask to enter it. The mercury of the gauge sinks, and when it is depressed one inch, the further supply of vapour will be stopped. The moment the vapour entered, the needle moved, and it now points to 65°. I can add another inch, and again determine the absorption; a third inch, and do the same. The absorptions effected by four inches, introduced in this way, are given in the following table. For the sake of comparison, the corresponding absorptions of olefiant gas are placed in the third column.

FIG. 86.

Sulphuric Ether.

Pressures in inches of mercury	Absorption	Corresponding absorption of olefiant gas
1	214	90
2	282	123
3	315	142
4	330	154

(405) For these pressures the absorption of radiant heat by the vapour of sulphuric ether is about two and two-third times the absorption by olefiant gas. There is, moreover, no proportionality between the quantity of vapour and the absorption.

(406) But reflections similar to those which we have already applied to olefiant gas are also applicable to sulphuric ether. Supposing we make our unit-measure small enough, the number of rays first destroyed will vanish in comparison with the total number, and probably, for a time, the absorption will be directly proportional to the density. To examine whether this is the case, the other portion of the apparatus, omitted in the general description, was made use of. K (Plate I.) is one of the small flasks already described, with a brass cap, which is closely screwed on to the stopcock *c'*. Between the cocks *c'* and *c*, the latter of which is connected with the experimental tube, is the chamber M, the capacity of which is accurately determined. The flask K is partially filled with ether, the air above the liquid and that dissolved in it being removed. The tube S S' and the chamber M being exhausted, the cock *c* is shut off; and *c'* being turned on, the chamber M is filled with pure ether vapour. By turning *c'* off and *c* on, this quantity of vapour is allowed to diffuse itself through the experimental tube, where its absorption is determined; successive measures are thus sent into the tube, and the effect produced by each is noted.

(407) In the following table, the unit-measure made use of had a volume of $\frac{1}{100}$th of a cubic inch.

Sulphuric Ether.

Unit-measure, $\frac{1}{100}$th of a cubic inch.

Measures	Absorption Observed	Absorption Calculated
1	5·0	4·6
2	10·3	9·2
4	19·2	18·4
5	24·5	23·0
6	29·5	27·0
7	34·5	32·2
8	38·0	36·8
9	44·0	41·4
10	46·2	46·2
11	50·0	50·6
12	52·8	55·2
13	55·0	59·8
14	57·2	64·4
15	59·4	69·0

(408) We here find that the proportion between density and absorption holds sensibly good for the first eleven measures, after which the deviation from proportionality gradually augments.

(409) No doubt, for smaller measures than $\frac{1}{100}$th of a cubic inch, the above law holds still more rigidly true; and in a suitable locality it would be easy to determine, with perfect accuracy, $\frac{1}{10}$th of the absorption produced by the first measure; this would correspond to $\frac{1}{1000}$th of a cubic inch of vapour. But, before entering the tube, the vapour had only the tension due to the temperature of the laboratory, namely 12 inches. This would require to be multiplied by 2·5 to bring it up to that of the atmosphere. Hence, the $\frac{1}{1000}$th of a cubic inch would, on being diffused through a tube possessing a capacity of 220 cubic inches, have a pressure of $\frac{1}{22} \times \frac{1}{25} \times \frac{1}{1000} = \frac{1}{500000}$th of an atmosphere. That the action of a transparent vapour so attenuated upon radiant heat should be at all measurable is simply astounding.

(410) These experiments with ether and olefiant gas

show that not only do gaseous bodies, at the ordinary tension of the atmosphere, offer an impediment to the transmission of radiant heat; not only are the interstitial spaces of such gases incompetent to allow the ethereal undulations free passage; but, also, that their density may be reduced vastly below that which corresponds to the atmospheric pressure, and still the door thus opened is not wide enough to let the undulations through. There is something in the constitution of the individual molecules, thus sparsely scattered, which enables them to destroy the calorific waves. The destruction, however, is merely one of form; there is no absolute loss. Through dry air the heat rays pass without sensibly warming it; through olefiant gas and ether vapour they cannot pass thus freely; but every wave withdrawn from the radiant beam produces its equivalent motion in the body of the absorbing gas, and raises its temperature. It is a case of transference, not of annihilation.

(411) Before changing the source of heat here made use of, let us direct our attention for a moment to the action of a few of the permanent gases on radiant heat. To measure the quantities introduced into the experimental tube, the mercury gauge of the air-pump was employed. In the case of carbonic oxide, the following absorptions correspond to the pressures annexed to them; the action of a full atmosphere of air, assumed to produce a deflection of one degree, being taken as unity:—

Carbonic Oxide.

Pressures in inches of mercury	Absorption	
	Observed	Calculated
0·5	2·5	2·5
1·0	5·6	5·0
1·5	8·0	7·5
2·0	10·0	10·0
2·5	12·0	12·5
3·0	15·0	15·0
3·5	17·5	17·5

As in former cases, the third column is calculated on the assumption that the absorption is directly proportional to the density of the gas; and we see that for seven measures, or up to a pressure of 3·5 inches, the proportionality holds strictly good. But for large quantities this is not the case; when, for instance, the unit measure is 5 inches, instead of half an inch, we obtain the following result:—

Pressures in inches	Absorption	
	Observed	Calculated
5	18	18
10	32·5	36
15	45	54

Carbonic acid, sulphide of hydrogen, nitrous oxide, and other gases, though differing in the energy of their absorption, and all of them exceeding carbonic oxide, exhibit, when small and large quantities are used, a similar deportment towards radiant heat.

(412) Thus, then, in the case of some gases, we find an almost absolute incompetence on the part of their atoms to intercept the ethereal waves, while the atoms of ether gases, struck by these same undulations, absorb their motion, and become themselves centres of heat. We have now to examine what gaseous bodies are competent to do in this latter capacity; we have to enquire whether these atoms and molecules, which can accept motion from the ether in such very different degrees, are not also characterised by their competency to *impart* motion to the ether in different degrees; or, to use the common language, having learned something of the power of different gases, as *absorbers* of radiant heat, we have now to enquire into their capacities as *radiators*.

(413) An arrangement is before you by means of which we can pursue this enquiry. P (fig. 87) is the thermo-electric pile, with its two conical reflectors; S is a double screen of polished tin; A is an argand burner,

consisting of two concentric perforated rings; C is a copper ball, which, during the experiments, is heated under redness; while the tube *t t* leads to a gasholder. When

FIG. 87.

the hot ball C is placed on the burner, it warms the air in contact with it; an ascending current is thus established, which, to some extent, acts upon the pile. To neutralise this action, a large Leslie's cube, L, filled with water, a few degrees above the air in temperature, is placed before the opposite face of the pile. The needle being thus brought to zero, the gas is forced, by a gentle water pressure, through the orifices of the burner; it meets the ball C, glides along its surface, and ascends, in a warm current, in front of the pile. The rays from the heated gas issue in the direction of the arrows, against the pile, and the consequent deflection of the galvanometer needle indicates the magnitude of the radiation.

(414) The results of the experiments are given in the second column of the following table; the numbers there recorded marking the extreme limit to which the needle swung, when the rays from the gas fell upon the pile:—

	Radiation	Absorption
Air	Insensible	Insensible
Oxygen	"	"
Nitrogen	"	"
Hydrogen	"	"
Carbonic oxide	12°	18·0°
Carbonic acid	18	25·0
Nitrous oxide	29	44·0
Olefiant gas	53	61·0

(415) In the second column of figures are placed the deflections due to the absorption of the gases here employed, at a common tension of 5 inches. A comparison of the two columns shows us that radiation and absorption go hand in hand; that the molecule which is competent to *intercept* a calorific beam, is competent, in a proportionate degree, to *generate* a calorific beam. That, in short, a capacity to accept motion from the ether, and to impart motion to it, by gaseous bodies, are correlative properties.

(416) And here, be it remarked, we are relieved from all considerations regarding the influence of cohesion on the results. In solids and liquids the particles are more or less in thrall, and cannot be considered as individually free. The difference in point of radiative and absorptive power, between alum and rocksalt, for example, might be fairly regarded as due to their character as aggregates, held together by crystallising force. But the difference between olefiant gas and atmospheric air cannot be explained in this way; it is a difference dependent on the individual molecules of these substances; and thus, our experiments with gases and vapours probe the question

of atomic constitution to a depth quite unattainable with solids and liquids.

(417) I have refrained, thus far, from giving you as full a tabular statement of the absorptive powers of gases and vapours as the experiments made with the apparatus already described would enable me to do; knowing that results, obtained with another apparatus, were in reserve which would better illustrate the subject. This second arrangement is the same in principle as the first; only two changes of importance being made in it. The first is that, instead of making a cube of boiling water the source of heat, a plate of copper is employed, against which a thin steady gas-flame from a Bunsen's burner is caused to play; the heated plate forms the back of our new front chamber, which latter can be exhausted independently, as before. The second alteration is the substitution of a tube of glass of the same diameter, and 2 feet 8 inches long, for the tube of brass s s′, Plate I. All the other parts of the apparatus remain as before. The gases were introduced in the manner already described into the experimental tube, and from the galvanometric deflection, consequent on the entrance of each gas, its absorption was calculated.

(418) The following table gives the relative absorptions of several gases, at a common pressure of one atmosphere. It may be remarked that the differences between air and the other gases would be still greater if the brass tube had been employed; but the use of it would have excluded the corrosive gases mentioned in the table.

Name	Absorption at 30 inches pressure
Air	1
Oxygen	1
Nitrogen	1
Hydrogen	1
Chlorine	39
Hydrochloric acid	62

Name	Absorption at 30 inches pressure
Carbonic oxide	90
Carbonic acid	90
Nitrous oxide	355
Sulphide of hydrogen	390
Marsh gas	403
Sulphurous acid	710
Olefiant gas	970
Ammonia	1195

(419) The most powerful and delicate tests yet applied have not enabled me to establish a difference between oxygen, nitrogen, hydrogen, and air. The absorption of these substances is exceedingly small—probably even smaller than I have assumed it. The more perfectly the above-named gases are purified, the more closely does their action approach to that of a vacuum. And who can say that the best drying apparatus is perfect? We cannot even say that sulphuric acid, however pure, may not yield a modicum of vapour to the gases passing through it, and thus make the absorption by those gases appear greater than it ought. Stopcocks also must be greased, and hence may contribute an infinitesimal impurity to the air passing through them. But however this may be, it is certain that if any further advance should be made in the purification of the more feebly acting gases, it will only serve to augment the enormous differences of absorption here exhibited.

(420) Ammonia, at the tension of an atmosphere, exerts an absorption at least 1,195 times that of air. If a metal screen be interposed between the pile and the experimental tube, the needle will move a little, but so little that you entirely fail to see it. What does this experiment prove? It proves that this ammonia which, within our glass tube, is as transparent to light as the air we breathe, is so opaque to the heat radiating from our source, that the addition of a plate of metal hardly augments its opacity.

There is, indeed, reason to believe that this light, transparent gas is really as black, at the present moment, to the calorific rays, as if the experimental tube were filled with ink, pitch, or any other impervious substance.

(421) With oxygen, nitrogen, hydrogen, and air, the action of a whole atmosphere is so small, that it would be quite useless to attempt to determine the action of a fractional part of an atmosphere. Could we, however, make such a determination, the difference between them and the other gases would come out still more forcibly than in the last table. In the case of the energetic gases, we know that the calorific rays are most copiously absorbed by the portion of gas which first enters the experimental tube; the quantities which enter last, producing, in many cases, a merely infinitesimal effect. If, therefore, instead of comparing the gases at a common pressure of one atmosphere, we were to compare them at a common pressure of an inch, we should doubtless find the difference between the least absorbent and the most absorbent gases greatly augmented. We have already learned that when the absorption is small the quantity absorbed is proportional to the amount of gas present. Assuming this to be true for air, and for the other feeble gases referred to; taking, that is, their absorption at 1 inch of pressure to be $\frac{1}{30}$th of that at 30 inches; we have the following comparative effects. It will be understood that in every case, except the first four, the absorption of 1 inch of the gas was determined by direct experiment:—

Name	Relative absorption at 1 inch pressure
Air	1
Oxygen	1
Nitrogen	1
Hydrogen	1
Chlorine	60
Bromine	160
Carbonic oxide	750

Name	Relative absorption at 1 inch pressure
Carbonic acid	972
Hydrobromic acid	1005
Nitric oxide	1590
Nitrous oxide	1860
Sulphide of hydrogen	2100
Ammonia	5460
Olefiant gas	6030
Sulphurous acid	6480

(422) What extraordinary differences in the constitution and character of the ultimate particles of various gases do the above results reveal! For every ray intercepted by air, oxygen, hydrogen, or nitrogen—ammonia intercepts 5,460; olefiant gas 6,030; while sulphurous acid destroys 6,480. With these results before us, we can hardly help attempting to visualise the molecules themselves, trying to discern, with the eye of intellect, the actual physical qualities on which these vast differences depend. These molecules are particles of matter, plunged in an elastic medium, accepting its motions and imparting theirs to it. Is the hope unwarranted, that we may ultimately make radiant heat such a *feeler* of atomic constitution, that we shall be able to infer, from their action upon it, the mechanism of the ultimate particles of matter themselves?

(423) Have we, even now, no glimpse of a relation between absorption and atomic constitution? You remember our experiments with gold, silver, and copper; you recollect how feebly they radiated, and how feebly they absorbed (§§ 340 and 341). We heated them by boiling water; that is to say, we imparted, by the contact of the water, motion to their atoms; but this motion was imparted with extreme slowness by the atoms to the ether in which they swung. That the atoms of these bodies glide through the ether with scarcely any resistance may also be inferred from the length of time which they require to cool in vacuo. But we have seen that when

the motion which the atoms possess, and which they are incompetent to transfer to the ether, is imparted, by contact, to a coat of varnish, or of chalk, or of lampblack, or even to flannel or velvet, these bodies, being good radiators, soon transfer the motion to the ether. The same we found true for glass and earthenware.

(424) In what respect do those good radiators differ from the metals referred to? In one profound particular —the metals *are elements*; the others *are compounds*. In the metals, the atoms vibrate singly; in the varnish, velvet, earthenware, and glass, they vibrate in groups. And now, in bodies as diverse from the metals as can possibly be conceived, we find the same significant fact making its appearance. Oxygen, hydrogen, nitrogen, and air, are elements, or mixtures of elements, and, both as regards radiation and absorption, their feebleness is declared. They swing in the ether, with scarcely any loss of moving force.

(425) It is impossible not to be struck by the position of chlorine and bromine in the last table. Chlorine is an extremely dense and also a coloured gas; bromine is a far more densely-coloured vapour; still we find them, as regards perviousness to the heat of our source, standing above every transparent compound gas in the table. The act of combination with hydrogen produces, in the case of each of these substances, a transparent compound; but the chemical act, which augments the transparency to light, augments the opacity to heat; hydrochloric acid absorbs more than chlorine; and hydrobromic acid absorbs more than bromine.

(426) Further, the element bromine is here in the liquid condition; a portion of it enclosed in this glass cell is of a thickness sufficient to extinguish utterly the flame of a lamp or candle. But when a candle is placed in front of the cell, and a thermo-electric pile behind it; the

prompt movement of the needle declares the passage of radiant heat through the bromine. This heat consists entirely of the obscure rays of the candle, for the light, as stated, is utterly cut off. Let us remove the candle, and put in its place a copper ball, heated not quite to redness. The needle at once flies to its stops, showing the transparency of the bromine to the heat emitted by the ball. It is impossible, I think, to close our eyes against this convergent evidence that the free atoms swing with ease in the ether, while when grouped in oscillating systems, they cause its waves to swell, imparting to it, when compounded into molecules, an amount of motion quite beyond their power to communicate, as long as they remained uncombined.*

(427) But it will occur to you, no doubt, that lampblack, which is an elementary substance, is one of the best absorbers and radiators in nature. Let us examine this substance a little. Ordinary lampblack contains many impurities; it has various hydrocarbons condensed within it, and these hydrocarbons are powerful absorbers and radiators. Lampblack, therefore, as hitherto applied, can hardly be considered an element at all. I have, however, had these hydrocarbons in great part removed, by carrying through red-hot lampblack a current of chlorine gas: but the substance has continued to be a powerful radiator and absorber. Well, what *is* lampblack? Chemists will tell you that it is an allotropic form of the diamond: here, in fact, is a diamond reduced to charcoal by intense heat. Now the allotropic condition has long been defined as due to a difference in the arrangement of a body's particles; hence, it is conceivable that this arrangement, which causes such a marked

* I should like to reserve my opinion as to the comparative strength of the radiation of the molecule as a whole, and that of its constituent atoms.

physical difference between lampblack and diamond, may consist of an atomic grouping, which causes the body to act on radiant heat as if it were a compound. Such an arrangement of an element, though exceptional, is quite conceivable; and it will afterwards be shown that this is actually and eminently the case, as regards an allotropic form of our highly ineffectual oxygen.

(428) But, in reality, lampblack is not so impervious as you might suppose it to be. Melloni has shown it to be transparent, in an unexpected degree, to radiant heat emanating from a low source, and the experiment now to be performed will corroborate his. This plate of rock-salt, by being held over a smoky lamp, has been so thickly coated with soot that it does not allow a trace of light from the most brilliant gas jet to pass through it. Between the smoked plate and this vessel of boiling water, which is to serve as our source of heat, is placed a screen. The thermo-electric pile is at the other side of the smoked plate. When the screen is withdrawn, instantly the needle moves from zero, its final and permanent deflection being 52°. I now cleanse the salt perfectly, and determine the radiation through the unsmoked plate,—it is 71°. But the value of the deflection 52°, expressed with reference to our usual unit, is 85, and the value of 71°, or the total radiation, is about 222. Hence, the radiation through the soot is to the whole radiation as

$$222 : 85 = 100 : 38$$

that is to say, 38 per cent. of the incident heat has been transmitted by the layer of lampblack.

(429) We shall have to deal subsequently with far more impressive illustrations of the diathermancy of opaque bodies than that here exhibited by lampblack. They may receive a passing notice here. Iodide of methyl is formed by the union of the element iodine with the radical methyl.

Exposure to light usually sets a portion of the iodine free and colours the liquid a rich brown. In a series of experiments on the radiation of heat through liquids, I compared, as regards their powers of transmission, a strongly coloured specimen of the iodide of methyl, with a perfectly transparent one; there was no difference between them. The iodine, which produced so marked an effect on light, did not sensibly affect the radiant heat. Here are the numbers which express the portion of the total radiation intercepted by the transparent and coloured liquids respectively:—

	Absorption per cent.
Iodide of methyl (transparent)	53·2
" " (strongly coloured with iodine) .	53·2

The source of heat, in this case, was a spiral of platinum wire raised to bright redness by an electric current. On looking through the coloured liquid, the incandescent spiral was visible. The colour was therefore intentionally deepened by adding iodine, until the solution was of sufficient opacity to cut off wholly the light of a brilliant jet of gas. The transparency of the liquid to the radiant heat was not sensibly affected by the addition of the iodine. The luminous heat was, of course, cut off; but this, as compared with the whole radiation, was so small as to be insensible in the experiments.

(430) It is known that iodine dissolves freely in the bisulphide of carbon, the colour of the solution in thin layers being a splendid purple; but in layers of moderate thickness it may be rendered perfectly opaque to light. A quantity of iodine was dissolved in the liquid, sufficient, when introduced into a cell 0·07 of an inch wide, to cut off the light of the most brilliant gas flame. Comparing the opaque solution with the transparent bisulphide, the following results were obtained:—

	Absorption
Bisulphide of carbon (opaque)	12·5
" " (transparent) . . .	12·5

Here the presence of a quantity of iodine, perfectly opaque to a brilliant light, was without measurable effect upon the heat emanating from our platinum spiral.

(431) The same liquid was placed in a cell 0·27 of an inch in width; that is to say, a solution opaque to light, at a thickness of 0·07, was employed in a layer of nearly four times this thickness. Here are the results:—

	Absorption
Bisulphide of carbon (transparent) . . .	18·8
„ „ (opaque)	19·0

The difference between both measurements lies within the limits of possible error.

(432) The light of the electric lamp has already been decomposed in your presence, the spectrum of the light being projected upon a white screen. For this purpose, a prism of transparent bisulphide of carbon was employed. The liquid is contained in a wedge-shaped flask with plane glass sides; it draws the colours very widely apart, and produces a more beautiful effect than could be obtained with a glass prism. I now project a little spectrum on this small screen, behind which is placed the thermo-electric pile, connected with the large galvanometer in front of the table. The spectrum, as you observe, is about 1½ inches wide and 2 inches long, its colours being rendered very vivid by concentration. If the screen were removed, the red and extra-red end of the spectrum would fall upon the pile behind, and doubtless produce a thermo-electric current. But we will allow no light to fall upon the instrument; my object being to show you that we have here a spectrum which you cannot see, and that you may entirely detach the non-luminous spectrum from the luminous one. Here, then, is a second prism, filled with the bisulphide of carbon, in which iodine has been dissolved. I remove the transparent prism, and put the opaque one exactly in

its place. The spectrum has disappeared; there is no longer a trace of light upon the screen; but a thermal spectrum is still there. The obscure rays of the electric lamp have traversed the opaque liquid, have been refracted like the luminous ones, and are now, though invisible, impinging upon the screen before you. This is proved by removing the screen: no light strikes the pile, but the heat falling upon it is competent to dash violently aside the needles of our large galvanometer.

(433) The action of gases upon radiant heat has been already illustrated with our glass experimental tube and our new source of heat. Let me now refer to the action of vapours, as examined with the same apparatus. Here are several glass flasks, each furnished with a brass cap, to which a stopcock can be screwed. Into each is poured a quantity of a volatile liquid, a flask being reserved for each liquid, so as to render the admixture of the vapours impossible. From each flask the air is carefully removed,—not only the air above the liquid, but the air dissolved in it, this latter bubbling freely away when the flask is exhausted. I now attach my flask to the exhausted experimental tube, and allow the vapour to enter, without permitting any ebullition to occur. The mercury column of the pump sinks, and when the required depression has been obtained, the supply of vapour is cut off. In this way, the vapours of the substances mentioned in the next table have been examined, at pressures of 0·1, 0·5, and 1 inch, respectively.

	Absorption of Vapours at the pressures		
	0·1	0·5	1·0
Bisulphide of carbon	15	47	62
Iodide of methyl	35	147	242
Benzol	66	182	267
Chloroform	85	182	236
Methylic alcohol	109	390	590
Amylene	182	535	823

	Absorption of Vapours at the pressures		
	0·1	0·5	1·0
Sulphuric ether	300	710	870
Alcohol	325	622	
Formic ether	480	870	1075
Acetic ether	590	980	1195
Propionate of ethyl	596	970	
Boracic acid	620		

(434) These numbers refer to the absorption of a whole atmosphere of dry air as their unit; that is to say, $\frac{1}{10}$th of an inch of bisulphide of carbon vapour does fifteen times the execution of 30 inches of atmospheric air; while $\frac{1}{10}$th of an inch of boracic ether vapour does 620 times the execution of a whole atmosphere of atmospheric air. Comparing air at a pressure of 0·01 with boracic ether at the same pressure, the absorption of the latter is probably more than 180,000 times that of the former.

(434*a*) It is easy to show in a general way the absorption of radiant heat by vapour. An open tube will answer the purpose when quantitative results are not sought. The tube even may be dispensed with, and the vapour discharged from a slit into the open air between the pile and the source. A few specimen results obtained in this rough way will suffice for illustration. Two cubes of boiling water were employed, and in the usual manner the needle was brought to zero. Dry air was then urged from a gas bag (a common bellows would answer the same purpose) through a U-tube containing fragments of glass, moistened with the liquid whose vapour was to be examined. The mixed air and vapour were discharged in the open air in front of the pile, and the extreme limit of the swing of the galvanometer needle was noted.

Vapour discharged in open air	Limit of swing of needle
Sulphuric ether	118°
Formic ether	117
Acetic ether	92

Vapour discharged in open air	Limit of swing of needle
Amylene	91
Bisulphide of carbon	61
Valeric ether	32
Benzol	31
Alcohol	31

The influence of volatility here forces itself upon the attention. The action of course depends on the amount of vapour discharged, a quantity directly dependent on the volatility of the liquid. It is in consequence of its greater volatility that bisulphide of carbon is here able to transcend the far more energetic alcohol.

APPENDIX TO CHAPTER X.

I GIVE here the method of calibrating the galvanometer which Melloni recommends, as leaving nothing to be desired as regards facility, promptness, and precision. His own statement of the method, translated from 'La Thermochrose,' p. 59, is as follows :—

FIG. 88.

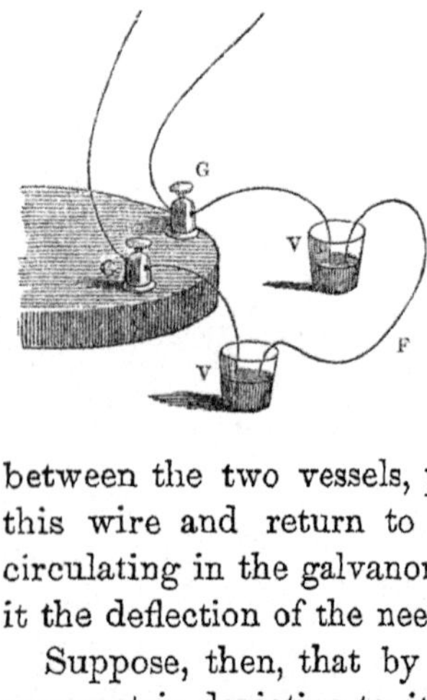

Two small vessels, V V, are half-filled with mercury, and connected separately, by two short wires, with the extremities G G of the galvanometer. The vessels and wires thus disposed make no change in the action of the instrument; the thermo-electric current being freely transmitted, as before, from the pile to the galvanometer. But if, by means of a wire F, a communication be established between the two vessels, part of the current will pass through this wire and return to the pile. The quantity of electricity circulating in the galvanometer will be thus diminished, and with it the deflection of the needle.

Suppose, then, that by this artifice we have reduced the galvanometric deviation to its fourth or fifth part; in other words, supposing that the needle, being at 10 or 12 degrees, under the action of a constant source of heat, placed at a fixed distance from the pile, descends to 2 or 3 degrees, when a portion of the current is diverted by the external wire; I say, that by causing the source to act from various distances, and observing in each case the *total* deflection, and the *reduced* deflection, we have all the data necessary to determine the ratio of the deflections of the needle, to the forces which produce these deflections.

To render the exposition clearer, and to furnish, at the same time, an example of the mode of operation, I will take the numbers relating to the application of the method to one of my thermo-multipliers.

The external circuit being interrupted, and the source of heat being sufficiently distant from the pile to give a deflection not exceeding 5 degrees of the galvanometer, let the wire be placed from V to V; the needle falls to 1·5°. The connection between the two vessels being again interrupted, let the source be brought near enough to obtain successively the deflections:—

5°, 10°, 15°, 20°, 25°, 30°, 35°, 40°, 45°.

Interposing after each the same wire between V and V, we obtain the following numbers:—

1·5°, 3°, 4·5°, 6·3°, 8·4°, 11·2°, 15·3°, 22·4°, 29·7°.

Assuming the force necessary to cause the needle to describe each of the first degrees of the galvanometer to be equal to unity, we have the number 5 as the expression of the force corresponding to the first observation. The other forces are easily obtained by the proportions:—

$$1{\cdot}5 : 5 = a : x = \frac{5}{1{\cdot}5} a = 3{\cdot}333.*$$

where a represents the deflection when the exterior circuit is closed. We thus obtain

5, 10, 15·2, 21, 28, 37·3

for the forces, corresponding to the deflections,

5°, 10°, 15°, 20°, 25°, 30°.

In this instrument, therefore, the forces are sensibly proportional to the arcs, up to nearly 15 degrees. Beyond this, the proportionality ceases, and the divergence augments as the arcs increase in size.

The forces belonging to the intermediate degrees are obtained with great ease, either by calculation or by graphical construction, which latter is sufficiently accurate for these determinations.

By these means we find,

Degrees . .	13°,	14°,	15°,	16°,	17°,	18°,	19°,	20°,	21°.
Forces . .	13,	14·1,	15·2,	16·3,	17·4,	18·6,	19·8,	21,	22·3.
Differences . .		1·1,	1·1,	1·1,	1·1,	1·2,	1·2,	1·2,	1·3.

* That is to say, one reduced current is to the total current to which it corresponds, as any other reduced current is to its corresponding total current.

Degrees . .	22°, 23°, 24°, 25°, 26°, 27°, 28°, 29°, 30°.
Forces . .	23·5, 24·9, 26·4, 28, 29·7, 31·5, 33·4, 35·3, 37·3.
Differences . .	1·4, 1·5, 1·6, 1·7, 1·8, 1·9, 1·9, 2.

In this table we do not take into account any of the degrees preceding the 13th, because the force corresponding to each of them possesses the same value as the deflection.

The forces corresponding to the first 30 degrees being known, nothing is easier than to determine the values of the forces corresponding to 35, 40, 45 degrees, and upwards.

The reduced deflections of these three arcs are,

15·3°, 22·4°, 29·7°.

Let us consider them separately; commencing with the first. In the first place, then, 15 degrees, according to our calculation, are equal to 15·2; we obtain the value of the decimal 0·3 by multiplying this fraction by the difference 1·1 which exists between the 15th and 16th degrees; for we have evidently the proportion

$$1 : 1{\cdot}1 = 0{\cdot}3 : x = 0{\cdot}3.$$

The value of the reduced deflection corresponding to the 35th degree, will not, therefore, be 15·3°, but 15·2° + 0·3° = 15·5°. By similar considerations we find 23·5° + 0·6° = 24·1°, instead of 22·4° and 36·7° instead of 29·7° for the reduced deflections of 40 and 45 degrees.

It now only remains to calculate the forces belonging to these three deflections, 15·5°, 24·1°, and 36·7°, by means of the expression 3·333 a; this gives us—

the forces, 51·7, 80·3, 122·3;
for the degrees, 35, 40, 45.

Comparing these numbers with those of the preceding table, we see that the sensitiveness of our galvanometer diminishes considerably when we use deflections greater than 30 degrees.

CHAPTER XI.

ACTION OF ODOROUS SUBSTANCES UPON RADIANT HEAT—ACTION OF OZONE UPON RADIANT HEAT—DETERMINATION OF THE RADIATION AND ABSORPTION OF GASES AND VAPOURS WITHOUT ANY SOURCE OF HEAT EXTERNAL TO THE GASEOUS BODY—DYNAMIC RADIATION AND ABSORPTION—RADIATION THROUGH THE EARTH'S ATMOSPHERE—INFLUENCE OF THE AQUEOUS VAPOUR OF THE ATMOSPHERE ON RADIANT HEAT—CONNECTION OF THE RADIANT AND ABSORBENT POWER OF AQUEOUS VAPOUR WITH METEOROLOGICAL PHENOMENA.

APPENDIX:—FURTHER DETAILS OF THE ACTION OF HUMID AIR.

(435) SCENTS and effluvia generally have long occupied the attention of observant men, and they have formed favourite illustrations of the 'divisibility of matter.' No chemist ever weighed the perfume of a rose; but in radiant heat we have a test more refined than the chemist's balance. The results brought before you in our last chapter would enable you to hear the assertion without surprise, that the quantity of volatile matter removed from a hartshorn bottle by any person in this room, by a single act of inhalation, would exercise a more potent action on radiant heat, than the whole body of oxygen and nitrogen which the room contains. Let us apply this test to other odours, and see whether they also, notwithstanding their almost infinite attenuation, do not exercise a measurable influence on radiant heat.

(436) We will operate in a very simple way. A number of small and equal squares of bibulous paper are rolled up so as to form little cylinders, each about two inches in length. Each paper cylinder is then moistened by dipping

one end of it into an aromatic oil; the oil creeps by capillary attraction through the paper, until the whole of the roll becomes moist. The roll is introduced thus into a glass tube, of a diameter which enables the paper cylinder to fill it without being squeezed, and between the drying apparatus and the experimental tube is placed the tube containing the scented paper. The experimental tube is now exhausted and the needle at zero. Turning this cock on, dry air passes gently through the folds of the saturated paper. The air takes up the perfume of the aromatic oil, and carries it forward into the experimental tube. The absorption of one atmosphere of dry air we assume to be unity; and any additional absorption which these experiments reveal, must be due to the scent which accompanies the air.

(437) The following table will give a condensed view of the absorption of the substances mentioned in it, with reference to the unit just mentioned.

Perfumes.

Name of perfume	Absorption
Patchouli	30
Sandal-wood	32
Geranium	33
Oil of cloves	34
Otto of roses	37
Bergamot	44
Neroli	47
Lavender	60
Lemon	65
Portugal	67
Thyme	68
Rosemary	74
Oil of laurel	80
Camomile flowers	87
Cassia	109
Spikenard	355
Aniseed	372

(438) The number of atoms of air here in the tube must be regarded as almost infinite, in comparison with

those of the odours; still the latter, thinly scattered as they are, intercept, in the case of patchouli, 30 times the quantity absorbed by the air; otto of roses does upwards of 36 times the execution of the air; thyme, 74 times; spikenard, 355 times; and aniseed, 372 times. It would be idle to speculate on the quantities of matter implicated in these results. Probably they would have to be multiplied by millions to bring them up to the pressure of ordinary air. Thus,—

> The sweet south
> That breathes upon a bank of violets,
> Stealing and giving odour,

owes its sweetness to an agent, which, though almost infinitely attenuated, may be more potent, as an intercepter of terrestrial radiation, than the entire atmosphere from 'bank' to sky.

(439) In addition to these experiments on the essential oils, others were made on aromatic herbs. A number of such were obtained from Covent Garden Market; they were dry, in the common acceptation of the term; that is to say, they were not green, but withered. Still, I fear the results obtained with them cannot be regarded as pure, on account of the probable admixture of aqueous vapour. The aromatic parts of the plants were stuffed into a glass tube eighteen inches long and a quarter of an inch in diameter. Previous to connecting them with the experimental tube, they were attached to a second air-pump, and dry air was carried over them for some minutes. They were then connected with the experimental cylinder, and treated as the essential oils; the only difference being that a length of eighteen inches, instead of two, was occupied by the herbs.

Thyme, thus examined, gave an action thirty-three times that of the air which passed over it.

Peppermint exercised thirty-four times the action of the air.

Spearmint exercised thirty-eight times the action of the air.

Lavender exercised thirty-two times the action of the air.

Wormwood exercised forty-one times the action of the air.

Cinnamon exercised fifty-three times the action of the air.

As already hinted, these results may be complicated with the action of aqueous vapour: its quantity, however, must have been infinitesimal.

(440) There is another substance of great interest to the chemist, to which we may apply the test of radiant heat, but the attainable quantities of it are so minute as almost to elude measurement. I mean that extraordinary substance, ozone. This body is known to be liberated at the oxygen electrode, when water is decomposed by an electric current. To investigate its action, three different decomposing cells were constructed. In the first, No. 1, the platinum plates used as electrodes had about four square inches of surface; the plates of the second (No. 2) had two square inches of surface; while the plates of the third (No. 3) had only one square inch of surface, each.

(441) My reason for using eloctrodes of different sizes was this:—On first applying radiant heat to the examination of ozone, I constructed a decomposing cell, in which, to diminish the resistance of the current, very large platinum plates were used. The oxygen thus obtained, which ought to have embraced the ozone, showed scarcely any of the reactions of this substance. It hardly discoloured iodide of potassium, and was almost without action on radiant heat. A second decomposing apparatus, with

smaller plates, was tried, and here the action, both on iodide of potassium and on radiant heat, was found very decided. Being unable to refer these differences to any other cause than the different magnitudes of the plates, I formally attacked the subject, by operating with the three cells above described. Calling the action of the main body of the electrolytic oxygen unity; that of the ozone which accompanied it, in the respective cases, is given in the following table :—

Number of Cell	Absorption
No. 1	20
No. 2	34
No. 3	47

(442) Thus, the modicum of ozone which accompanied the oxygen, and in comparison to which it is a vanishing quantity, exerted, in the case of the first pair of plates, an action twenty times that of the oxygen itself, while with the third pair of plates the ozone was forty-seven times more energetic than the oxygen. The influence of the size of the plates, or, in other words, of the *density* of the current where it enters the liquid, on the production of ozone, is rendered strikingly manifest by these experiments.

(443) Portions of the plates of cell No. 2 were then cut away, so as to make them smaller than those of No. 3. The reduction of the plates was accompanied by an increase of the action upon radiant heat; the absorption rose at once from 34 to

65.

The reduced plates of No. 2 here transcend those of No. 3, which, in the first experiments, gave the largest action.

The plates of No. 3 were next reduced, so as to make

them smallest of all. The ozone now generated by No. 3 effected an absorption of

85.

Thus, we see that the action upon radiant heat advances as the size of the electrodes is diminished.

Heat is known to be very destructive of ozone; and suspecting the development of heat at the small electrodes of the cell last made use of, I surrounded the cell with a mixture of pounded ice and salt. Kept thus cool, the absorption of the ozone generated rose to

136.

(444) There is a perfect correspondence between these results, and those of MM. de la Rive, Soret, and Meidinger, though there is no resemblance between our respective modes of experiment. Such an agreement is calculated to augment our confidence in radiant heat, as an investigator of molecular condition.*

* M. Meidinger commences his paper by showing the absence of agreement between theory and experiment in the decomposition of water, the difference showing itself very decidedly in a deficiency of oxygen *when the current was strong*. On heating his electrolyte, he found that this difference disappeared, the proper quantity of oxygen being then liberated. He at once surmised that the defect of oxygen might be due to the formation of ozone; but how did the substance act to produce the diminution of the oxygen? If the defect were due to the great density of the ozone, the destruction of this substance, by heat, would restore the oxygen to its true volume. Strong heating, however, which destroyed the ozone, produced no alteration of volume, hence M. Meidinger concluded that the effect which he observed was not due to the ozone which remained mixed with the oxygen itself. He finally concluded, and justified his conclusion by satisfactory experiments, that the loss of oxygen was due to the formation, in the water, of peroxide of hydrogen by the ozone; the oxygen being thus withdrawn from the tube to which it belonged. He also, as M. de la Rive had previously done, experimented with electrodes of different sizes, and found the loss of oxygen much more considerable when a small electrode was used than with a large one; whence he inferred that the formation of ozone was facilitated by *augmenting the density of the current at the place where electrode and electrolyte meet.* The same conclusion is deduced from the above

(445.) The quantities of ozone involved in the foregoing experiments must be perfectly unmeasurable by ordinary means. Still, its action upon radiant heat is so energetic as to place it beside olefiant gas, or boracic ether, as an absorbent—bulk for bulk, it might transcend either. No *elementary gas* that I have examined behaves at all like ozone. In its swing through the ether it must powerfully disturb the medium. If it be oxygen, it must be oxygen atoms, packed into groups. I sought to decide the question whether it is oxygen, or a compound of hydrogen, in the following way. Heat destroys ozone. If it were oxygen only, heat would convert it into the common gas; if it were the hydrogen compound, which some chemists considered it to be, heat would convert it into oxygen, plus aqueous vapour. The gas alone, admitted into the experimental tube, would give the neutral action of oxygen, but the gas, plus the aqueous vapour, would probably give a greater action. The dried electrolytic gas was first caused to pass through a glass tube heated to redness, and thence without drying direct into the experimental tube. Secondly, after heating, it was dried before entering the experimental tube. Hitherto I have not been able to establish, with certainty, a difference between the dried and undried gas. If, therefore, the act of heating develope aqueous vapour, the experimental means employed have not yet enabled me to detect it. For the present, therefore, I hold the belief, that ozone is produced by the packing of the atoms of elementary oxygen into oscil-

experiments on radiant heat. No two things could be more diverse than the two modes of proceeding. M. Meidinger sought for the oxygen which had disappeared, and found it in the liquid; I examined the oxygen actually liberated, and found that the ozone mixed with it augments in quantity, as the electrodes diminish in size. It may be added, that since the perusal of M. Meidinger's paper I have repeated his experiments with my own decomposition cells, and found that those which gave me the greatest absorption, also showed the greatest deficiency in the amount of oxygen liberated.

lating groups; and that heating dissolves the bond of union, and allows the atoms to swing singly, thus disqualifying them for either intercepting or generating the motion, which, as systems, they are competent to intercept and generate.*

(446) Your attention is now to be directed to a series of facts which surprised and perplexed me, when they were observed. I permitted, on one occasion, a quantity of alcohol vapour, sufficient to depress the mercury gauge 0·5 of an inch, to enter the experimental tube; it produced a deflection of 72°. While the needle pointed to this high figure, and before pumping out the vapour, I allowed dry air to stream into the tube, and happened, as it entered, to keep my eye upon the galvanometer.

(447) The needle, to my astonishment, sank speedily to zero, and went to 25° on the opposite side. The entry of the ineffective air not only neutralised the absorption previously observed, but left a considerable balance in favour of the face of the pile turned towards the source. A repetition of the experiment brought the needle down from 70° to zero, and sent it to 38° on the opposite side. In like manner, a very small quantity of the vapour of sulphuric ether produced a deflection of 30°; on allowing dry air to fill the tube, the needle descended speedily to zero, and swung to 60° at the opposite side.

My first thought, on observing these extraordinary effects, was, that the vapours had deposited themselves in opaque films on the plates of rock-salt, and that the dry air, on entering, had cleared these films away, and allowed the heat from the source free transmission.

* The foregoing conclusion regarding the constitution of ozone was described at a time when the most eminent authorities regarded ozone as consisting of single atoms, and ordinary oxygen of groups of atoms. Chemical investigations have since independently established the view suggested by the above experiments on radiant heat.

(448) But a moment's reflection dissipated this supposition. The clearing away of such a film could, at best, but restore the state of things existing prior to the entrance of the vapour. It might be conceived able to bring the needle again to 0°, but it could not possibly produce the negative deflection. Nevertheless, I dismounted the tube, and subjected the plates of salt to a searching examination. No such deposit as that surmised was observed. The salt remained perfectly transparent while in contact with the vapour. How, then, are the effects to be accounted for?

(449) We have already made ourselves acquainted with the thermal effects produced when air is permitted to stream into a vacuum. We know that the air is warmed by its collision against the sides of the receiver. Can it be that the heat thus generated, imparted by the air to the alcohol and ether vapours, and radiated by them against the pile, was more than sufficient to make amends for the absorption? The *experimentum crucis* at once suggests itself here. If the effects observed be due to the heating of the air, on entering the partial vacuum in which the vapour was diffused, we ought to obtain the same effects, when the sources of heat hitherto made use of are entirely abolished. We are thus led to the consideration of the novel and, at first sight, utterly paradoxical problem —to determine the radiation and absorption of a gas or vapour *without any source of heat external to the gaseous body itself*.

(450) Let us, then, erect our apparatus, and abandon our two sources of heat. Here is our glass tube, stopped at one end by a plate of glass, for we do not now need the passage of the heat through this end; and at the other end by a plate of rock-salt. In front of the salt is placed the pile, connected with its galvanometer. Though there is now no special source of heat acting upon the pile, the

needle does not come quite to zero; indeed the walls of this room, and the people who sit around, are so many sources of heat, to neutralise which, and thus to bring the needle accurately to zero, I must slightly warm the defective face of the pile. This is done without any difficulty by a cube of lukewarm water, placed at a distance; the needle is now at zero.

(451) The experimental tube being exhausted, air is permitted to enter, till the tube is filled. This air is at present warm; every one of its atoms is oscillating; and if the atoms possessed any sensible power of communicating their motion to the luminiferous ether, we should have, from each atom, a train of waves impinging on the face of the pile. But you observe scarcely any motion of the galvanometer, and hence may infer that the quantity of heat radiated by the air is exceedingly small. The deflection produced is 7°.

(452) But these 7° are not really due to the radiation of the air. To what then? I open one of the ends of the experimental tube, and place a bit of black paper as a lining within it; the paper merely constitutes a ring, which covers the interior surface of the tube for a length of 12 inches. Let us now close the tube and repeat the last experiment. The air is now entering; but mark the needle—it has already flown through an arc of 70°. You see here exemplified the influence of this bit of paper lining; it is warmed by the air, and it radiates against the pile in this copious way. *The interior surface of the tube itself must do the same*, though in a less degree, and to the radiation from this surface, and not from the air itself, the deflection of 7° which we have just obtained is, I believe, to be ascribed.

(453) Removing the bit of lining from the tube, instead of air we will permit nitrous oxide to stream into it; the needle swings to 28°, thus showing the superior radiative

power of this gas over that of air. On working the pump, the gas within the experimental tube is chilled; and into it the pile pours its heat, a swing of 20° in the opposite direction being the consequence.

(454) Instead of nitrous oxide, I allow olefiant gas to enter the exhausted tube. We have already learned that this gas is highly gifted with the power of absorption and radiation. Its atoms are now warmed, and every one of them asserts its power; the needle swings through an arc of 67°. Let it waste its heat, and let the needle come to zero. On pumping out, the chilling of the gas within the tube produces a deflection of 40° on the side of cold. We have certainly here a key to the solution of the enigmatical effects, observed with the alcohol and ether vapour.

(455) For the sake of convenience we may call the heating of the gas on entering the vacuum *dynamic heating*; its radiation may be called *dynamic radiation*, and its absorption, when it is chilled by pumping out, *dynamic absorption*. These terms being understood, the following table explains itself. In each case, the extreme limit to which the needle swung, on the entry of the gas into the experimental tube, is recorded.

Dynamic Radiation of Gases.

Name	Limit of first impulsion
Air	7°
Oxygen	7
Hydrogen	7
Nitrogen	7
Carbonic oxide	19
Carbonic acid	21
Nitrous oxide	31
Olefiant gas	63

(456) We observe that the order of the radiative powers, determined in this novel way, is the same as that already obtained from a totally different mode of experiment. It

must be borne in mind that the discovery of dynamic radiation is quite recent, and that the conditions of perfect accuracy have not yet been developed; it is, however, certain, that the mode of experiment is susceptible of the highest degree of precision.

(457) Let us now turn to our vapours, and while dealing with them I shall endeavour to unite two effects which, at first sight, might appear utterly incongruous. We have already learned that a polished metal surface emits an extremely feeble radiation; but that, when the same surface is coated with varnish, the radiation is copious. In the communication of motion to the ether of space,* the atoms of the metal need a mediator, and this they find in the varnish. *You may varnish a metallic surface by a*

FIG. 89.

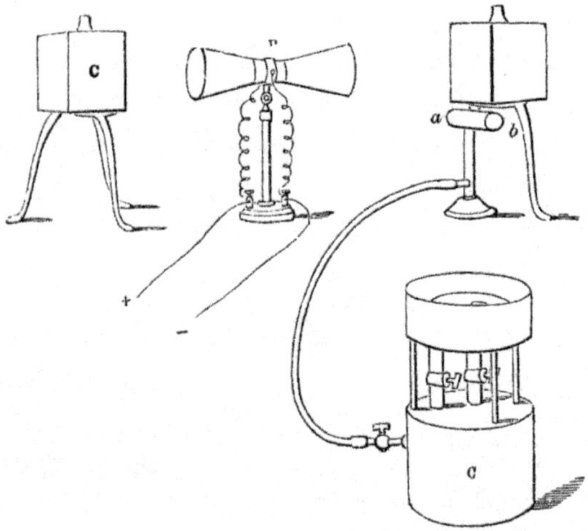

film of a powerful gas. The arrangement before you enables me to cause a thin stream of olefiant gas to pass

* If we could change either the name given to the interstellar medium, or that given to certain volatile liquids by chemists, it would be an advantage. It is difficult to avoid confusion in the use of the same term for objects so utterly diverse.

from the gasholder G (fig. 89) through a slit tube *a b*, over the heated surface of the cube C. At present no gas issues, and the radiation from C is neutralised by that from C′; but now I pour the gas from G over the cube C; and though the surface is actually cooled by the passage of the gas, for the gas has to be warmed by the metal, the effect is to augment considerably the radiation. As soon as the gas begins to flow, the needle begins to move, and reaches an amplitude of 45°.

(458) We have here varnished a metal by a gas, but a more interesting and subtle effect is *the varnishing of one gaseous body by another*. This flask contains acetic ether, a volatile, and, as you know, a highly absorbent substance. I attach the flask to the experimental tube, and permit the vapour to enter it, until the mercury column has been depressed half an inch. There is now vapour, under half an inch of pressure, in the tube. I intend to use that vapour as my varnish; the element oxygen, instead of the element gold, silver, or copper, being the substance to which this varnish is to be applied. At the present moment, the needle is at zero; permitting dry oxygen to enter the tube, the gas is dynamically heated, and we have seen its incompetence to radiate its heat; but now it comes into contact with the acetic ether vapour, and, communicating its heat-motion to the vapour by direct collision, the latter is able to send on the motion to the pile. Observe the needle—it is caused to swing through an arc of 70° by the radiation from the vapour molecules. It is not necessary to insist upon the fact, that in this experiment the vapour bears precisely the same relation to the oxygen as the varnish to the metal, in our former experiments.

(459) Let us wait a little, and allow the vapour to pour away the heat: it is the discharger of the calorific motion generated by the oxygen—the needle is again at zero. On

working the pump, the vapour within the tube is chilled, and the needle swings to nearly 45° on the other side of zero. In this way, the dynamic radiation and absorption of the vapours mentioned in the following table have been determined; air, however, instead of oxygen, being the substance employed to heat the vapour. The limit of the first swing of the needle is noted as before.

Dynamic Radiation and Absorption of Vapours.

	Deflections	
	Radiation	Absorption
1. Bisulphide of carbon	14°	6°
2. Iodide of methyl	20	8
3. Benzol	30	14
4. Iodide of ethyl	34	16
5. Methylic alcohol	36	18
6. Chloride of amyl	41	23
7. Amylene	48	26
8. Alcohol	50	28
9. Sulphuric ether	64	34
10. Formic ether	69	38
11. Acetic ether	70	43

(460) We have here used eleven different kinds of vapour, as varnish for our air, and we find that the dynamic radiation and absorption augment exactly in the order established by experiments with external sources of heat. We also see how beautifully dynamic radiation and absorption go hand in hand, the one augmenting and diminishing with the other.

(461) The smallness of the quantities of matter concerned in some of these actions on radiant heat has been often referred to; and I wish now to describe an experiment, which shall furnish a more striking example of this kind than any hitherto brought before you. The absorption of boracic ether vapour (see page 328) exceeds that of any other substance hitherto examined; and its dynamic radiation may be presumed to be commensurate. Let us exhaust the experimental tube as perfectly as pos-

sible, and introduce into it a quantity of boracic ether vapour, sufficient to depress the mercury column $\frac{1}{10}$th of an inch. The barometer stands to-day at 30 inches; hence, the pressure of the ether vapour now in our tube is $\frac{1}{300}$th of an atmosphere.

On sending dry air into the tube, the vapour is warmed, and the dynamic radiation produces the deflection 56°.

By working the pump, I reduce the residue of air within it to a pressure of 0·2 of an inch, or $\frac{1}{150}$th of an atmosphere. A portion of the boracic ether vapour remains of course in the tube, the pressure of this residue being the $\frac{1}{150}$th part of that of the vapour, when it first entered the tube. When dry air is permitted to enter, the dynamic radiation of the residual vapour produces a deflection of 42°.

We will again work the pump till the pressure of the air within it is 0·2 of an inch; the quantity of ether vapour now in the tube being $\frac{1}{150}$th of that present in the last experiment. The dynamic radiation of this residue gives a deflection of 20°.

Two additional experiments, conducted in the same way, gave deflections of 14° and 10° respectively. The question now is, what was the tenuity of the boracic ether vapour when this last deflection was obtained? The following table contains the answer to this question.

Dynamic Radiation of Boracic Ether.

Pressure in parts of an atmosphere	Deflection
$\frac{1}{300}$	56°
$\frac{1}{150} \times \frac{1}{300} = \frac{1}{45000}$	42
$\frac{1}{150} \times \frac{1}{150} \times \frac{1}{300} = \frac{1}{6750000}$	20
$\frac{1}{150} \times \frac{1}{150} \times \frac{1}{150} \times \frac{1}{300} = \frac{1}{1012500000}$	14
$\frac{1}{150} \times \frac{1}{150} \times \frac{1}{150} \times \frac{1}{150} \times \frac{1}{300} = \frac{1}{151875000000}$	10

(462) The air itself, warming the interior of the tube, produces, as we have seen, a deflection of 7°; hence the entire deflection of 10° was not due to the radiation of the

vapour. Deducting 7°, it would leave a remainder of 3°. But supposing we entirely omit the last experiment, we can then have no doubt that at least half the deflection 14° is due to the residue of boracic ether vapour; this we find, by strict measurement, would have to be multiplied by one thousand millions, to bring it up to the pressure of ordinary atmospheric air.

(463) Another reflection here presents itself, which is worthy of our consideration. We have measured the dynamic radiation of olefiant gas, by allowing the gas to enter our tube, until the latter was quite filled. Let us consider the state of the warm radiating column of olefiant gas in this experiment. It is manifest, that those portions of the column most distant from the pile must radiate *through the gas in front of them*, and, in this forward portion of the column of gas, a large quantity of the rays emitted by its hinder portion will be absorbed. In fact, it is quite certain that if we made our column sufficiently long, the frontal portions would act as a perfectly impenetrable screen to the radiation from the hinder ones. Thus, by cutting off that part of the gaseous column most distant from the pile, we might diminish only in a very small degree the amount of radiation which reaches the pile.

(464) Let us now compare the dynamic radiation of a vapour with that of olefiant gas. In the case of our vapour, we use only 0·5 of an inch of pressure, hence the radiating molecules of the vapour are much wider apart than those of the olefiant gas, under 60 times the pressure; and, consequently, the radiation of the hinder portions of the column of vapour will have a comparatively open door, through which to reach the pile. These considerations render it manifest that, in the case of the vapour, *a greater length of tube* is available for radiation than in the case of olefiant gas. This leads further to the conclusion, that if

we shorten the tube, we shall diminish the radiation, in the case of the vapour, more considerably than in that of the gas. Let us now bring our reasoning to the test of experiment.

(465) We have found the dynamic radiation of the following four substances, when the radiating column was 2 feet 9 inches long, to be represented by the annexed deflections:—

Olefiant gas	63°
Sulphuric ether vapour	64
Formic ether ,, ...	69
Acetic ether ,, ...	70

olefiant gas giving here the least dynamic radiation.

(466) Experiments made, in precisely the same manner, with a tube 3 inches long, or $\frac{1}{11}$th of the former length, gave the following deflections:—

Olefiant gas	39°
Sulphuric ether vapour	11
Formic ether ,, ...	12
Acetic ether ,, ...	15

The verification of our reasoning is therefore complete. It is proved, that in the long tube the dynamic radiation of the vapour exceeds that of the gas, while in a short one the dynamic radiation of the gas exceeds that of the vapour. The result proves, if proof were needed, that though diffused in air, the vapour molecules are really the centres of radiation.

(467) Up to the present point, I have purposely omitted a reference to the most important vapour of all, as far as our world is concerned—the vapour of water. This vapour, as you know, is always diffused through the atmosphere. The clearest day is not exempt from it: indeed, in the Alps, the purest skies are often the most treacherous, the blue deepening with the amount of

aqueous vapour in the air. It is needless, therefore, to remind you, that when aqueous vapour is spoken of, nothing visible is meant; it is not fog; it is not cloud; it is not mist of any kind. These are formed of vapour which has been condensed to water; but the true vapour, with which we have to deal, is an impalpable transparent gas. It is diffused everywhere throughout the atmosphere, though in very different proportions.

(468) To prove the existence of aqueous vapour in the air by which we are now surrounded, a copper vessel, filled an hour ago with a mixture of pounded ice and salt, is placed in front of the table. The surface of the vessel was then black, but it is now white—furred all over with hoar-frost—produced by the condensation, and subsequent congelation upon its surface, of the aqueous vapour. This white substance can be scraped off; as the frozen vapour is removed, the black surface of the vessel reappears; and now a sufficient quantity is collected to form a respectable snowball. Let us go one step farther. I place this snow in a mould, and squeeze it before you into a cup of ice; and thus, without quitting this room, we have experimentally illustrated the manufacture of glaciers, from beginning to end. On the plate of glass used to cover the vessel, the vapour is not congealed, but it is condensed so copiously, that when the plate is held edgeways, the water runs off it in a stream.

(469) The quantity of this vapour is small. Oxygen and nitrogen constitute about 99½ per cent. of our atmosphere; of the remaining 0·5, about 0·4½ is aqueous vapour; * the rest is carbonic acid. Had we not been

* The known tenuity of the aqueous vapour of the atmosphere caused Prof. Magnus, when he made his first experiments on the vapour of the air, to say, 'that it was evident beforehand that such vapour could exert no sensible action.' Had he approached the subject, as we have done, through the foregoing experiments, so cautious a philosopher would not, I think, have made this statement.

already acquainted with the action of almost infinitesimal quantities of matter on radiant heat, we might well despair of being able to establish a measurable action, on the part of the aqueous vapour of our atmosphere. Indeed, I quite neglected the action of this substance for a time, and could hardly credit my first result, which made the action of the aqueous vapour of our laboratory fifteen times that of the air in which it was diffused. This, however, by no means expresses the true relation between aqueous vapour and dry air.

(470) To illustrate this point, our first arrangement, shown in Plate I. has been resumed, a brass tube being employed, and two sources of heat, acting on the opposite faces of the pile. The experiment with dry air is repeated by permitting it to enter the experimental cylinder. The needle does not move sensibly. If close to it, you would observe a motion through about one degree. Could we get our air quite pure, its action would be even less than this. Let us again pump out, and allow the air of this room to enter the experimental cylinder direct, without permitting it to pass through the drying apparatus. The needle, you observe, moves as the air enters, and the final deflection is 48°. The needle will point steadily to this figure, as long as the sources of heat remain constant, and as long as the air continues in the tube. These 48° correspond to an absorption of 72; that is to say, the aqueous vapour contained in the atmosphere of this room to-day exerts an action on the radiant heat 72 times as powerful as that of the air itself.

(471) This result is obtained with perfect ease, still not without due care. In comparing dry with humid air, it is perfectly essential that the substances be pure. You may work for months with an imperfect drying apparatus, and fail to obtain air which shows this almost total absence of action on radiant heat. An amount of organic impurity

too small to be seen by the eye, is sufficient to augment fiftyfold the action of the air. Knowing the effect which an almost infinitesimal amount of matter, in certain cases, can produce, you are better prepared for such facts than I was, when they first forced themselves on my attention. The experimental result which we have just obtained will, if true, have so important an influence on the science of meteorology, that, before it is admitted, it ought to be subjected to the closest scrutiny. First of all, then, look at this piece of rock-salt brought in from the next room, where it has stood for some time near a tank, but not in contact with visible moisture. The salt is wet; it is a hygroscopic substance, and freely condenses moisture upon its surface. Here, also, is a polished plate of the substance, which is now quite dry: I breathe upon it, and instantly its affinity for moisture causes the vapour of my breath to overspread the surface, in a film which exhibits beautifully the colours of thin plates.* Now we know, from a former table (page 303) how opaque a solution of rock-salt is to the calorific rays, and hence arises the question whether, in the above experiment with undried air, we may not in reality be measuring the action of a thin stratum of such a solution, deposited on our plates of salt, instead of the pure action of the aqueous vapour of the air.

(472) If we operate incautiously, and, more particularly, if it be our actual intention to wet the plates of salt, we may readily obtain the deposition of moisture. This is a point on which any competent experimenter will soon instruct himself; but the essence of good experimenting consists in the exclusion of circumstances which would

* Receiving the beam from the electric lamp upon the polished plate of salt, so as to reflect the light on to a screen; and placing a lens in front of the salt, so as to produce an image of its polished surface on the screen; on breathing against the salt through a glass tube, rings of vivid iridescence instantly flash forth, which may be seen by hundreds at once. The order of the colours is that of Newton's rings.

render the pure and simple questions which we intend to put to Nature, impure and composite ones. The first way of replying to the doubt here raised is to examine our plates of salt; if the experiments have been properly conducted, no trace of moisture is found upon the surface. To render the success of this experiment more certain, we will slightly alter the arrangement of our apparatus. Hitherto we have had the thermo-electric pile and its two reflectors entirely *outside* the experimental cylinder. Taking this right-hand reflector from the pile, and removing this plate of rock-salt, I push the reflector into the experimental cylinder. The hollow reflecting cone is 'sprung' at its base *a b* (fig. 90), (this is our former arrangement, Plate I., with the single exception that one

FIG. 90.

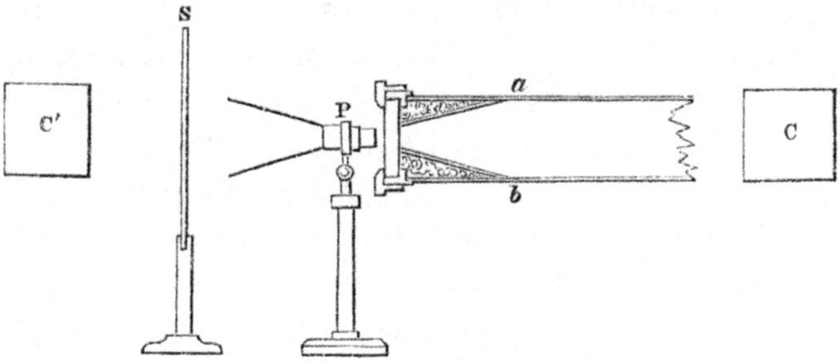

of the reflectors of the pile P is now within the tube) so that it is held tightly by its own pressure against the inner surface of the cylinder. The space between the outer surface of the reflector and the inner surface of the tube is filled with fragments of fused chloride of calcium, which are prevented from falling out by a little screen of wire-gauze, and then the plate of salt is reattached. Against the inner surface of the salt the narrow end of the reflector now abuts. Bringing the face of the pile close up to the plate, though not into actual contact with it, our arrangement is complete.

(473) In the first place, it is to be remarked, that the plate of salt nearest to the source of heat is never moistened, unless the experiments are of the roughest character. Its proximity to the source enables the heat to chase away every trace of humidity from its surface. The distant plate is the one in danger, and now we have the circumferential portions of this plate kept perfectly dry by the chloride of calcium. No moist air can at all reach the rim of the plate; while upon its central portion, measuring about a square inch in area, *we have converged our entire radiation.* On *à priori* grounds, we should conclude that it is quite impossible for a film of moisture to collect there; and this conclusion is justified by fact. Testing, as before, the dried and the undried air of this room, we find, as in the former instance, that the latter produces seventy times the effect of the former. The needle is now deflected, by the absorption of the undried air; allowing this air to remain in the tube, we will unscrew the plate of salt, and examine its surface. We may even use a lens for this purpose, taking care, however, that the breath does not strike the plate. It was carefully polished when attached to the tube; it is perfectly polished now. Glass, or rock-crystal, could not show a surface more exempt from any appearance of moisture. When a dry handkerchief placed over my finger is drawn along the surface, no trace is left behind. There is not the slightest deposition of moisture; still, we see that absorption has taken place. This experiment is conclusive against the hypothesis that the effects observed are due to a film of brine, instead of to aqueous vapour.

(474) The doubt may, however, linger, that although we are unable to detect the film of moisture, it may still be there. This doubt is answered in the following way: I detach the experimental tube from the front chamber, and remove the two plates of rock-salt; the tube is now *open*

at both ends, and my aim will be to introduce dry and moist air into this open tube, and to compare their effects upon the radiation. And here, as in all other cases, the practical tact of the experimenter must come into play. The source on the one hand, and the pile on the other, are now freely exposed to the air; a very slight agitation, acting upon either, would disturb, and might indeed altogether mask, the effect we seek. The air, then, must be introduced into the open tube, without producing any commotion, either near the source or near the pile. The length of the experimental tube is now 4 feet 3 inches; at C (fig. 91) is a cock connected with an india-rubber bag containing common air, and subjected

FIG. 91.

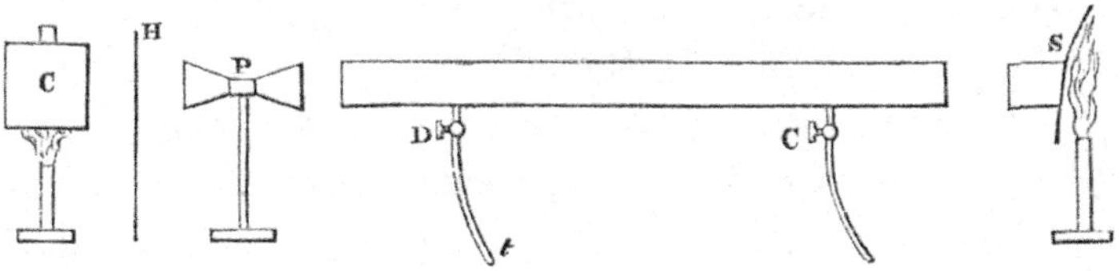

by a weight to gentle pressure; at D is a second cock, connected by a flexible tube, *t*, with an air-pump. Between the cock C, and the india-rubber bag, our drying tubes are introduced; and when that cock is opened, the air is forced gently through the drying tubes into the experimental cylinder. The air-pump is slowly worked at the same time, the dry air being thereby drawn towards D. The distance of C from the source S is 18 inches, and the distance of D from the pile P is 12 inches: the compensating cube C, and the screen H, serve the same purposes as before. By thus isolating the central portion of the tube, we can displace dry air by moist, or moist air by dry, without permitting any agitation to reach either the source or the pile.

(475) At present, the tube is filled with the common air

of the laboratory, and the needle of the galvanometer points steadily to zero. I allow air to pass through the drying apparatus, and to enter the open tube at c, the pump being worked as already described. Mark the effect. When the dry air enters, the needle commences moving, and the direction of its motion shows that more heat is now passing than before. The substitution of dry air for the air of the laboratory has rendered the tube more transparent to the rays of heat. The final deflection thus obtained is 45°. Here the needle steadily remains, and beyond this point it cannot be moved by any further drawing in of dry air.

(476) Let us now shut off the supply of dry air, and cease working the pump; the needle sinks, but with great slowness, indicating a correspondingly slow diffusion of the aqueous vapour of the adjacent air into the dry air of the tube. If the pump be worked, the removal of the dry air is hastened, and the needle sinks more speedily,—it now points to zero. The experiment may be made a hundred times in succession without any deviation from this result; on the entrance of the dry air, the needle invariably goes up to 45°, showing augmented transparency; on the entrance of the undried air, the needle sinks to 0°, showing augmented absorption.

(477) But the atmosphere to-day is not saturated with moisture; hence, if saturated, we might expect to find a greater action. I remove the drying apparatus, and put in its place a U-tube, filled with fragments of glass moistened by distilled water. Through this tube air is forced from the india-rubber bag, the pump being worked as before. We are now displacing the humid air of the laboratory by still more humid air, and see the consequence. The needle moves in a direction which indicates augmented opacity, the final deflection being 15°.

(478) Here, then, we have substantially the same result

as that obtained when our tube was stopped with plates of rock-salt; the action, therefore, cannot be referred to a film of moisture deposited upon the surface of the plates. And be it remarked that there is not the slightest caprice or uncertainty in these experiments when properly conducted. They have been executed at different times and seasons; the tube has been dismounted and remounted; the suggestions of eminent men who have seen the experiments, and whose object it was to test the results, have been complied with; but no deviation from the effects just recorded has been observed. The entrance of each kind of air is invariably accompanied by its characteristic action; the needle is under the most complete control: in short, no experiments hitherto made with solid and liquid bodies are more certain in their execution than the foregoing experiments on dry and humid air.

(479) We can easily estimate the percentage of the entire radiation absorbed by the common air, between the points C and D. Introducing this tin screen between the experimental cylinder and the pile, one of the sources of heat is entirely shut off. The deflection produced by the other source indicates the *total radiation*. This deflection corresponds to about 780 of the units which have been hitherto adopted; one unit being the quantity of heat necessary to move the needle from 0° to 1°. The deflection of 45° corresponds to 62 units; out of 780, therefore, 62, in this instance, have been absorbed by the moist air. The following statement gives us the absorption per hundred:—

$$780 : 100 = 62 : 7{\cdot}9$$

An absorption of nearly 8 per cent. was, therefore, effected by the atmospheric vapour which occupied the tube between C and D. Air *perfectly saturated* gives a still greater absorption.

(480) This absorption took place, notwithstanding the partial *sifting* of the heat, in its passage from the source to C, and from D to the pile. The moist air, moreover, was, probably, only in part displaced by the dry. In other experiments with a tube 4 feet long, and polished within, it was found that the atmospheric vapour, on a day of average dryness, absorbed over 10 per cent. of the radiation from our source. Regarding the earth as a source of heat, no doubt *at least* 10 *per cent. of its heat is intercepted within ten feet of the surface.** This single fact suggests the enormous influence which this newly-developed property of aqueous vapour must have in the phenomena of meteorology.

(480*a*) But we have not yet disposed of all objections. It has been intimated to me that the air of our laboratory might be impure; the suspended carbon particles of the London air have also been referred to, as a possible cause of the absorption ascribed to aqueous vapour. The results, however, were obtained, when the apparatus was removed from the laboratory—they are obtainable in this room. Air, moreover, was brought from the following localities in impervious bags:—Hyde Park, Primrose Hill, Hampstead Heath, Epsom Downs (near the Grand Stand); a field near Newport, Isle of Wight; St. Catharine's Down, Isle of Wight; the seabeach near Black Gang Chine. *The aqueous vapour of the air from all these localities, examined in the usual way, exerted an absorption seventy times that of the air in which the vapour was diffused.*

(481) Again, I experimented thus. The air of the laboratory was dried and purified, until its absorption fell below unity; this purified air was then led through a U-tube, filled with fragments of perfectly clean glass moistened

* Under some circumstances, the absorption, I have reason to believe, considerably exceeds this amount.

with distilled water. Its neutrality, when dry, showed that all prejudicial substances had been removed from it, and in passing through the U-tube, it could take up nothing but the pure vapour of water. The vapour thus carried into the experimental tube produced an action ninety times greater than that of the air which carried it.

(482) But fair and philosophic criticism does not end even here. The tube with which these experiments were made is polished within, and it might be surmised that the vapour of the humid air had, on entering, deposited itself upon the interior surface of the tube, thus diminishing its reflective power, and producing an effect apparently the same as absorption. To this it may in the first place be replied that the amount of heat intercepted is proportional to the quantity of air present. This is shown by the following table, which gives the absorption, by humid air, at pressures varying from 5 to 30 inches of mercury:—

Humid Air.

Pressure in inches	Absorption: Observed	Absorption: Calculated
5	16	16
10	32	32
15	49	48
20	64	64
25	82	80
30	98	96

(483) The third column of this table is calculated on the assumption that the absorption is proportional to the quantity of vapour in the tube, and the agreement of the calculated and observed results shows this to be the case, within the limits of the experiment. It cannot be supposed that effects so regular as these, and agreeing so completely with those obtained with small quantities of other vapours, and even with small quantities of the per-

manent gases, can be due to the condensation of the vapour on the interior surface. When, moreover, five inches of air were in the tube, less than one-sixth of the vapour necessary to saturate the space was present. The driest day would make no approach to this dryness. That condensation, especially condensation which should destroy, by its action upon the inner reflector, quantities of heat so accurately proportional to the quantities of matter present, should here occur is scarcely to be thought of.

(484) My desire, however, was to take this important question quite out of the domain of mere reasoning, however strong this might be. It was therefore resolved to abandon not only the plates of rock-salt but also the experimental tube itself, and to displace one portion of the free atmosphere by another. With this view, the following arrangement was made: C (fig. 92), a cube of boiling water, is our source of heat. Y is a hollow brass cylinder set upright, 3·5 inches wide, and 7·5 inches high. P is the thermo-electric pile, and C′ a compensating cube, between which and P is an adjusting screen, to regulate the amount of heat falling on the posterior surface of the pile. The whole arrangement was surrounded by a hoarding, the space within which was divided into compartments by sheets of tin, and these spaces were stuffed loosely with paper or horsehair. These precautions, which required time to be learned, were necessary to prevent the formation of local air-currents, and also to intercept the irregular action of the external air. The effect to be measured here is very small, and hence the necessity of removing all causes of disturbance, which could possibly interfere with its clearness and purity.

(485) A rose-burner *r* was placed at the bottom of the cylinder Y, and from it a tube passed to an india-rubber bag containing air. The cylinder Y was first filled with

fragments of rock-crystal, moistened with distilled water. On subjecting the india-rubber bag to pressure, the air was gently forced up among the fragments of quartz, and having there charged itself with vapour, it was discharged in the space between the cube C and the pile. Previous to this the needle stood at zero; but on the

FIG. 92.

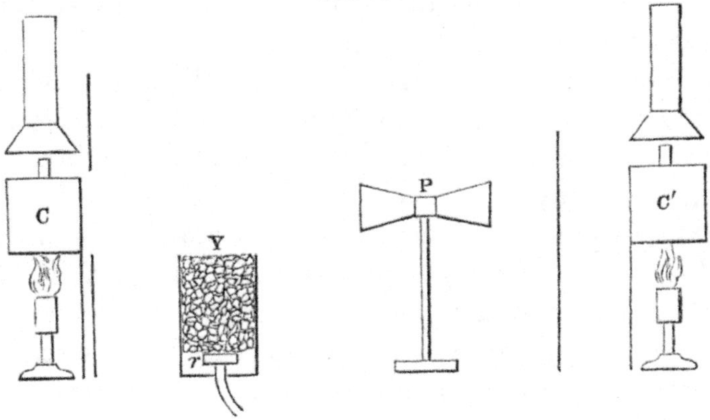

emergence of the saturated air from the cylinder, the needle moved and took up a final deflection of 5 degrees. The direction of the deflection showed that the opacity of the space, between the source C and the pile, was augmented by the saturated air.

(486) The quartz fragments were now removed, and the cylinder filled with fragments of fresh chloride of calcium, through which the air was gently forced, exactly as in the last experiment. Now, however, in passing through the chloride of calcium, it was, in great part, robbed of its aqueous vapour; and the air, thus dried, displaced the common air between the source and the pile. The needle moved, declaring a permanent deflection of 10 degrees; the direction of the deflection showed that the transparency of the space was augmented by the presence of the dry air. By properly timing the discharges of the

air, the swing of the needle could be augmented to 15 or 20 degrees. Repetition showed no deviation from this result; the saturated air always augmented the opacity, the dry air always augmented the transparency, of the space between the source and the pile. Not only, therefore, have the plates of rock-salt been abandoned, but also the experimental tube itself; and the results are all perfectly concurrent, as regards the action of aqueous vapour upon radiant heat.

(486*a*) Many remarkable corroborations of these views have been published by that excellent meteorologist, Col. Richard Strachey, of the Royal Engineers. And his testimony is all the more valuable, as it is based on observations made long before the property of aqueous vapour here developed was known to have an existence. From his important paper, published in the Philosophical Magazine for July 1866, I extract a single representative series of observations, made between the 4th and the 25th of March 1850; during which period 'the sky remained remarkably clear, while great variations in the quantity of vapour took place.' The first column of figures gives the tension of aqueous vapour, and the second the fall of the thermometer from 6.40 p.m. to 5.40 a.m.

Tension of vapour	Fall of thermometer
0·888 inches	6·0°
0·849 "	7·1
0·805 "	8·3
0·749 "	8·5
0·708 "	10·3
0·659 "	12·6
0·605 "	12·1
0·554 "	13·1
0·435 "	16·5

The general result is here unmistakable. In clear nights the fall of the thermometer, which is an expression of the energy of the radiation, is determined by the amount of transparent aqueous vapour in the air. The presence of the vapour checks the loss by radiation, while its removal favours radiation and promotes the nocturnal chill.

(487) We shall subsequently add another powerful proof to those here given. Were the subject less important, I should not have dwelt upon it so long. It was thought right to remove as far as possible every objection, so that meteorologists might apply, without the faintest misgiving, the results of experiment. The applications of these results to their science must be innumerable; and here I cannot but regret that the incompleteness of my knowledge prevents me from making the proper applications myself. I would, however, ask your permission to refer to such points as can now be called to mind, with which the facts just established appear to be more or less intimately connected.

(488) And, first, it is to be remarked, that the vapour which absorbs heat thus greedily, radiates it copiously. This fact must come powerfully into play in the tropics. We know that the sun raises from the equatorial ocean enormous quantities of vapour, and that immediately under him, in the region of calms, the rain, due to the condensation of the vapour, descends in deluges. Hitherto, this has been ascribed to the chilling which accompanies the expansion of ascending air; and no doubt this, as a true cause, must produce its proportional effect. But the radiation from the vapour itself must also be influential. Imagine a column of saturated air ascending from the equatorial ocean; for a time, the vapour entangled in this air is surrounded by air almost fully saturated. The ascending vapour radiates, but it radiates into the surrounding vapour; and to the radiation from any vapour, the same vapour, as proved by Kirchhoff, is particularly opaque. Hence, for a time, the radiation from our ascending column is intercepted, and in great

part returned, by the surrounding vapour; condensation under such circumstances is rendered difficult. But the quantity of aqueous vapour in the air diminishes speedily as we ascend; the decrement of its tension, as proved by the observations of Hooker, Strachey, and Welsh, is much more speedy than that of the air; and finally, our vaporous column finds itself elevated beyond the protecting screen which, during the first portion of its ascent, was spread above it. It is now in the presence of pure space, and into space it pours its heat, without stoppage or requital. To the loss of heat thus endured, the condensation of the vapour, and its torrential descent to the earth, must certainly be in part ascribed.

(489) Similar remarks apply to the formation of cumuli in our own latitudes; they are the heads of columns of vapour, which rise from the earth's surface, and are precipitated as soon as they reach a certain elevation. Thus, the visible cloud forms the capital of an invisible pillar of saturated air. Certainly the top of such a column, raised above the lower vapour-screen which clasps the earth, and offering itself to space, must be chilled by radiation; in this action alone we have to some extent a physical cause for the generation of clouds.

(490) Mountains act as condensers, partly by the coldness of their own masses; which they owe to their elevation. Above them spreads no vapour-screen of sufficient density to intercept their heat, which consequently gushes unrequited into space. When the sun is withdrawn, this loss is shown by the quick descent of the thermometer. The descent is not due to radiation from the air, but to radiation from the earth, or from the thermometer itself. Thus, the difference between a thermometer which, properly confined, gives the true temperature of the night air, and one which is permitted to radiate freely towards space, must be greater at high elevations than at low ones. This

conclusion is entirely confirmed by observation. On the Grand Plateau of Mont Blanc, for example, MM. Martins and Bravais found the difference between two such thermometers to be 24° Fahr.; when a difference of only 10° was observed at Chamouni.

(491) But mountains also act as condensers, by the deflection upwards of moist winds, and the consequent expansion of the air. The chilling thus produced is the same as that which accompanies the direct ascent of a column of warm air into the atmosphere; the elevated air performs work, and its heat is correspondingly consumed. But, in addition to these causes, we must take into account the radiant power of the moist air, when thus tilted upwards. It is thereby lifted beyond the protection of the aqueous layer which lies close to the earth, and therefore pours its heat freely into space, thus effecting its own condensation. No doubt, I think, can be entertained, that the extraordinary energy of water as a radiant, *in all its states of aggregation*, must play a powerful part in a mountain region. As vapour, it pours its heat into space, and promotes condensation; as liquid, it pours its heat into space, and promotes congelation; as snow, it pours its heat into space, and thus converts the surfaces on which it falls into more powerful condensers than they otherwise would be. Of the numerous wonderful properties of water, not the least important is this extraordinary power which it possesses, of discharging the motion of heat upon the interstellar ether.

(492) A freedom of escape, similar to that from bodies of vapour at great elevations, would occur at the earth's surface generally, were the aqueous vapour removed from the air above it, for the great body of the atmosphere is a practical vacuum, as regards the transmission of radiant heat. The withdrawal of the sun from any region over which the atmosphere is dry, must be followed by quick

refrigeration. The moon would be rendered entirely uninhabitable by beings like ourselves through the operation of this single cause; with a radiation, uninterrupted by aqueous vapour, the difference between her monthly maxima and minima must be enormous. The winters of Thibet are almost unendurable, from the same cause. Witness how the isothermal lines dip from the north into Asia, in winter, as a proof of the low temperature of this region. Humboldt has dwelt upon the 'frigorific power' of the central portions of this continent, and controverted the idea that it was to be explained by reference to the elevation; there being vast expanses of country, not much above the sea-level, with an exceedingly low temperature. But not knowing the influence which we are now studying, Humboldt, I imagine, omitted the most potent cause of the cold. The refrigeration at night is extreme when the air is dry. The removal, for a single summer night, of the aqueous vapour from the atmosphere which covers England, would be attended by the destruction of every plant which a freezing temperature could kill. In Sahara, where 'the soil is fire and the wind is flame,' the cold at night is often painful to bear. Ice has been formed in this region at night. In Australia, also, the *diurnal range* of temperature is very great, amounting, commonly, to between 40 and 50 degrees. In short, it may be safely predicted, that wherever the air is *dry*, the daily thermometric range will be great. This, however, is quite different from saying that where the air is *clear*, the thermometric range will be great. Great clearness to light is perfectly compatible with great opacity to heat; the atmosphere may be charged with aqueous vapour while a deep blue sky is overhead, and on such occasions the terrestrial radiation would, notwithstanding the 'clearness,' be intercepted.

(493) And here we are led to an easy explanation of a fact which evidently perplexed Sir John Leslie. This celebrated experimenter constructed an instrument which he named an *æthrioscope*, the function of which was to determine the radiation against the sky. It consisted of two glass bulbs united by a vertical glass tube, so narrow that a little column of liquid was supported in the tube by its own adhesion. The lower bulb D (fig. 93) was protected by a metallic envelope, and gave the temperature of the air; the upper bulb B was blackened, and was surrounded by a metallic cup C, which protected the bulb from terrestrial radiation.

FIG. 93.

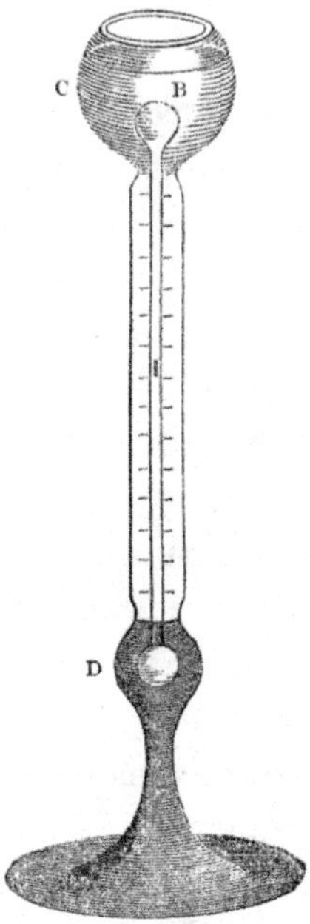

(494) 'This instrument,' says its inventor, 'exposed to the open air in clear weather, will at all times, both during the day and the night, indicate an impression of cold shot downwards from the higher regions. . . . The sensibility of the instrument is very striking, for the liquor incessantly falls and rises in the stem with every passing cloud. But the cause of its variations does not always appear so obvious. Under a fine blue sky the *æthrioscope* will sometimes indicate a cold of 50 millesimal degrees; yet on other days, *when the air seems equally bright*, the effect is hardly 30°.' This anomaly is simply due to the difference in the quantity of aqueous vapour present in the atmosphere. Indeed, Leslie himself connects the effect with aqueous vapour in these words: 'The pressure of hygrometric moisture in the air probably affects the instrument.' It is not, however, the 'pressure'* that is effective; the presence of invisible

* Possibly the word 'pressure' is a misprint for 'presence.'

vapour intercepted the radiation from the æthrioscope, while its absence opened a door for the escape of this radiation into space. As regards experiments on terrestrial radiation, a new definition will have to be given for 'a clear day;' it is manifest, for example, that in experiments with the pyrheliometer,* two days of equal visual clearness may give totally different results. We are also enabled to account for the fact that the radiation from this instrument is often intercepted, when no cloud is seen. Could we, however, make the constituents of the atmosphere, its vapour included, objects of vision, we should see sufficient to account for this result.

(495) Another interesting point, on which this subject has a bearing, is the theory of *serein*. 'Most authors,' writes Melloni, 'attribute to the cold resulting from the radiation of the air, the excessively fine rain which sometimes falls in a clear sky, during the fine season, a few moments after sunset.' 'But,' he continues, 'as no fact is yet known which directly proves the emissive power of pure and transparent elastic fluids,† it appears to me more conformable,' &c. &c. If the difficulty here urged against the theory of *serein* be its only one, the theory will stand, for transparent elastic fluids are now proved to possess the power of radiation which the theory assumes. It is not, however, to radiation from the *air* that the chilling can be ascribed, but to radiation from the body itself, whose condensation produces the *serein*.

(496) Let me add the remark, that as far as I can at present judge, aqueous vapour and liquid water absorb the same class of rays; this is another way of stating that the colour of pure water is shared by its vapour. In virtue of aqueous vapour, the atmosphere is therefore a blue medium. It has been remarked, that the colour

* The instrument is described in Chapter XIII.

† This statement indicates the state of the science of thermotics in reference to the gaseous form of matter when these researches were begun.

of the firmamental blue, and of distant hills, *deepens* with the amount of aqueous vapour in the air ; but the substance which produces a *variation* of depth must be effective as an *origin* of colour. Whether the azure of the sky—the most difficult question of meteorology,—is to be thus accounted for, I will not at present stop to enquire.*

* In connection with the investigation of the radiation and absorption of heat by gases and vapours, it gives me pleasure to refer to the prompt and intelligent aid rendered me by Mr. Becker, of the firm of Elliott's, West Strand. Mr. Becker is well acquainted with the apparatus necessary for those experiments.

The fear of being led too far from my subject causes me to withhold all speculation as to the cause of atmospheric polarisation. I may, however, remark, that the polarisation of heat was illustrated by means of the mica piles with which Professor (now Principal) J. D. Forbes first succeeded in establishing the fact of polarisation.

NOTE.—In a paper recently published by him, Professor Magnus controverts the views maintained in the foregoing chapter regarding the action of aqueous vapour. I shall give the experiments and reasonings of my eminent friend (by which I am not convinced) due consideration as soon as I can command the time.

NOTE, 1870.—The fifteenth chapter of this volume is devoted to the colour and polarization of the light of the sky.

APPENDIX TO CHAPTER XI.

EXTRACTS FROM A DISCOURSE 'ON RADIATION THROUGH THE EARTH'S ATMOSPHERE.'

'NOBODY ever obtained the idea of a line from Euclid's definition of it—"length without breadth." The idea is obtained from a real physical line, drawn by a pen or pencil, and therefore possessing width; this idea being afterwards brought, by a process of abstraction, more nearly into accordance with the conditions of the definition. So, also, with regard to physical phenomena; we must help ourselves to a conception of the invisible, by means of proper images derived from the visible, afterwards purifying our conceptions to the needful extent. Definiteness of conception, even though at some expense to delicacy, is of the greatest utility in dealing with physical phenomena. Indeed, it may be questioned whether a mind trained in physical research can at all enjoy peace, without having made clear to itself some possible way of conceiving those operations which lie beyond the boundaries of sense, and in which sensible phenomena originate.

'When we speak of radiation through the atmosphere, we ought to be able to affix definite physical ideas, both to the term atmosphere and the term radiation. It is well known that our atmosphere is mainly composed of the two elements, oxygen and nitrogen. These elementary atoms may be figured as small spheres, scattered thickly in the space which immediately surrounds the earth. They constitute about 99½ per cent. of the atmosphere. Mixed with these atoms, we have others of a totally different character; we have the molecules, or atomic groups, of carbonic acid, of ammonia, and of aqueous vapour. In these substances diverse atoms have coalesced, forming little systems of atoms. The molecule of aqueous vapour, for example, consists of two atoms of hydrogen, united to one of oxygen; and they mingle, as little triads, among the monads of oxygen and nitrogen which constitute the great mass of the atmosphere.

'These atoms and molecules are separate, but they are embraced by a common medium. Within our atmosphere exists a second and a finer atmosphere, in which the atoms of oxygen and nitrogen hang like suspended grains. This finer atmosphere unites not only atom with atom, but star with star; and the light of all suns, and of all stars, is in reality a kind of music, propagated through this interstellar air. This image must be clearly seized, and then we have to advance a step. We must not only figure our atoms suspended in this medium, but vibrating in it. In this motion of the atoms consists what we call their heat. "What is heat in us," as Locke has perfectly expressed it, "is in the body heated nothing but motion." Well, we must figure this motion communicated to the medium in which the atoms swing, and sent in ripples through it, with inconceivable velocity, to the bounds of space. Motion in this form, unconnected with ordinary matter, but speeding through the interstellar medium, receives the name of Radiant Heat; and, if competent to excite the nerves of vision, we call it Light.

'Aqueous vapour was defined to be an invisible gas. Vapour was permitted to issue horizontally with considerable force from a tube connected with a small boiler. The track of the cloud of condensed steam was vividly illuminated by the electric light. What was seen, however, was not vapour, but vapour condensed to water. Beyond the visible end of the jet, the cloud resolved itself into true vapour. A lamp was placed under the jet, at various points; the cloud was cut sharply off at that point, and when the flame was placed near the efflux orifice, the cloud entirely disappeared. The heat of the lamp completely prevented precipitation. This same vapour was condensed and congealed on the surface of a vessel containing a freezing mixture, from which it was scraped, in quantities sufficient to form a small snowball. The beam of the electric lamp, moreover, was sent through a large receiver placed on an air-pump. A single stroke of the pump caused the precipitation of the aqueous vapour within, which became beautifully illuminated by the beam; while, upon a screen behind, a richly-coloured halo, due to diffraction by the little cloud within the receiver, flashed forth.

'The waves of heat speed from our earth through the atmosphere towards space. These waves dash in their passage against the atoms of oxygen and nitrogen, and against the molecules of aqueous vapour. Thinly scattered as these latter are, we might naturally think meanly of them, as barriers to the waves of heat. We might imagine that the wide spaces between the vapour molecules would be an open door for the passage of the undulations; and that if those waves were at all intercepted, it would be by the substances which form 99½ per cent. of the whole atmosphere. Three or four years ago, however, it was found by the speaker that this small modicum of aqueous vapour intercepted fifteen times the quantity of heat stopped by the whole of the air in which it was diffused. It was afterwards found that the dry air then experimented with was not perfectly pure; and that the purer the air became, the more it approached the character of a vacuum, and the greater, by comparison, became the action of the aqueous vapour. The vapour was found to act with 30, 40, 50, 60, 70 times the energy of the air in which it was diffused; and no doubt was entertained that the aqueous vapour of the air which filled the Royal Institution theatre, during the delivery of the discourse, absorbed 90 or 100 times the quantity of radiant heat which was absorbed by the main body of the air of the room. Looking at the single atoms, for every 200 of oxygen and nitrogen there is about 1 of aqueous vapour. This 1 is 80 times more powerful than the 200; and hence, comparing a single atom of oxygen or nitrogen with a single atom of aqueous vapour, we may infer that the action of the latter is 16,000 times that of the former.

'No doubt can exist of the extraordinary opacity of this substance to the rays of obscure heat; particularly such rays as are emitted by the earth, after being warmed by the sun. Aqueous vapour is a blanket, more necessary to the vegetable life of England than clothing is to man. Remove for a single summer-night the aqueous vapour from the air which overspreads this country, and you would assuredly destroy every plant capable of being destroyed by a freezing temperature. The warmth of our fields and gardens would pour itself unrequited into space, and the sun would rise upon an island held fast in the iron grip of frost. The aqueous vapour constitutes a local dam, by which the temperature at the earth's surface is deepened: the dam, however, finally overflows, and we give to space all that we receive from the sun.

'The sun raises the vapours of the equatorial ocean; they rise, but for a time a vapour screen spreads above and around them. But the higher they rise, the more they come into the presence of pure space; and when, by their levity, they have penetrated the vapour screen, which lies close to the earth's surface, what must occur?

'It has been said that, compared atom for atom, the absorption of an atom of aqueous vapour is 16,000 times that of air. Now the power to absorb and the power to radiate are perfectly reciprocal and proportional. The atom of aqueous vapour will therefore radiate with 16,000 times the energy of an atom of air. Imagine, then, this powerful radiant in the presence of space, and with no screen above it to check its radiation. Into space it pours its heat, chills itself, condenses, and the tropical torrents are the consequence. The expansion of the air, no doubt, also refrigerates it; but in accounting for deluges, the chilling of the vapour by its own radiation must play a most important part. The rain quits the ocean as vapour; returns to it as water. How are the vast stores of heat, set free by the change from the vaporous to the liquid condition, disposed of? Doubtless, in great part, they are wasted by radiation into space. Similar remarks apply to the cumuli of our latitudes. The warmed air, charged with vapour, rises in columns, so as to penetrate the vapour screen which hugs the earth; in the presence of space, the head of each pillar wastes its heat by radiation, condenses to a cumulus, which constitutes the visible capital of an invisible column of saturated air. Numberless other meteorological phenomena receive their solution by reference to the radiant and absorbent properties of aqueous vapour.'

The radiant power of a vapour is proportional to its absorbent power. Experiments on the dynamic radiation of dried and undried air prove the superiority of the latter as a radiator. The following experiment, performed by Dr. Frankland in the theatre of the Royal Institution, showed the effect to a large audience. A charcoal chauffer, 14 inches high and 6 inches in diameter, was placed in front of a thermo-electric pile, and at a distance from it of two feet. The radiation from the chauffer itself was intercepted by a metallic screen. The deflection due to the radiation from the ascending column of hot carbonic acid was then carefully neutralised by a constant source of heat, radiating against the opposite face of the pile. A current of steam was then forced vertically through the chauffer. The deflection of the galvanometer was prompt and powerful. When the current of steam was interrupted, the needle returned to zero. When, instead of a current of steam, a current of air was forced through the chauffer, the slight effect produced showed the pile to be chilled instead of warmed. In this experiment Dr. Frankland compared aqueous vapour, not with air, but with the more powerful carbonic acid, and demonstrated the superiority of the vapour as a radiator.*

The following remarkable passage from Hooker's 'Himalayan Journals,' 1st edit. vol. ii. p. 407, also bears upon the present subject: 'From a multitude of desultory observations I conclude that, at 7,400 feet, 125·7°, or 67° above the temperature of the air, is the average effect of the sun's rays on a black bulb thermometer. . . . These results, though greatly above those obtained at Calcutta, are not much, if at all, above what may be observed on the plains of India. The effect is much increased by elevation. At 10,000 feet, in December, at 9 a.m., I saw the mercury mount to 132°, while the temperature of shaded snow hard by was 22°. At 13,100 feet, in January, at 9 a.m., it has stood at 98°, with a difference of 68·2°, and at 10 a.m. at 114°, with a difference of 81·4°, *whilst the radiating thermometer on the snow had fallen at sunrise to* 0·7°.'

These enormous differences between the shaded and the unshaded air, and between the air and the snow, are, no doubt, due to the comparative absence of aqueous vapour at these elevations. The air is incompetent to check either the solar or the terrestrial radiation, and hence the maximum heat in the sun and the maximum cold in the shade must stand very wide apart. The difference between Calcutta and the plains of India is accounted for in the same way.

Dr. Livingstone, in his 'Travels in South Africa,' has given

* Phil. Mag. vol. xxvii. p. 326.

some striking examples of the difference in nocturnal chilling when the air is dry and when laden with moisture. Thus he finds in South Central Africa during the month of June, 'the thermometer early in the mornings at from 42° to 52°; at noon, 94° to 96°,' or a mean difference of 48° between sunrise and midday. The range would probably have been found still greater had not the thermometer been placed in the shade of his tent, which was pitched under the thickest tree he could find. He adds, moreover, 'the sensation of cold after the heat of the day was very keen. The Balonda at this season never leave their fires till nine or ten in the morning. As the cold was so great here, it was probably frosty at Linyanti; I therefore feared to expose my young trees there.' *

Dr. Livingstone afterwards crosses the continent and reaches the river Zambesi at the beginning of the year. Here the thermometric range is reduced from 48° to 12°. He thus describes the change he felt on entering the valley of the river: 'We were struck by the fact, that as soon as we came between the range of hills which flank the Zambesi, the rains felt warm. At sunrise the thermometer stood at from 82° to 86°; at midday, in the coolest shade, namely, in my little tent, under a shady tree, at 96° to 98°; and at sunset at 86°. This is different from anything we experienced in the interior.' †

Proceeding towards the mouth of the river, on January 16 he makes the following additional observation: 'The Zambesi is very broad here (at Zumbo), but contains many inhabited islands. We slept opposite one on the 16th, called Shibanga. The nights are warm, the temperature never falling below 80°; it was 91° even at sunset. One cannot cool the water by a wet towel round the vessel. . . .' ‡

In Central Australia the daily range of the thermometer is still greater. The following extract is from a paper by Mr. W. S. Jevons 'On some Data concerning the Climate of Australia and New Zealand': '. . . In the interior of the continent of Australia the fluctuations of temperature are immensely increased. The heat of the air, as described by Captain Sturt, is fearful during summer; thus, in about lat. 30° 50′ S., and lon. 141° 18′ E., he writes: "The thermometer every day rose to

* Livingstone's Travels, p. 484. † Ibid. p. 575. ‡ Ibid. p. 589.

112° or 116° in the shade, whilst in the direct rays of the sun from 140° to 150°." Again, "at a quarter past three p.m. on January 21 (1845), the thermometer had risen to 131° in the shade, and to 154° in the direct rays of the sun." . . . In the winter the thermometer was observed as low as 24°, giving an extreme range of 107°.

'The fluctuations of temperature were often very great and sudden, and were severely felt. On one occasion (October 25), the temperature rose to 110° during the day, but a squall coming on, it fell to 38° at the following sunrise; it thus varied 72° in less than twenty-four hours. . . . Mitchell, on his last journey to the N.W. interior, had very cold frosty nights. On May 22, the thermometer stood at 12° in the open air. . . . Still, in the daytime, the air was warm, and the daily range of temperature was enormous. Thus, on June 2, the thermometer rose from 11° at sunrise to 67° at four p.m.; or through a range of 56°. On June 12, the range was 53°, and on many other days nearly as great.'

Even at Sydney the average daily range of the thermometer is 21°, whilst at Greenwich the average range is only 17°. 'It thus appears that even close to the ocean the mean daily range of the Australian climate is very considerable. It is least in the autumn and greatest during the cloudless days of spring.' After giving a table of the seasonal variation of the rainfall in Australia, Mr. Jevons remarks that 'it is plainly shown that the most rainy season of the year on the east coast is the autumn, that is, the three months, March, April, May. The spring season appears the driest, summer and winter being intermediate.'

Without quitting Europe, we find places where, while the day temperature is very high, the hour before sunrise is intensely cold. I have often experienced this in the post-wagens of Germany; and I am informed that the Hungarian peasants, if exposed at night, take care, even in hot weather, to protect themselves by heavy cloaks against the nocturnal chill. The observations of MM. Bravais and Martins on the Grand Plateau of Mont Blanc have been already referred to. M. Martins has recently added to our knowledge by making observations on the heating of the soil at great elevations, and finds on the summit of the Pic du Midi the heat of the soil exposed to the sun, above that of the air, to be twice as great as in the valley at the base of the mountain. 'The immense heating of the soil,' writes M. Martins, 'compared with that of the air on high mountains, is the more remarkable, since, during the nights, the cooling by radiation is there much greater than in the plain.' The observations of the Messrs. Schlagentweit furnish, if I mistake not, many illustrations of the action of aqueous vapour; and I do not doubt, that the more this question is tested, the more clearly will it appear that the radiant and absorbent powers of this substance enable it to play a most important part in the phenomena of meteorology.

CHAPTER XII.

ABSORPTION OF HEAT BY VOLATILE LIQUIDS—ABSORPTION OF HEAT BY THE VAPOURS OF THOSE LIQUIDS AT A COMMON PRESSURE—ABSORPTION OF HEAT BY THE SAME VAPOURS WHEN THE QUANTITIES OF VAPOUR ARE PROPORTIONAL TO THE QUANTITIES OF LIQUID—COMPARATIVE VIEW OF THE ACTION OF LIQUIDS AND THEIR VAPOURS UPON RADIANT HEAT—PHYSICAL CAUSE OF OPACITY AND TRANSPARENCY—INFLUENCE OF TEMPERATURE ON THE TRANSMISSION OF RADIANT HEAT—CHANGES OF POSITION THROUGH CHANGES OF TEMPERATURE—RADIATION FROM FLAMES—INFLUENCE OF OSCILLATING PERIOD ON THE TRANSMISSION OF RADIANT HEAT—EXPLANATION OF CERTAIN RESULTS OF MELLONI AND KNOBLAUCH.

(497) THE natural philosophy of the future will certainly for the most part consist in the investigation of the relations subsisting between the ordinary matter of the universe and the wonderful ether in which this matter is immersed. Regarding the motions of the ether itself, the optical investigations of the last half century leave nothing to be desired; but regarding the atoms and molecules, whence issue the undulations of light and heat, and their relations to the medium in which they move, and by which they are set in motion, these investigations teach us little. To come closer to the origin of the ethereal waves—to obtain, if possible, some experimental hold of the oscillating atoms themselves—has been the main object of those researches on the radiation and absorption of heat by gases and vapours, which, in brief outline, have been sketched before you.

(498) These enquiries have made known the enormous differences existing between different gaseous molecules, as

regards their power of emitting and absorbing radiant heat. When a gas is condensed to a liquid, the molecules approach and grapple with each other, by forces which are insensible as long as the gaseous state is maintained. But though thus condensed and enthralled, the all-pervading ether still surrounds the molecules. If, then, the power of radiation and absorption depend upon them individually, we may expect that the deportment towards radiant heat of the free molcule will maintain itself after that molecule has relinquished its freedom and formed part of a liquid. If, on the other hand, the state of aggregation be of paramount importance, we may expect to find, on the part of liquids, a deportment altogether different from that of their vapours. Which of these views corresponds with the truth of nature, we have now to enquire.

(499) Melloni examined the diathermancy of various liquids, but he employed for this purpose the flame of an oil-lamp, covered by a glass chimney. His liquids, moreover, were contained in glass cells; hence, the radiation was profoundly modified before it entered the liquid at all, glass being impervious to a considerable part of the emission. Melloni moreover did not occupy himself with the questions of molecular physics, which to us are of paramount interest. In the examination of the question now before us, it was my wish to interfere as little as possible with the primitive emission, and an apparatus was therefore devised in which a layer of liquid, of any thickness, could be enclosed between two polished plates of rock-salt.

(500) The apparatus consists of the following parts:—A B C (fig. 94) is a plate of brass, 3·4 inches long, 2·1 inches wide, and 0·3 of an inch thick. Into it, at its corners, are rigidly fixed four upright pillars, furnished at the top with screws, for the reception of the nuts *q r s t*. D E F is a second plate of brass, of the same size as the former, and

pierced with holes at its four corners, so as to enable it to slip over the four columns of the plate A B C. Both these plates are perforated by circular apertures, *m n* and *o p*,

FIG. 94.

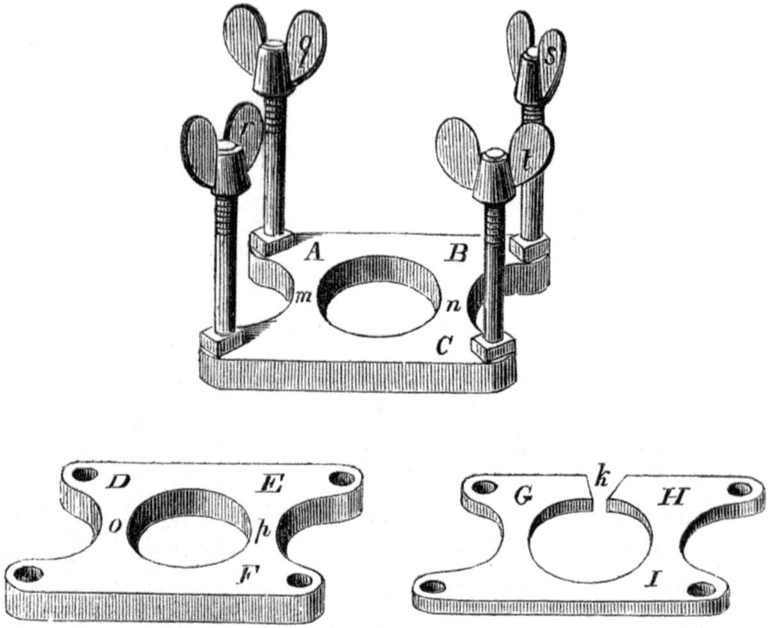

1·35 inch in diameter. G H I is a third plate of brass, of the same area as D E F, and, like it, having its centre and its corners perforated. The plate G H I is intended to separate the two plates of rock-salt which are to form the walls of the cell, and its thickness determines that of the liquid layer. The separating plate G H I was ground with the utmost accuracy, and the surfaces of the plates of salt were polished with extreme care, with a view to rendering the contact between the salt and the brass water-tight. In practice, however, it was found necessary to introduce washers of thin letter-paper between the plates of salt and the separating plate.

(501) In arranging the cell for experiment, the nuts *q r s t* are unscrewed, and a washer of india-rubber is first placed on A B C. On this washer is placed one of the plates of rock-salt. On the plate of rock-salt is laid the washer of letter-paper, and on this again the separating plate G H I. A second washer of paper is placed on this plate, then comes the second plate of salt, on which another india-rubber washer is laid. The plate D E F is finally slipped over the columns, and the whole arrangement is tightly screwed together by the nuts *q r s t*.

(502) Thus, when the plates of rock-salt are in position, a circular space, as wide as the plate G H I is thick, is enclosed between them, and the space can be filled with any liquid through the orifice *k*. The use of the india-rubber washers is to relieve the crushing pressure which would be applied to the plates of salt, if they were in actual contact with the brass; and the use of the paper washers is, as already explained, to render the cell liquid-tight. After each experiment, the apparatus is unscrewed, the plates of salt are removed and thoroughly cleansed; the cell is then remounted, and in two or three minutes all is ready for a new experiment.

(503) My next necessity was a perfectly steady source of heat, of sufficient intensity to penetrate the most absorbent of the liquids to be subjected to examination. This was found in a spiral of platinum wire, rendered incandescent by an electric current. The frequent use of this source led to the construction of the lamp shown in fig. 95. A is a globe of glass three inches in diameter, fixed upon a stand, which can be raised and lowered. At the top of the globe is an opening, into which a cork is fitted, and through the cork pass two wires, the ends of which are united by the platinum spiral *s*. The wires are carried down to the binding screws *a b*, which are fixed in the foot of the stand, so that when the instrument is attached to the

battery, no strain is ever exerted on the wires which carry the spiral. The ends of the thick wire to which the spiral is attached are also of stout platinum, for when it was attached to copper wires unsteadiness was introduced through oxidation. The heat issues from the incandescent spiral by the opening *d*, which is an inch and a half in

FIG. 95.

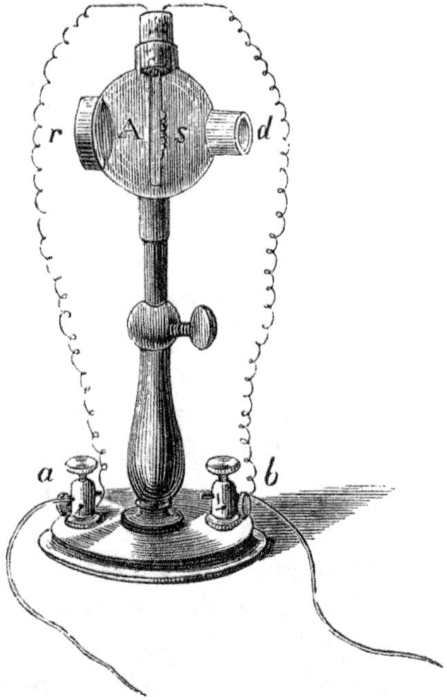

diameter. Behind the spiral, finally, is a metallic reflector, *r*, which augments the flux of heat without sensibly changing its quality. In the open air the red-hot spiral is a capricious source of heat, but surrounded by its glass globe its steadiness is admirable.*

(504) The whole experimental arrangement will be immediately understood from the sketch given in fig. 96. A

* I have had also lamps constructed in which the spiral was placed in vacuo, its rays passing to external space through a plate of rock-salt. Their steadiness is perfect.

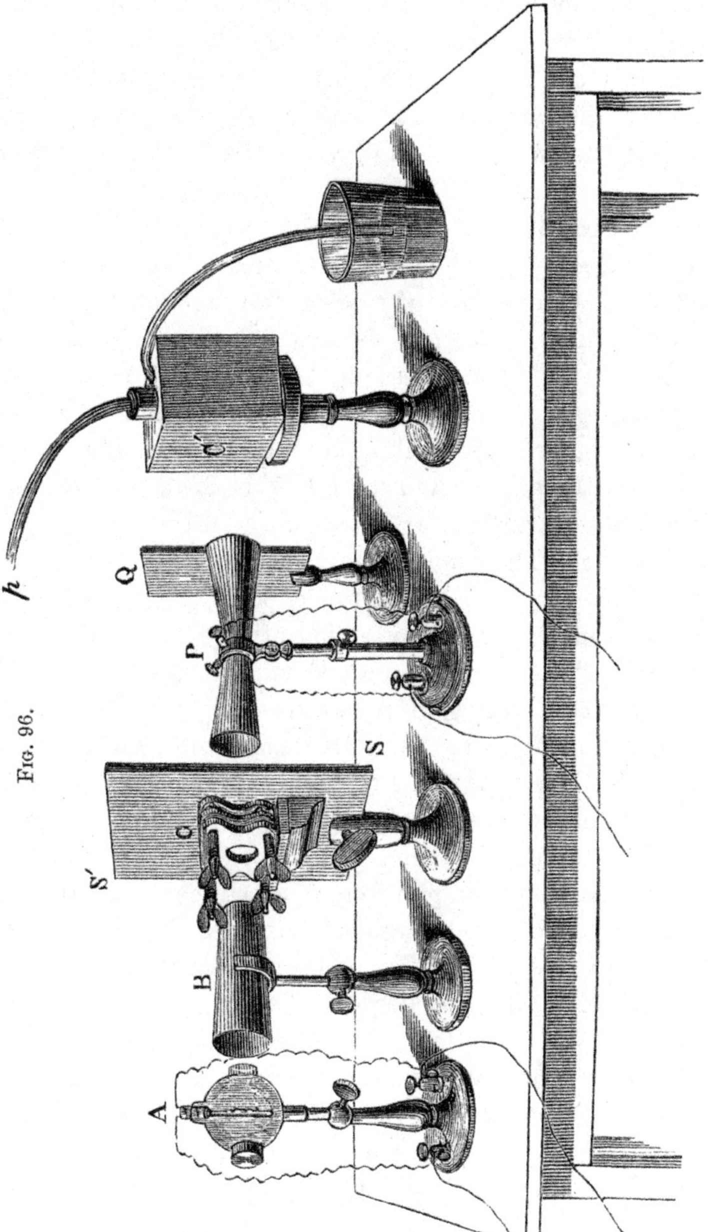

FIG. 96.

is the platinum lamp just described, heated by a current from a Grove's battery of five cells. Means were devised to render this lamp perfectly constant throughout the day. In front of the spiral, and with an interior reflecting surface, is the tube B, through which the heat passes to the rock-salt cell C. This cell is placed on a little stage, soldered to the back of the perforated screen S S′, so that the heat, after having crossed the cell, passes through the hole in the screen, and afterwards impinges on the thermo-electric pile P. The pile is placed at some distance from the screen S S′, so as to render the temperature of the cell C itself of no account. C′ is the compensating cube, containing water kept boiling by steam from the pipe *p*. Between the cube C′ and the pile P is the screen Q, which regulates the amount of heat falling on the posterior face of the pile. The whole arrangement is here exposed, but, in practice, the pile P and the cube C′ are carefully protected from the capricious action of the surrounding air.

(505) The experiments are thus performed. The empty rock-salt cell C being placed on its stage, a double silvered screen, (not shown in the figure) is first introduced between the end of the tube B and the cell C′; the heat of the spiral being thus totally cut off, and the pile subjected to the action of the cube C alone. By means of the screen Q′ the heat received by the pile from C, is reduced until the total heat to be adopted throughout the series of experiments is obtained: say, that it is sufficient to produce a galvanometric deflection of 50 degrees. The double screen used to intercept the radiation from the spiral is then gradually withdrawn, until this radiation completely neutralises that from the cube C′, and the needle of the galvanometer points steadily to zero. The position of the double screens, once fixed, remains subsequently unchanged. The rays in the first instance pass from the spiral through the *empty* rock-salt cell. A small funnel, supported by a suit-

able stand, dips into the aperture which leads into the cell, and through this the liquid is poured. The introduction of the liquid destroys the previous equilibrium, the galvanometer needle moves, and finally assumes a steady deflection. From this deflection we can immediately calculate the quantity of heat absorbed by the liquid, and express it in hundredths of the entire radiation.

(506) The experiments were executed with eleven different liquids, employing each liquid in five different thicknesses. The results are collected together in the following table:—

ABSORPTION OF HEAT BY LIQUIDS. SOURCE OF HEAT: PLATINUM SPIRAL RAISED TO BRIGHT REDNESS BY A VOLTAIC CURRENT.

Liquid	Thickness of liquid in parts of an inch				
	0·02	0·04	0·07	0·14	0·27
Bisulphide of carbon	5·5	8·4	12·5	15·2	17·3
Chloroform	16·6	25·0	35·0	40·0	44·8
Iodide of methyl	36·1	46·5	53·2	65·2	68·6
Iodide of ethyl	38·2	50·7	59·0	69·0	71·5
Benzol	43·4	55·7	62·5	71·5	73·6
Amylene	58·3	65·2	73·6	77·7	82·3
Sulphuric ether	63·3	73·5	76·1	78·6	85·2
Acetic ether	—	74·0	78·0	82·0	86·1
Formic ether	65·2	76·3	79·0	84·0	87·0
Alcohol	67·3	78·6	83·6	85·3	89·1
Water	80·7	86·1	88·8	91·0	91·0

(507) Here, for a thickness of 0·02 of an inch we find the absorption varying from a minimum of 5·5 per cent. in the case of bisulphide of carbon, to a maximum of 80·7 per cent. in the case of water. The bisulphide therefore transmits 94·5 per cent., while the water—a liquid equally transparent to light—transmits only 19·3 per cent. of the entire radiation. At all thicknesses, water, it will be observed, asserts its predominance. Next to it, as an absorbent, stands alcohol; a body which also resembles it chemically.

(508) As liquids, then, those bodies are shown to possess very different capacities of intercepting the heat emitted by our radiating source; and we have next to enquire whether these differences continue, after the molecules have been released from the bond of cohesion and reduced to the state of vapour. We must, of course, test the vapours by waves of the same period as those applied to the liquids, and this our mode of experiment renders easy of accomplishment. The heat generated in a wire by a current of a given strength being invariable, it was only necessary, by means of a tangent compass and rheocord, to keep the current constant from day to day, in order to obtain, both as regards quantity and quality, an invariable source of heat.

(509) The liquids from which the vapours were derived were placed in small long flasks, a separate flask being devoted to each. The air above the liquid, and within it, being first carefully removed by an air-pump, the flask was attached to the experimental tube, in which the vapours were to be examined. This tube was of brass, 49·6 inches long, and 2·4 inches in diameter, its two ends being stopped by plates of rock-salt. Its interior surface was polished. With the single exception that the source of heat was a red-hot platinum spiral, instead of a cube of hot water, the arrangement was that figured in Plate I. At the commencement of each experiment, the brass tube being thoroughly exhausted, and the radiation from the spiral being neutralised by that from the compensating cube, the needle stood at zero. The cock of the flask containing the volatile liquid was then carefully turned on, and the vapour allowed slowly to enter the experimental tube. When a pressure of 0·5 of an inch was obtained, the vapour was cut off, and the permanent deflection of the needle noted. Knowing the total heat, the absorption in 100ths of the entire radiation could be

at once deduced from the deflection. The following table contains the results:—

RADIATION OF HEAT THROUGH VAPOURS. SOURCE: RED-HOT PLATINUM SPIRAL. PRESSURE, 0·5 OF AN INCH.

	Absorption per cent.
Bisulphide of carbon	4·7
Chloroform	6·5
Iodide of methyl	9·6
Iodide of ethyl	17·7
Benzol	20·6
Amylene	27·5
Alcohol	28·1
Formic ether	31·4
Sulphuric ether	31·9
Acetic ether	34·6
Total heat	100·0

(510) We are now in a condition to compare the action of a series of volatile liquids, with that of the vapours of those liquids, upon radiant heat. Commencing with the substance of the lowest absorptive energy, and proceeding to the highest, we have the following orders of absorption:—

Liquids	Vapours
Bisulphide of carbon.	Bisulphide of carbon.
Chloroform.	Chloroform.
Iodide of methyl.	Iodide of methyl.
Iodide of ethyl.	Iodide of ethyl.
Benzol.	Benzol.
Amylene.	Amylene.
Sulphuric ether.	Alcohol.
Acetic ether.	Formic ether.
Formic ether.	Sulphuric ether.
Alcohol.	Acetic ether.
Water.	

(511) Here, as far as amylene, the order of absorption is the same for both liquids and vapours. But from amylene downwards, though strong liquid absorption is, in a general way, paralleled by strong vapour absorption, the order of both is not the same. There is not the slightest

doubt that, next to water, alcohol is the most powerful absorber in the list of liquids; but there is just as little doubt that the position which it occupies in the list of vapours is the correct one. This has been established by reiterated experiments. Acetic ether, on the other hand, though certainly the most energetic absorber in the state of vapour, falls behind both formic ether and alcohol in the liquid state. Still, on the whole, it is perfectly impossible to contemplate these results, without arriving at the conclusion that the act of absorption is, in the main, *molecular*, and that the molecules maintain their power as absorbers and radiators when they change their state of aggregation. Should any doubt, however, linger as to the correctness of this conclusion, it will speedily disappear.

(512) A moment's reflection will show that the comparison here instituted is not a strict one. We have taken the liquids at a common thickness, and the vapours at a common volume and pressure. But if the layers of liquid employed were turned, bodily, into vapour, the volumes obtained would *not* be the same. Hence, the quantities of matter traversed by the radiant heat are not proportional to each other in the two cases, and to render the comparison strict, they ought to be proportional. It is easy, of course, to make them so; for the liquids being examined at a constant volume, their specific gravities give us the relative quantities of matter traversed by the radiant heat, and from these, and the vapour-densities, we can immediately deduce the corresponding volumes of the vapour. Dividing, in fact, the specific gravities of our liquids by the densities of their vapours, we obtain the following series of vapour volumes, whose weights are proportional to the masses of liquid employed.

TABLE OF PROPORTIONAL VOLUMES.

Bisulphide of carbon	0·48
Chloroform	0·36
Iodide of methyl	0·46
Iodide of ethyl	0·36
Benzol	0·32
Amylene	0·26
Alcohol	0·50
Sulphuric ether	0·28
Formic ether	0·36
Acetic ether	0·29
Water	1·60

(513) Introducing the vapours, in the volumes here indicated, into the experimental tube, the following results were obtained:—

RADIATION OF HEAT THROUGH VAPOURS. QUANTITY OF VAPOUR PROPORTIONAL TO THAT OF LIQUID.

Name of Vapour	Pressure in parts of an inch	Absorption per cent.
Bisulphide of carbon	0·48	4·3
Chloroform	0·36	6·6
Iodide of methyl	0·46	10·2
Iodide of ethyl	0·36	15·4
Benzol	0·32	16·8
Amylene	0·26	19·0
Sulphuric ether	0·28	21·5
Acetic ether	0·29	22·2
Formic ether	0·36	22·5
Alcohol	0·50	22·7

(514) Arranging both liquids and vapours in the order of their absorption, we now obtain the following result:—

Liquids	Vapours
Bisulphide of carbon.	Bisulphide of carbon.
Chloroform.	Chloroform.
Iodide of methyl.	Iodide of methyl.
Iodide of ethyl.	Iodide of ethyl.
Benzol.	Benzol.
Amylene.	Amylene.
Sulphuric ether.	Sulphuric ether.
Acetic ether.	Acetic ether.
Formic ether.	Formic ether.
Alcohol.	Alcohol.
Water.	*

* Aqueous vapour, unmixed with air, condenses so readily that it cannot be directly examined in our experimental tube.

(515) Here the discrepancies revealed by our former series of experiments entirely disappear, and it is proved that for heat of the same quality, the order of absorption for liquids and their vapours is the same. We may, therefore, safely infer that the position of a vapour, as an absorber or a radiator, is determined by that of the liquid from which it is derived. Granting the validity of this inference, the position of *water* fixes that of *aqueous vapour*. But we have found that, for all thicknesses, water exceeds the other liquids in the energy of its absorption. Hence, if no single experiment on the vapour of water existed, we should be compelled to conclude, from the deportment of its liquid, that, weight for weight, aqueous vapour transcends all others in absorptive power. Add to this the direct and multiplied experiments, by which the action of this substance on radiant heat has been established, and we have before us a body of evidence sufficient, I trust, to set this question for ever at rest, and to induce the meteorologist to apply the result, without misgiving, to the phenomena of his science.

(516) We must now prepare the way for the consideration of an important question. A pendulum swings at a certain definite rate, which depends upon the length of the pendulum. A spring will oscillate at a rate which depends upon the weight and elastic force of the spring. If we coil a wire into a long spiral, and attach a bullet to the end, the bullet may be caused to oscillate up and down, at a rate which depends upon its weight, and upon the elasticity of the spiral. A musical string, in like manner, has its determinate rate of vibration, which depends upon its length, weight, and tension. A beam which bridges a gorge has also its own rate of oscillation; and we can often, by timing our movements on such a beam, so accumulate the impulses as to endanger its safety. Soldiers, in crossing pontoon bridges, tread irregularly, lest the

motion imparted to the pontoons should accumulate to a dangerous extent. The step of a person carrying water on his head in an open pail sometimes coincides with the oscillation of the water from side to side of the vessel, until, impulse being added to impulse, the liquid finally splashes over the rim. The water carrier instinctively alters his step, and thus reduces the liquid to comparative tranquillity. You have heard a particular pane of glass respond to a particular note of an organ; if you open a piano, and sing into it, some one string will also respond. Now, in the case of the organ the pane responds, because its period of vibration happens to coincide with the period of the sonorous waves that impinge upon it; and in the case of the piano, that string responds whose period of vibration coincides with the period of the vocal chords of the singer. In each case, there is an accumulation of the effect, similar to that observed when you stand upon a plank-bridge, and time your impulses to its rate of vibration. In the case of the singing flame already referred to, you had the influence of period exemplified in a very striking manner. It responded to the voice, only when the pitch of the voice corresponded to its own. A higher and a lower note were equally ineffective to put the flame in motion.

(517) These ordinary mechanical facts will help us to an insight of the more subtle phenomena of light and radiant heat. I have shown you the transparency of lampblack, and the far more wonderful transparency of iodine, to the purely thermal rays; and we have now to enquire why iodine stops light and allows heat to pass. The sole difference between light and radiant heat is one of period. The waves of the one are short and of rapid recurrence, while those of the other are long, and of slow recurrence. The former are intercepted by the iodine, and the latter are allowed to pass. Why? There can, I think, be only

one answer to this question—that the intercepted waves are those whose periods coincide with the periods of oscillation possible to the atoms of the dissolved iodine. The waves transfer their motion to the atoms which synchronise with them. Supposing waves of any period to impinge upon an assemblage of molecules of any other period, it is, I think, physically certain that a *tremor* of greater or less intensity will be set up among the molecules; but for the motion to *accumulate*, so as to produce sensible absorption, coincidence of period is necessary. Briefly defined, therefore, transparency is synonymous with *discord*, while opacity is synonymous with *accord*, between the periods of the waves of ether and those of the molecules of the body on which they impinge. The opacity, then, of our solution of iodine to light shows that its atoms are competent to vibrate in all periods which lie within the limits of the visible spectrum; while its transparency to the extra-red undulations demonstrates the incompetency of its atoms to vibrate in unison with the longer waves.

(518) The term 'quality,' as applied to radiant heat, has been already defined; the ordinary test of quality being the power of radiant heat to pass through diathermic bodies. If the heat of two beams be transmitted by the selfsame substance in different proportions, the two beams are said to be of different qualities. Strictly speaking, this question of quality is one of period; and if the heat of one source be more or less copiously transmitted than the heat of another source, it is because the waves of ether excited by the one are different in length and period from those excited by the other. When we raise the temperature of our platinum spiral, we alter the quality of its heat. As the temperature is raised, shorter and ever shorter waves mingle in the radiation. Dr. Draper, in a very beautiful investigation, has shown that

when platinum first appears luminous, it emits only red rays; but as its temperature augments, orange, yellow, and green are successively added to the radiation; and when the platinum is so intensely heated as to emit white light, the decomposition of that light gives all the colours of the solar spectrum.

(519) Almost all the vapours which we have hitherto examined are transparent to light, while all of them are, in some degree, opaque to obscure rays. This proves the incompetence of the molecules of these vapours to vibrate in visual periods, and their competence to vibrate in the slower periods of the waves which fall beyond the red of the spectrum. Conceive, then our platinum spiral to be gradually raised from a state of obscure to a state of luminous heat; the change would manifestly tend to produce *discord* between the radiating platinum and the molecules of our vapours. And the higher we raise the temperature of our platinum, the more decided will be the discord. On *à priori* grounds, then, we should infer, that the raising of the temperature of the platinum spiral ought to augment the power of its rays to pass through our list of vapours. This conclusion is entirely verified by the experiments recorded in the following tables:—

RADIATION THROUGH VAPOURS. SOURCE OF HEAT: PLATINUM SPIRAL BARELY VISIBLE IN THE DARK.

Name of Vapour	Absorption per cent.
Bisulphide of carbon	6·5
Chloroform	9·1
Iodide of methyl	12·5
Iodide of ethyl	21·0
Benzol	25·4
Amylene	35·8
Sulphuric ether	43·4
Formic ether	45·2
Acetic ether	49·6

(520) With the same platinum spiral raised to a white heat, the following results were obtained:—

RADIATION THROUGH VAPOURS. SOURCE OF HEAT: WHITE-HOT PLATINUM SPIRAL.

Name of Vapour	Absorption per cent.
Bisulphide of carbon	2·9
Chloroform	5·6
Iodide of methyl	7·8
Iodide of ethyl	12·8
Benzol	16·5
Amylene	22·6
Formic ether	25·1
Sulphuric ether	25·9
Acetic ether	27·2

(521) With the same spiral, brought still nearer to its point of fusion, the following results were obtained with four of the vapours :—

RADIATION THROUGH VAPOURS. SOURCE: PLATINUM SPIRAL AT AN INTENSE WHITE HEAT.

Name of Vapour	Absorption
Bisulphide of carbon	2·5
Chloroform	3·9
Formic ether	21·3
Sulphuric ether	23·7

(522) Placing the results obtained with the respective sources side by side, the influence of the vibrating period on the transmission comes out in a very decided manner:—

ABSORPTION OF HEAT BY VAPOURS.

Name of Vapour	Source: Platinum Spiral			
	Barely visible	Bright red	White-hot	Near fusion
Bisulphide of carbon	6·5	4·7	2·9	2·5
Chloroform	9·1	6·3	5·6	3·9
Iodide of methyl	12·5	9·6	7·8	
Iodide of ethyl	21·3	17·7	12·8	
Benzol	26·4	20·6	16·5	
Amylene	35·8	27·5	22·7	
Sulphuric ether	43·4	31·4	25·9	23·7
Formic ether	45·2	31·9	25·1	21·3
Acetic ether	49·6	34·6	27·2	

(523) The gradual augmentation of penetrative power, as the temperature is augmented, is here very manifest. By raising the spiral from a barely visible to an intense white heat, we reduce the proportionate absorption, in the case of bisulphide of carbon and chloroform, to less than one-half. At barely visible redness, moreover, 56·6 and 54·8 per cent. pass through sulphuric and formic ether respectively; while of the intensely white-hot spiral, 76·3 and 78·7 per cent. pass through the same vapours.* Thus, by augmenting the temperature of the solid platinum, we introduce into the radiation waves of shorter period, which, being in discord with the periods of the vapours, pass more easily through them.

(524) Running the eye along the numbers which express the absorptions of sulphuric and formic ether in the last table, we find that, for the lowest heat, the absorption of the latter exceeds that of the former; for a bright red heat they are nearly equal, but the formic still retains a slight predominance; at a white heat, however, the sulphuric slips in advance, and at the heat near fusion its predominance is decided. I have tested this result in various ways, and by multiplied experiments, and placed it beyond doubt. We may at once infer from it that the capacity of the molecule of formic ether to enter into rapid vibration is less than that of sulphuric, and thus we obtain a glimpse of the inner character of these bodies. By augmenting the temperature of the spiral, we produce vibrations of quicker periods, and the more of these that are introduced, the more opaque, in comparison with formic ether, does sulphuric ether become. The atom of oxygen which formic ether possesses, in excess of sulphuric, renders it more sluggish as a vibrator. Experiments made with a source of 100° C., establish more decidedly

* The *transmission* is found by subtracting the absorption from 100.

the preponderance of the formic ether for vibrations of slow period.

RADIATION THROUGH VAPOURS. SOURCE: LESLIE'S CUBE, COATED WITH LAMPBLACK. TEMPERATURE, 212° FAHR.

Name of Vapour	Absorption per cent.
Bisulphide of carbon	6·6
Iodide of methyl	18·8
Chloroform	21·6
Iodide of ethyl	29·0
Benzol	34·5
Amylene	47·1
Sulphuric ether	54·1
Formic ether	60·4
Acetic ether	69·9

For heat issuing from this source, the absorption by formic ether is 6·3 per cent. in excess of that by sulphuric.

(525) But in this table we notice another case of reversal. In all the experiments with the platinum spiral thus far recorded, chloroform showed itself less energetic, as an absorber, than iodide of methyl; but here chloroform shows itself to be decidedly the more powerful of the two. This result has been placed beyond doubt, by repeated experiments. To the radiation emitted by lampblack, heated to 212°, chloroform is certainly more opaque than iodide of methyl.

(526) We have hitherto occupied ourselves with the radiation from heated solids: let us now pass on to the examination of the radiation from flames. The first experiments were made with a steady jet of gas, issuing from a small circular burner, the flame being long and tapering. The top and bottom of the flame were excluded, and its most brilliant portion was chosen as the source. The results obtained are recorded in the following table:—

RADIATION OF HEAT THROUGH VAPOURS. SOURCE: A HIGHLY LUMINOUS JET OF GAS.

Name of Vapour	Absorption	White-hot Spiral
Bisulphide of carbon	9·8	2·9
Chloroform	12·0	5·6
Iodide of methyl	16·5	7·8
Iodide of ethyl	19·5	12·8
Benzol	22·0	16·5
Amylene	30·2	22·7
Formic ether	34·6	25·9
Sulphuric ether	35·7	25·1
Acetic ether	38·7	27·2

(527) It is interesting to compare the heat emitted by the white-hot carbon with that emitted by the white-hot platinum; and to facilitate the comparison, beside the results given in the last table are placed those recorded in a former one. The emission from the flame is thus proved to be far more powerfully absorbed than the emission from the spiral. Doubtless, however, the carbon, in reaching incandescence, passes through lower stages of temperature, and in those stages emits heat more in accord with the vapours. It is also mixed with the vapour of water and carbonic acid, both of which contribute their quota to the total radiation. It is therefore probable that the greater absorption of the heat emitted by the flame is due to the slower periods of the substances, which are unavoidably mixed with the white-hot carbon.

(528) The next source of heat employed was the flame of a Bunsen's burner,* the temperature of which is known to be very high. The flame was of a pale blue colour, and emitted a very feeble light. The following results were obtained:—

* Described in Chapter II.

RADIATION OF HEAT THROUGH VAPOURS. SOURCE: PALE-BLUE FLAME OF BUNSEN'S BURNER.

Name of Vapour	Absorption
Chloroform	6·2
Bisulphide of carbon	11·1
Iodide of ethyl	14·0
Benzol	17·9
Amylene	24·2
Sulphuric ether	31·9
Formic ether	33·3
Acetic ether	36·3

(529) The total heat radiated from the flame of Bunsen's burner is much less than that radiated when the incandescent carbon is present in the flame. The moment the air is permitted to mix with the luminous flame, the radiation falls so considerably, that the diminution is at once detected, even by the hand or face brought near the flame. Comparing the two last tables, we see that the radiation from Bunsen's flame is, on the whole, less powerfully absorbed than that from the luminous gas jet. In some cases, as in that of formic ether, they come very close to each other; in the case of amylene, and a few other substances, they differ more markedly. But an extremely interesting case of reversal here shows itself. Bisulphide of carbon, instead, of being first, stands decidedly below chloroform. With the luminous jet, the absorption of bisulphide of carbon is to that of chloroform as 100: 122, while with the flame of Bunsen's burner the ratio is 100: 56; the removal of the lampblack from the flame more than doubles the relative transparency of the chloroform. We have here, moreover, another instance of the reversal of formic and sulphuric ether. For the luminous jet, the sulphuric ether is decidedly the more opaque; for the flame of Bunsen's burner, it is excelled in opacity by the formic.

(530) The main radiating bodies in the flame of a Bunsen's burner are, no doubt, aqueous vapour and carbonic acid. Highly-heated nitrogen is also present, which may produce a sensible effect. But the main source of the radiation is, no doubt, the aqueous vapour and the carbonic acid. I wished to separate these two constituents, and to study them separately. The radiation of aqueous vapour could be obtained from a flame of pure hydrogen, while that of carbonic acid could be obtained from an ignited jet of carbonic oxide. To me the radiation from the hydrogen flame possessed a peculiar interest; for notwithstanding the high temperature of such a flame, I thought it likely that the accord between its periods of vibration and those of the cool aqueous vapour of the atmosphere would still be such as to cause the atmospheric vapour to exert a special absorbent power upon the radiation. The following experiments test this surmise.

RADIATION THROUGH ATMOSPHERIC AIR. SOURCE: A HYDROGEN FLAME.

	Absorption
Dry air	0
Undried air	17·2

Thus, in a polished tube 4 feet long, the aqueous vapour of our laboratory air absorbed 17 per cent. of the radiation from the hydrogen flame. A platinum spiral, raised by electricity to a degree of incandescence not greater than that attainable by plunging a wire into the hydrogen flame, being used as a source of heat, the undried air of the laboratory was found to absorb

5·8 per cent.

of its radiation, or one-third of the quantity absorbed in the case of the flame of hydrogen.

(531) The plunging of a spiral of platinum wire into the flame reduces its temperature; but at the same time introduces vibrations, which are not in accord with those

of aqueous vapour; the absorption, by ordinary undried air, of heat emitted by this composite source amounted to

8·6 per cent.

On humid days, the absorption of the heat emitted by a hydrogen flame exceeds even the above large figure. Employing the same experimental tube and a new burner, the experiments were repeated some days subsequently, with the following result:—

RADIATION THROUGH AIR. SOURCE: HYDROGEN FLAME.

	Absorption
Dry air	0
Undried air	20·3

(532) The physical causes of transparency and opacity have been already pointed out; and we may infer from the foregoing powerful action of atmospheric vapour on the radiation from the hydrogen flame, that accord reigns between the oscillating molecules of the flame at a temperature of 5898° Fahr. and the molecules of aqueous vapour at a temperature of 60° Fahr. The enormous temperature of the hydrogen flame increases the amplitude, but does not change the rate of oscillation.

(533) We must devote a moment's attention, in passing, to the word 'amplitude' here employed. The *pitch* of a note depends solely on the number of aërial waves which strike the ear in a second. The loudness, or *intensity* of a note depends upon the distance within which the separate atoms of air vibrate. This distance is called the *amplitude* of the vibration. When we pull a harp-string very gently aside, and let it go, it disturbs the air but little; the amplitude of the vibrating air-atoms is small, and the intensity of the sound feeble. But if we pull the string vigorously aside, on letting it go, we have a note of the same pitch as before, but, as the amplitude of vibration is greater, the sound is more intense.

While, then, the wave-length, or period of recurrence, is independent of the amplitude, it is this latter which determines the loudness of the sound.

(534) The same holds good for light and radiant heat. Here the individual ether particles vibrate to and fro across the line of propagation; and the extent of their excursion is called the amplitude of the vibration. We may, as in the case of sound, have the same wave-length with very different amplitudes, or, as in the case of water, we may have high waves and low waves, with the same distance between crest and crest. Now, while the colour of light, and the quality of radiant heat, depend entirely upon the length of the ethereal waves, the intensity of the light and heat is determined by the amplitude. And, inasmuch as it has been shown, that the periods of vibration of a hydrogen flame coincide with those of cool aqueous vapour, we are compelled to conclude that the enormous temperature of the flame is not due to the rapidity, but to the extraordinary amplitude of its molecular vibration.

(535) The other component of the flame of Bunsen's burner is carbonic acid, and the radiation of this substance is immediately obtained from a flame of carbonic oxide. Of the radiation from this source, the small amount of carbonic acid diffused in the air of our laboratory absorbed 13·8 per cent. This high absorption proves that the vibrations of the molecules of carbonic acid, within the flame, are synchronous with the vibrations of those of the carbonic acid of the atmosphere. The temperature of the flame, however, is 5508° Fahr., while that of the atmosphere is only 60°. But if the high temperature is incompetent to change the rate of oscillation, we may expect cold carbonic acid, when used in large quantities, to be highly opaque to the radiation from the carbonic oxide flame. Here follow the results of experiments executed to test this conclusion:—

RADIATION THROUGH DRY CARBONIC ACID. SOURCE: CARBONIC OXIDE FLAME.

Pressure in inches	Absorption
1·0	48·0
2·0	55·5
3·0	60·3
4·0	65·1
5·0	68·6
10·0	74·3

For the rays emanating from the heated solids employed in our former researches, carbonic acid proved to be one of the most feeble absorbers; but here, when the waves sent into it emanate from molecules of its own substance, its absorbent energy is enormous. The thirtieth of an atmosphere of the gas cuts off half the entire radiation; while at a pressure of 4 inches, 65 per cent. of the radiation is intercepted.

(536) The energy of olefiant gas, both as an absorbent and a radiant, is now well known. For the solid sources of heat just referred to, its power is incomparably greater than that of carbonic acid; but for the radiation from the carbonic oxide flame, the power of olefiant gas is feeble, when compared with that of carbonic acid. This is proved by the experiments recorded in the following table:—

RADIATION THROUGH DRY OLEFIANT GAS AND DRY CARBONIC ACID. SOURCE: CARBONIC OXIDE FLAME.

Pressure in inches	Olefiant gas absorption	Carbonic acid absorption
1·0	23·2	48·0
2·0	34·7	55·5
3·0	44·0	60·3
4·0	50·6	65·1
5·0	55·1	68·6
10·0	65·5	74·3

(537) Beside the absorption by olefiant gas, I have placed that by carbonic acid derived from the last table. The superior power of the acid is very decided, and most

so in the smaller pressures; at a pressure of an inch it is twice that of the olefiant gas. The substances approach each other more closely, as the quantity of gas augments. Here, in fact, both of them approach perfect opacity, and as they draw near to this common limit, their absorptions, as a matter of course, approximate.

(538) These experiments prove that the presence of an infinitesimal quantity of carbonic acid gas might be detected, by its action on the rays emitted by a carbonic oxide flame. The action, for example, of the carbonic acid expired by the lungs is very decided. An india-rubber bag was filled from the lungs; it contained, therefore, both the aqueous vapour and the carbonic acid of the breath. The air from the bag was then conducted through a drying apparatus, the moisture being thus removed, and the neutral air and active carbonic acid permitted to enter the experimental tube. The following results were obtained:—

Air from the Lungs containing CO_2. Source: Carbonic Oxide Flame.

Pressure in inches	Absorption
1	12·0
3	25·0
5	33·3
30	50·0

(539) Thus, the tube filled with the dry exhalation from the lungs intercepted 50 per cent. of the entire radiation from a carbonic oxide flame. It is quite manifest that we have here a means of testing, with surpassing delicacy, the amount of carbonic acid emitted under various circumstances from the lungs.

(540) The application of radiant heat to the determination of the carbonic acid of the breath has been illustrated by my late assistant, Mr. Barrett. The deflection produced by the breath, freed from its moisture, but

retaining its carbonic acid, was first determined. Carbonic acid, artificially prepared, was then mixed with perfectly dry air, in such proportions that its action upon the radiant heat was the same as that of the carbonic acid of the breath. The percentage of the former being known, immediately gives that of the latter. I here give the results of three chemical analyses, determined by Dr. Frankland, as compared with three physical analyses performed by my late assistant.

PERCENTAGE OF CARBONIC ACID IN HUMAN BREATH.

By chemical analysis	By physical analysis
4·311	4·00
4·66	4·56
5·33	5·22

(541) The agreement between the results is very fair. Doubtless, with greater practice a closer agreement could be attained. We thus find, in the quantity of ethereal motion which it is competent to intercept, an accurate and practical measure of the amount of carbonic acid expired from the human lungs.

(542) Water at moderate thickness is a very transparent substance; that is to say, the periods of its molecules are in discord with those of the visible spectrum. It is also highly transparent to the extra-violet rays; so that we may safely infer from the deportment of this substance, its incompetence to enter into rapid molecular vibration. When, however, we once quit the visible spectrum for the rays beyond the red, the opacity of the substance begins to show itself; for such rays, indeed, its absorbent power is unequalled. The synchronism of the periods of the water molecules with those of the extra-red waves is thus demonstrated. We have already seen that undried atmospheric air manifests an extraordinary opacity to the radiation from a hydrogen flame, and from this deportment we inferred the synchronism of the cold vapour of the

air, and the hot vapour of the flame. But if the periods of a vapour be the same as those of its liquid, we ought to find water highly opaque to the radiation from a hydrogen flame. Here are the results obtained with five different thicknesses of the liquid:—

RADIATION THROUGH WATER. SOURCE: HYDROGEN FLAME.

	Thickness of liquid				
	0·02 inch	0·04 inch	0·07 inch	0·14 inch	0·27 inch
Transmission per cent.	5·8	2·8	1·1	0·5	0·0

(543) Through a layer of water 0·36 of an inch thick, Melloni found a transmission of 11 per cent. for the heat of an Argand lamp. Here we employ a source of higher temperature, and a layer of water only 0·27 of an inch, and find the whole of the heat intercepted. A layer of water 0·27 of an inch in thickness is perfectly opaque to the radiation from a hydrogen flame, while a layer about one-tenth of the thickness employed by Melloni, cuts off more than 97 per cent. of the entire radiation. Hence we may infer the coincidence in vibrating period between cold water and aqueous vapour heated to a temperature of 5898° Fahr. (3259° C.)

(544) From the opacity of water to the radiation from aqueous vapour, we may infer the opacity of aqueous vapour to the radiation from water, and hence conclude that the very act of nocturnal refrigeration which causes the condensation of water on the earth's surface, gives to terrestrial radiation that particular character which renders it most liable to be intercepted by our atmosphere, and thus prevented from wasting itself in space.

(545) This is a point which deserves a moment's further consideration. I find that olefiant gas contained in a polished tube 4 feet long, absorbs about 80 per cent. of the radiation from an obscure source. A layer of the same gas 2 inches thick absorbs 33 per cent., a layer

1 inch thick absorbs 26 per cent., while a layer $\frac{1}{100}$th of an inch in thickness absorbs 2 per cent. of the radiation. Thus the absorption increases, and the quantity transmitted diminishes, as the thickness of the gaseous layer is augmented. Let us now consider for a moment the effect upon the earth's temperature of a shell of olefiant gas, surrounding our planet at a little distance above its surface. The gas would be transparent to the solar rays, allowing them, without sensible hindrance, to reach the earth. Here, however, the luminous heat of the sun would be converted into non-luminous terrestrial heat; at least 26 per cent. of this heat would be intercepted by a layer of gas one inch thick, and in great part returned to the earth. Under such a canopy, trifling as it may appear, and perfectly transparent to the eye, the earth's surface would be maintained at a stifling temperature.

(546) A few years ago, a work possessing great charms of style and ingenuity of reasoning, was written to prove that the more distant planets of our system are uninhabitable. Applying the law of inverse squares to their distances from the sun, the diminution of temperature was found to be so great, as to preclude the possibility of human life in the more remote members of the solar system. But in those calculations the influence of an atmospheric envelope was overlooked, and this omission vitiated the entire argument. It is perfectly possible to find an atmosphere which would act the part of a *barb* to the solar rays, permitting their entrance towards the planet, but preventing their withdrawal. For example, a layer of air two inches in thickness, and saturated with the vapour of sulphuric ether, would offer very little resistance to the passage of the solar rays, but I find that it would cut off fully 35 per cent. of the planetary radiation. It would require no inordinate thickening of the layer of vapour to double this absorption; and it is perfectly evident that,

with a protecting envelope of this kind, permitting the heat to enter, but preventing its escape, a comfortable temperature might be obtained on the surface of the most distant planet.

(547) Dr. Miller was the first to infer from the inability of the rays of burning hydrogen to pass through glass screens, that the vibrating periods of the flame must be extra-red; and that consequently, the oscillating periods of the lime-light must be more rapid than those of the oxyhydrogen flame to which it owes its incandescence.* As pointed out by Dr. Miller, the lime-light furnishes a case of exalted refrangibility. The same remark applies to a platinum wire plunged into a hydrogen flame. We have, in this case also, a conversion of unvisual periods into visual ones. This shortening of the periods must augment the discord between the radiating source and our series of liquids (§ 506), whose periods are slow, and hence augment their transparency to the radiation. The conclusion was tested and verified by experiments on layers of the liquids of two different thicknesses.

RADIATION THROUGH LIQUIDS. SOURCES: 1. HYDROGEN FLAME; 2. HYDROGEN FLAME AND PLATINUM SPIRAL.

Name of liquid	Transmission: Thickness of liquid 0·04 inch: Flame only	Transmission: Thickness of liquid 0·04 inch: Flame and spiral	Transmission: Thickness of liquid 0·07 inch: Flame only	Transmission: Thickness of liquid 0·07 inch: Flame and spiral
Bisulphide of carbon	77·7	87·2	70·4	86·0
Chloroform	54·0	72·8	50·7	69·0
Iodide of methyl	31·6	42·4	26·2	36·2
Iodide of ethyl	30·3	36·8	24·2	32·6
Benzol	24·1	32·6	17·9	28·8
Amylene	14·9	25·8	12·4	24·3
Sulphuric ether	13·1	22·6	8·1	22·0
Acetic ether	10·1	18·3	6·6	18·5
Alcohol	9·4	14·7	5·8	12·3
Water	3·2	7·5	2·0	6·4

* After referring to the researches of Professor Stokes on 'degraded' refrangibility, Dr. Miller says:—'Heat of low refrangibility may, however, be converted into that of higher refrangibility: for example, a jet of mixed oxygen and hydrogen gases furnishes a heat nearly as intense as any which

The transmission in each case is shown to be considerably augmented by the introduction of the platinum wire.

(548) Direct experiments on the radiation from a hydrogen-flame completely verify the inference of Dr. Miller. I had constructed for me a complete rock-salt train, capable of being substituted for the ordinary glass train of the electric lamp. A double rock-salt lens placed in the camera rendered the rays parallel; they then passed through a slit, and a second rock-salt lens placed without the camera produced, at an appropriate distance, an image of the slit. Behind this lens was placed a rock-salt prism, while laterally stood the linear thermo-electric pile already described (§ 309). Within the camera of the electric lamp was placed a burner with a single aperture, the flame issuing from it occupying the position usually taken up by the coal points. This burner was connected with a T-piece, from which two pieces of india-rubber tubing were carried, the one to a large hydrogen-holder, the other to the gas-pipe of the laboratory. It was thus in my power to have, at will, either the gas-flame or the hydrogen-flame. When the former was employed, a visible spectrum was produced, which enabled me to fix the thermo-electric pile in its proper position. To obtain the latter, it was only necessary to turn on the hydrogen until it reached the gas-flame and was ignited; then to turn off the gas and leave the hydrogen-flame behind. In this way, indeed, the one flame could be substituted for the other without opening the door of the camera, or producing any change

art can command, yet it does not emit rays which have the power of traversing glass in any considerable quantity even though a lens be employed for their concentration. Upon introducing a cylinder of lime into the jet of burning gases, though the amount of heat is not thus increased, the light becomes too bright for the unprotected eye to endure, and the thermic rays acquire the property of traversing glass, as is shown by their action upon a thermometer the bulb of which is placed in the focus of the lens.'—Chemical Physics, 1855, p. 210.

in the positions of the source, the lenses, the prism, and the pile.

(549) The spectrum of the luminous gas-flame being cast upon the brass screen (which, to render the colours more visible, was covered with tin-foil), the pile was gradually moved until the deflection of the galvanometer became a maximum. To reach this it was necessary to pass to some distance beyond the red of the spectrum; the deflection then observed was

30°.

When the pile was moved in either direction from this position, the deflection diminished.

(550) The hydrogen-flame was now substituted for the gas-flame; the visible spectrum disappeared, and the deflection fell to

12°.

Hence, as regards rays of this particular refrangibility, the emission from the luminous gas-flame was two-and-a-half times that from the hydrogen-flame.

(551) The pile was again moved to and fro, the movement in both directions being accompanied by a diminished deflection. Twelve degrees, therefore, was the maximum deflection for the hydrogen-flame; and the position of the pile, determined previously by means of the luminous flame, proves that this deflection was produced by extra-red undulations. I moved the pile a little forwards, so as to reduce the deflection from 12° to 4°, and then, in order to ascertain the refrangibility of the rays which produced this small deflection, relighted the gas. The face of the pile was found invading the red. When the pile was caused to pass successively through positions corresponding to the various colours of the spectrum, and to its extra-violet rays, no measurable deflection was produced by the hydrogen-flame.

(552) It is thus conclusively proved that the radiation

from a hydrogen-flame, as far as it is capable of measurement by our delicate arrangement, is extra-red. The other constituents of the radiation are so feeble as to be thermally insensible.

(553) And here we find ourselves in a position to offer solutions of various facts, which have hitherto stood out as enigmas in researches upon radiant heat. It was for a time generally supposed that the power of heat to penetrate diathermic substances augmented as the temperature of the source became more elevated. Knoblauch contended against this notion, showing that the heat emitted by a platinum wire plunged in an alcohol flame was less absorbed, by certain diathermic substances, than the heat of the flame itself, and justly arguing that the temperature of the spiral could not be higher than that of the body from which it derived its heat. A plate of transparent glass being introduced between his incandescent platinum spiral and his thermo-electric pile, the deflection of his needle fell from 35° to 19°; while, when the source was the flame of alcohol, without the spiral, the deflection fell from 35° to 16°. This proved the radiation from the flame to be intercepted more powerfully than that from the spiral; or, in other words, that the heat emanating from the body of highest temperature possessed the least penetrative power. Melloni afterwards corroborated this experiment.

(554) Transparent glass allows the rays of the visible spectrum to pass freely through it; but it is well known to be highly opaque to the radiation from obscure sources; or to waves of long period. A plate 0·1 of an inch thick intercepts all the rays from a source of 100° C., and transmits only 6 per cent. of the heat emitted by copper raised to 400° C. Now the products of an alcohol flame are aqueous vapour and carbonic acid, whose waves have been proved to be of slow period; of the particular character, consequently, most powerfully intercepted by glass.

But by plunging a platinum wire into such a flame, we virtually convert its heat into heat of higher refrangibility; we change the long periods into shorter ones, and thus establish the discord between the periods of the source and the periods of the diathermic glass, which, as before defined, is the physical cause of transparency. On purely *à priori* grounds, therefore, we might infer that the introduction of the platinum spiral would augment the penetrative power of the heat. With a plate of glass Melloni, in fact, found the following transmissions for the flame and the spiral:—

For the flame	For the platinum
41·2	52·8

The same remarks apply to the transparent selenite examined by Melloni. This substance is highly opaque to the extra-red undulations; but the radiation from an alcohol flame is mainly extra-red, and hence the opacity of the selenite to this radiation. The introduction of the platinum spiral shortens the periods and augments the transmission. Thus, with a specimen of selenite, Melloni found the transmissions to be as follows:—

Flame	Platinum
4·4	19·5

(555) So far the results of Melloni coincide with those of Knoblauch; but the Italian philosopher pursues the matter further, and shows that Knoblauch's results, though true for the particular substances examined by him, are not true of diathermic media generally. Melloni shows that in the case of *black* glass and *black* mica, a striking inversion of the effect is observed: through these substances the radiation from the flame is more copiously transmitted than that from the platinum. For black glass he found the following transmissions:—

From the flame	From the platinum
52·6	42·8

And for a plate of black mica the following transmissions:—

From the flame	From the platinum
62·8	52·5

(556) These results were left unexplained by Melloni, but the solution is now easy. The black glass and the black mica owe their blackness to the carbon incorporated in them, and the opacity of this substance to light, as already remarked, proves the accord of its vibrating periods with those of the visible spectrum. But it has been shown that carbon is, in a considerable degree, pervious to the waves of long period; that is to say, to such waves as are emitted by a flame of alcohol. The case of the carbon is therefore precisely antithetical to that of the transparent glass, the former transmitting the heat of long period, and the latter that of short period most freely. Hence it follows that the introduction of the platinum wire, by converting the long periods of the flame into short ones, augments the transmission through the transparent glass and selenite, and diminishes it through the opaque glass and mica.

CHAPTER XIII.

DISCOVERY OF DARK SOLAR RAYS—HERSCHEL AND MÜLLER'S EXPERIMENTS—RISE OF INTENSITY WITH TEMPERATURE—HEAT OF ELECTRIC SPECTRUM—RAY-FILTERS: SIFTING THE ELECTRIC LIGHT—TRANSMUTATION OF RAYS—THERMAL IMAGE RENDERED LUMINOUS—COMBUSTION AND INCANDESCENCE BY DARK RAYS—FLUORESCENCE AND CALORESCENCE—DARK SOLAR RAYS—DARK LIME-LIGHT RAYS—FRANKLIN'S EXPERIMENT ON COLOURS—ITS ANALYSIS AND EXPLANATION.

(557) ON a former occasion I promised to make known to you the progress of recent enquiry as regards the subject of invisible radiation. A hope was expressed that I should be able to sift in your presence the composite emission of the electric lamp; to detach its rays of darkness from its rays of light, and to show you the power of those dark rays when they are properly intensified and concentrated.

(558) The hour now before us shall be devoted to an attempt to redeem this promise and realise this hope. And in the first place it is necessary that we should have distinct notions regarding these dark rays, or obscure rays, or invisible rays—all these adjectives have been applied to them. We have defined light as wave motion; we have learned that the different colours of light are due to waves of different lengths; and we have also learned that side by side with the visible rays emitted by luminous sources, we have an outflow of invisible rays. This accurately expressed means that together with those waves which cross the humours of the eye, impinge upon the retina,

and excite the sense of vision, there are others which either do not reach the retina at all, or which, if they do, are not gifted with the power of producing that specific motion in the optic nerve which results in vision. Whether, and in what degree, the dark rays of the electric light reach the retina, shall be decided subsequently; but no matter what may be the cause of their inefficacy, whether it be due to their being quenched in the humours of the eye, or to a specific incompetence on their part to arouse the retina, all rays which fail to excite vision are called dark, obscure, or invisible rays: while all rays that can excite vision are called visible, or luminous rays.

(559) It must be confessed that there is a defect in the terms employed; for we cannot see light. In interstellar space we should be plunged in darkness, though the waves from all suns and all stars might be speeding through it. We should see the suns and we should see the stars themselves, but the moment we ceased to face a star, the moment we turned our backs upon it, its light would become darkness, though the ether all around us might be agitated by its waves. We cannot see the ether or its motions, and hence, strictly speaking, it is a misuse of language to speak of its waves or rays being visible or invisible. This form of expression, however, has taken root; its convenience has brought it into general use, and understanding by the terms visible and invisible rays, wave motions which are respectively competent and incompetent to excite the optic nerve, no harm can result from the employment of the terms.

(560) To the detection of those dark rays in the emission of the sun reference has been already made, and their existence in the emission of that source which comes next to the sun in power—the electric light, has also been demonstrated. The discoverer of the dark rays of the sun was, as you have been already informed, Sir William

Herschel. His means of observation were far less perfect than those now at our command; but, like Newton, he could extract from nature great results with very poor appliances. He caused thermometers to pass through the various colours of the solar spectrum, and noted the temperature corresponding to each colour. He pushed his thermometers beyond the extreme red of the spectrum, and found that the radiation, so far from terminating with the visible spectrum, rose to its maximum energy beyond the red. The experiment proved that side by side with its luminous rays the sun emitted others of lower refrangibility, which, although they possessed high calorific power, were incompetent to excite the sense of vision.

(561) Now the rise of the thermometric column, when the instrument is placed in any colour of the spectrum, may be represented by a straight line. For example, if a line

FIG. 97.

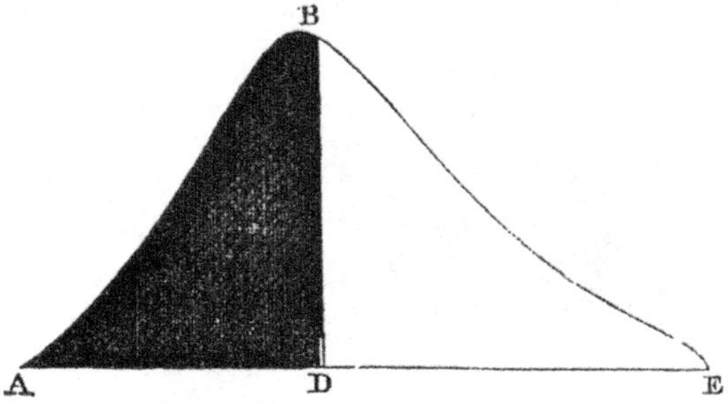

of a certain length be taken to represent a rise of one degree, a line of twice that length will represent a rise of two degrees, while a line of half the length would represent a rise of half a degree. In order to show the distribution of heat in the spectrum of the sun, Sir William Herschel adopted this device of representing temperatures by lines. Drawing a horizontal line A E, fig. 97, to represent the length of the spectrum, and erecting at its various points perpendiculars to represent the heat of the spectrum at those points, on uniting the ends of those perpendiculars, he obtained a curve, which exhibits at a glance the distribution of heat in the solar spectrum. The letter E marks a point in the blue of the spectrum where the heat first became sensible; from E to D, which marks the limit of the red, the temperature steadily increased, as shown by the increased height of the curve. At D the visible spectrum ceased, but an invisible one extended beyond D to A, where it vanished. According, then, to the observations of Sir William Herschel, the white space BDE represents the thermal value of the visible, while the black space ABD represents the thermal value of the invisible radiation of the sun.

(562) With the more perfect apparatus subsequently devised by Melloni, Professor Müller of Freiburg examined

FIG. 98.

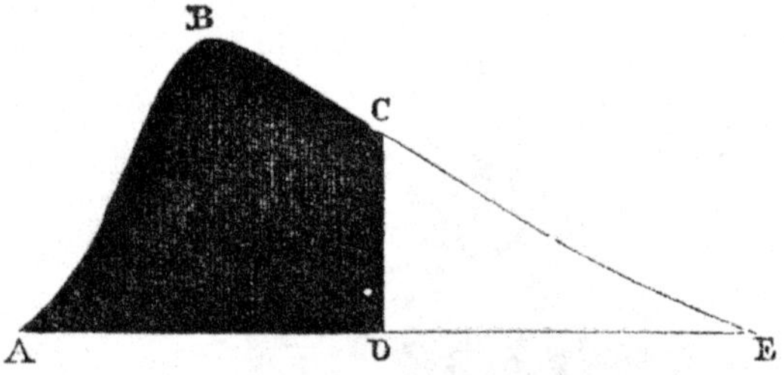

the distribution of heat in the solar spectrum. The results of his observations are rendered graphically in fig. 98, where the area DCE represents the visible, and ABCD the invisible radiation.

(563) Before proceeding to our own measurements, it is desirable to make a few remarks upon the generation and intensification of rays, visible and invisible. A solid body at the ordinary temperature of our air has its molecules in motion: but it emits rays of too low a refrangibility, or,

in other words, it generates undulations which are too long, and of too slow recurrence to excite vision. Conceive its temperature gradually augmented. With the increased temperature more rapid vibrations are introduced among the molecules of the body; and at a certain temperature the vibrations are sufficiently rapid to affect the eye as light. The body glows, and first of all, as proved by Dr. Draper, the light is a pure red. As the temperature heightens, orange, yellow, green and blue are introduced in succession.

(564) The vibrations corresponding to these successive colours are essentially new vibrations. But simultaneously with the introduction of each new and more rapid vibration, we have an intensification of *all those vibrations* which preceded it. The vibration executed when our ball was at the temperature of the air, continues to be executed when, the ball is white hot. But while the period remains thus constant, the amplitude, on which the intensity of the radiation depends, is enormously increased. For this reason, the rays emitted by an obscure body can never approach the intensity of the obscure rays of the same refrangibility emitted by a highly luminous one.

(565) Let me rivet this subject upon your attention by a numerical example of the rise in the intensity of a special vibration, while more rapid ones are being introduced. A spiral of platinum wire was placed in this camera, and in front of the camera a slit. A voltaic current was sent through the spiral, but not in sufficient strength to make it glow. By means of lenses and prisms of pure rock-salt, and by other suitable devices, an invisible spectrum of the rays emitted by the platinum-wire was obtained. A thin slice of this spectrum was permitted to fall upon the face of the linear thermo-electric pile already described. The band of the spectrum was so narrow and the radiation so weak, that the deflection of the galvano-

meter was in the first instance only one degree. Without altering the position of any portion of the apparatus, the current was gradually strengthened; raising the temperature of the wire, causing it to glow, and finally raising it to an intense white heat. When this occurred a brilliant light-spectrum was projected on the screen to which the pile was attached, but the pile itself was outside of the spectrum. It received invisible rays alone, and throughout the experiment it continued to receive those particular vibrations which first affected it. The rate of vibration being determined by the position of the pile, as this position remained throughout unchanged, the vibration was unchanged also.

(566) The following column of numbers shows the rise of intensity of the particular obscure rays falling on the pile, as the platinum spiral passed through its various degrees of incandescence up to white heat.

Appearance of spiral	Radiation of obscure band
Dark	1
Dark	6
Faint red	10
Dull red	13
Red	18
Full red	27
Orange	60
Yellow	93
Full white	122

Thus we prove that as the new and more rapid vibrations are introduced, the old ones become more intense, until at a white heat the obscure rays of a special refrangibility reach an intensity 122 times that possessed by them at the commencement. This abiding and augmentation of the dark rays when the bright ones are introduced may be expressed by the phrase *persistance of rays.*

(567) What has been here demonstrated regarding an incandescent platinum spiral is also true of the electric

light. Side by side with this outflow of intensely luminous rays, we have a corresponding outflow of obscure ones. The carbon points, like the platinum spiral, may be raised from a state of obscure warmth to a brilliancy almost equal to that of the sun, and as this occurs, the obscure radiation also rises enormously in intensity. The investigation of the distribution of heat in the spectrum of the electric light will yield us important results, and will fitly prepare the way for those experiments on invisible rays to which I shall subsequently direct your attention.

(568) The thermo-electric pile employed is this beautiful instrument already referred to as constructed by Ruhmkorff. It consists, as you know, of a single row of elements properly mounted and attached to a double brass screen. It has in front two silvered edges, which, by means of a screw, can be caused to close upon the pile, so as to render its face as narrow as desirable, reducing it to the width of the finest hair, or, indeed, shutting it off altogether. By means of a small handle and long screw, the plate of brass and the pile attached to it can be moved gently to and fro, and thus the vertical slit of the pile can be caused to traverse the entire spectrum, or to pass beyond it in both directions. The width of the spectrum was in each case equal to the length of the face of the pile.

(569) To produce a steady spectrum of the electric light, I employed a regulator devised by M. Foucault and constructed by Duboscq, the constancy of which is admirable. A complete rock-salt train was constructed, the arrangement of which has been already indicated. In the front orifice of the camera which surrounds the electric lamp was placed a lens of transparent rock-salt, intended to reduce to parallelism the divergent rays proceeding from the carbon-points. The parallel beam was permitted to pass through a narrow slit, in front of which was

placed another rock-salt lens, the position of this lens being so arranged that a sharply-defined image of the slit was obtained at a distance beyond it equal to that at which the spectrum was to be formed. Immediately behind this lens was placed a pure rock-salt prism (sometimes two of them). The beam was thus decomposed, a brilliant horizontal spectrum being cast upon the screen which bore the thermo-electric pile. By turning the handle already referred to, the face of the pile could be caused to traverse the spectrum, an extremely narrow band of light or radiant heat falling upon it at each point of its march.* A sensitive galvanometer was connected with the pile, and from its deflection the heating-power of every part of the spectrum, visible and invisible, was inferred.

(570) Two modes of moving the instrument were practised, the description of one of which will be sufficient here. The face of the pile was brought to the violet end of the spectrum, where the heat is insensible, and then moved, as I now move it, through all the colours to the red; then past the red up to the position of maximum heat, and afterwards beyond this position until the heat of the invisible spectrum gradually faded away. The following Table contains a series of measurements executed in this manner. The motion of the pile is measured by turns of its handle, every turn corresponding to the shifting of the face of the instrument through a space of one millimetre, or $\frac{1}{25}$th of an inch. At the beginning, where the increment of heat was slow and gradual, the readings were taken at every two turns of the handle; on quitting the red, where the heat suddenly increases, the intervals were only half a turn, while near the maximum, where the changes were most sudden, the intervals were reduced

* The width of the linear pile was 0·03 of an inch.

to a quarter of a turn, which corresponded to a translation of the pile through $\frac{1}{100}$th of an inch. Intervals of one and of two turns were afterwards resumed until the heating-power ceased to be distinct. At every halting-place the deflection of the needle was noted. Calling the maximum effect in each series of experiments 100, the column of figures in the following Table expresses the heat of all the other parts of the spectrum:—

DISTRIBUTION OF HEAT IN SPECTRUM OF ELECTRIC LIGHT.

Movement of pile	Calorific intensity, in 100th of the maximum
Before starting (pile in the blue)	0
Two turns forward (green entered)	2
,,	5
,,	8
,, (red entered)	21
,, (extreme red)	45
Half turn forward	60
,,	74
,,	85
,,	96
,,	99
Quarter turn forward, *maximum*	100
,,	97
Half turn forward	78
,,	62
,,	45
,,	36
Two turns forward	14
,,	9
,,	7
,,	5
,,	3
,,	2
,,	2

(571) Here, as already stated, we begin in the blue, and pass first through the visible spectrum. Quitting this at the place marked ('extreme red'), we enter the invisible calorific spectrum and reach the position of maximum heat, from which, onwards, the thermal power falls till it practically disappears.

(572) More than a dozen series of such measurements were executed, each series giving its own curve. On superposing the different curves a very close agreement was found to exist between them. The annexed figure (fig. 99), which is the mean of several, expresses, with a close approximation to accuracy, this distribution of heat in the spectrum of the electric light from fifty cells of Grove. The space A B C D represents the invisible, while C D E represents the visible radiation. We here see the gradual augmentation of thermal power, from the blue end of the spectrum to the red. But in the region of dark rays beyond the red the curve shoots suddenly upwards in a steep and massive peak,—a kind of Matterhorn of heat—which quite dwarfs by its magnitude the portion of the diagram representing the visible radiation.

(573) The sun's rays before reaching the earth have to pass through our atmosphere, the aqueous vapour of which exercises a powerful absorption on the invisible calorific rays. From this, apart from other considerations, it would follow that the ratio of the invisible to the visible radiation in the case of the sun must be less than in the case of the electric light. Experiment, we see, justifies this conclusion; for whereas fig. 98 shows the invisible radiation of the sun to be about twice the visible, fig. 99 shows the invisible radiation of the electric light to be nearly eight times the visible. If we cause the beam from the electric lamp to pass through a layer of water of suitable thickness, we place its radiation in approximately the same condition as that of the sun; and on decomposing the beam after it has been thus sifted, we obtain a distribution of heat closely resembling that observed in the solar spectrum.

(574) The curve representing the distribution of heat in the electric spectrum falls most steeply on that side of the maximum which is most distant from the red. On

FIG. 99.

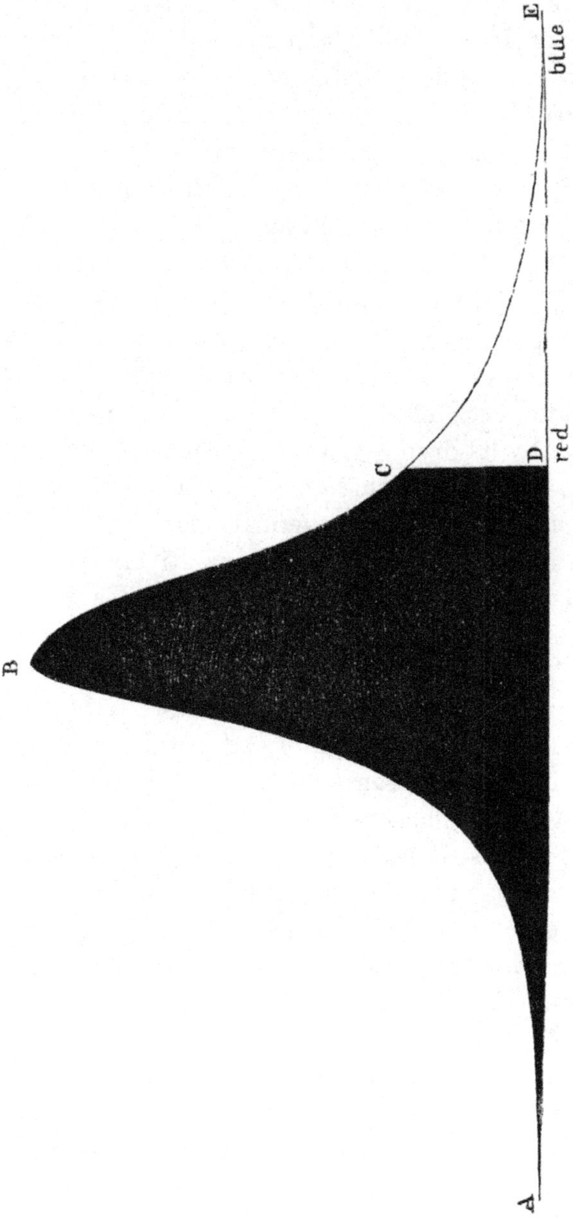

SPECTRUM OF ELECTRIC LIGHT.

both sides, however, we have a *continuous* falling off. I have made numerous experiments to ascertain whether there is any interruption of continuity in the calorific spectrum; but all the measurements hitherto executed with artificial sources reveal a gradual and continuous augmentation of heat from the point where it first becomes sensible up to the maximum.

(575) Sir John Herschel has shown that this is not the case with the radiation from the sun when analysed by a flint-glass prism. Permitting the solar spectrum to fall upon a sheet of blackened paper, over which had been spread a wash of alcohol, this eminent philosopher determined by its drying-power the heating-power of the spectrum. He found that the wet surface dried in a series of spots representing thermal maxima separated from each other by spaces of comparatively feeble calorific intensity. No such maxima and minima were observed in the spectrum of the electric light, nor in the spectrum of a platinum wire raised to a white heat by a voltaic current. Prisms and lenses of rock-salt, of crown glass, and of flint glass were employed in these cases. In other experiments the beam intended for analysis was caused to pass through layers of water and other liquids of various thicknesses. Gases and vapours of various kinds were also introduced into the path of the beam. In all cases there was a general lowering of the calorific power, but the descent of the curve on both sides of the maximum was unbroken.*

(576) The rays from an obscure source cannot, as already remarked, compete in point of intensity with the obscure rays of a luminous source. No body heated under incandescence could emit rays of an intensity comparable to those of the maximum region of the electric spectrum.

* At a future day I hope to subject this question to a more severe examination.

If, therefore, we wish to produce intense calorific effects by invisible rays, we must choose those emitted by an intensely luminous source. The question then arises, how are the invisible calorific rays to be isolated from the visible ones?

(577) The interposition of an opaque screen suffices to cut off the visible spectrum of the electric light, and leaves us the invisible calorific rays to operate upon at our pleasure. Sir William Herschel experimented thus when he sought, by concentrating them, to render the invisible rays of the sun visible. But to form a spectrum in which the invisible rays shall be completely separated from the visible ones, a narrow slit or a small aperture is necessary; and this circumstance renders the amount of heat separable by prismatic analysis very limited. If we wish to ascertain what the intensely concentrated invisible rays can accomplish, we must devise some other mode of detaching them from their visible companions. We must, in fact, discover a substance which shall filter the composite radiation of a luminous source by stopping the visible rays and allowing the invisible ones free transmission.

(578) The main object of these researches was, as already intimated, to make radiant heat an explorer of molecular condition, and the marked difference between elementary and compound bodies which the experiments reveal is, in my estimation, a point destined to be fruitful in important consequences. As soon as this difference came clearly out in the case of gases, liquids were looked to, and the action of such as I was able to examine fell in surprisingly with the previously observed deportment of gaseous bodies. Could we then obtain a *black* elementary body thoroughly homogeneous, and with all its parts in perfect optical contact, we should probably find it an effectual filter for the radiation of the sun or of the electric light. While cutting off the visible radiation, the black element would, probably, allow the invisible to pass.

(570) Carbon in the state of soot is black, but its parts are not optically continuous. In black glass the continuity is far more perfect, and hence the result established by Melloni, that black glass possesses a considerable power of transmission. Gold in ruby glass, or in a state of jelly prepared by Mr. Faraday, is exceedingly transparent to the invisible calorific rays, but it is not black enough to quench entirely the visible ones. The densely-brown liquid bromine is better suited to our purpose; for, in thicknesses sufficient to quench the light of our brightest flames, this element displays extraordinary diathermancy. Iodine cannot be applied in the solid condition, but it dissolves freely in various liquids, the solution in some cases being intensely dark. Here, however, the action of the element may be masked by that of its solvent. Iodine, for example, dissolves freely in alcohol; but alcohol is so destructive of the extra-red rays, that it would be entirely unfit for experiments the object of which is to retain these rays, while quenching the visible ones. The same remark applies in a greater or less degree to many other solvents of iodine.

(580) The deportment of bisulphide of carbon, both as a vapour and a liquid, suggests the thought that it would form a most suitable solvent. It is extremely diathermic, and there is hardly another substance able to hold so large a quantity of iodine in solution. Experiments already recorded (§ 506) prove that, of the rays emitted by a red-hot platinum spiral, 94·5 per cent. is transmitted by a layer of the liquid 0·02 of an inch in thickness, the transmission through layers 0·07 and 0·27 of an inch thick being 87·5 and 82·5 respectively. Another experiment with a layer of greater thickness will exhibit the deportment of the transparent bisulphide towards the far more intense radiation of the electric light.

(581) This cylindrical cell, 2 inches in length and 2·8

inches in diameter, with its ends stopped by plates of perfectly transparent rock-salt, was placed empty in front of an electric lamp; the radiation from the lamp, after having crossed the cell, fell upon a thermo-electric pile, and produced a deflection of

73°.

Leaving the cell undisturbed, the transparent bisulphide of carbon was poured into it: the deflection fell to

72°.

A repetition of the experiment gave the following results.

	Deflection
Through empty cell	74°
Through bisulphide	73

Taking the values of these deflections from a Table of calibration and calculating the transmission, that through the empty cell being 100, we obtain the following results:—

	Transmission
From the first experiment . . .	94·9 per cent.
From the second experiment . . .	94·6 ,,
Mean . .	94·8

Hence the introduction of the bisulphide lowers the transmission only from 100 to 94·8.*

(582) A *perfect* solvent of the iodine would be entirely neutral to the total radiation; and the bisulphide of carbon is shown by the foregoing experiment to approach very near perfection. We have in it a body capable of transmitting with little loss the total radiation of the electric light. Our object is now to filter this total, by the introduction into the bisulphide of a substance competent to quench the visible and transmit the invisible rays. That iodine does this with marvellous sharpness it is now my business to prove.

* The diminution of the reflection from the sides of the cell by the introduction of the bisulphide is not here taken into account.

(583) A rock-salt cell, filled with the *transparent* bisulphide of carbon, was placed in front of the camera which contained the white-hot platinum spiral. The transparent liquid was then drawn off and its place supplied by the solution of iodine. The deflections observed in the respective cases are as follows :—

RADIATION FROM WHITE-HOT PLATINUM.

Through transparent liquid	Through opaque liquid
73·9°	73·8°
73·0	72·9

(584) *All* the luminous rays passed through the transparent bisulphide, *none* of them passed through the solution of iodine. Still we see what a small effect is produced by their withdrawal. The actual proportion of luminous to obscure rays, as calculated from the above observations, may be thus expressed :—

Dividing the radiation from a platinum wire raised to a dazzling whiteness by an electric current into twenty-four equal parts, one of those parts is luminous, and twenty-three obscure.

(585) A bright gas flame was substituted for the platinum spiral, the top and bottom of the flame being shut off, and its most brilliant portion chosen as the source of rays. The result of forty experiments with this source may be thus expressed :—

Dividing the radiation from the most brilliant portion of a flame of coal gas into twenty-five equal parts, one of those parts is luminous and twenty-four obscure.

(586) I next examined the ratio of obscure to luminous rays in the electric light. A battery of fifty cells was employed, and the rock-salt lens was used to render the rays from the coal points parallel. To prevent the deflection from reaching an inconvenient magnitude, the parallel rays were caused to pass through a circular aperture 0·1 of an inch in diameter, and were sent alternately

through the transparent bisulphide and the opaque solution. It is not easy to obtain perfect steadiness on the part of the electric light; but three experiments carefully executed gave the following deflections :—

RADIATION FROM ELECTRIC LIGHT.

	Through transparent CS^2	Through opaque solution
Experiment No. I.	72·0°	70·0°
Experiment No. II.	76·5	75·0
Experiment No. III.	77·5	76·5

Calculating from these measurements the proportion of luminous to obscure heat, the result may be thus expressed.

Dividing the radiation from the electric light produced by a Grove's battery of fifty cells, into ten equal parts, one of those parts is luminous and nine obscure.

The results hitherto obtained with various sources, radiating through iodine, may be thus tabulated :—

RADIATION THROUGH DISSOLVED IODINE.

Source	Absorption	Transmission
Dark spiral	0	100
Lampblack at 212° Fahr. . .	0	100
Red-hot spiral	0	100
Hydrogen flame	0	100
Oil flame	3	97
Gas flame	4	96
White-hot spiral	4·6	95·4
Electric light	10	90

(587) Subsequent experiments with a battery of fifty cells made the transmission in the case of the electric light 89, and the absorption 11. Considering the transparency of the iodine for heat emitted by all sources heated barely up to incandescence, as exhibited in the above table, it may be inferred that the absorption of 11 per cent. represents the calorific intensity of the *luminous rays* alone. By the method of filtering, therefore, we make the invisible radiation of the electric light

eight times the visible. Computing, by means of a proper scale, the area of the spaces A B C D, C D E (fig. 99), the former, which represents the invisible emission, is found to be 7·7 times the latter. Prismatic analysis, therefore, and the method of filtering yield almost exactly the same result.

(588) It is plain from the description of the experiments that the foregoing results refer to the action of the iodine dissolved in the bisulphide of carbon. The transmission of 100, for example, does not indicate that the solution itself, but that the iodine in the solution is perfectly diathermic to the radiation from the first four sources.

(589) Having thus, in the solution of iodine, found a means of almost perfectly detaching the obscure from the luminous heat-rays of electric light, we are able to operate at will upon the former. I place a rock-salt lens in this camera so as to form a small image of the coal-points. A battery of forty cells being employed, the track of the cone of rays emergent from the lamp is plainly seen in the air, and their point of convergence therefore easily fixed. Fixing the cell containing the opaque solution in front of the lamp, the luminous cone is entirely cut off, but the intolerable temperature of the focus, when the hand is placed there, shows that the calorific rays are still transmitted. Placing successively in the dark focus thin plates of tin and zinc; they are speedily fused; matches are ignited, gun-cotton is exploded, and brown paper set on fire. With a battery of sixty of Grove's cells, all these results are readily obtained with the ordinary glass lenses of Duboscq's electric lamp. It is extremely interesting to observe in the middle of the air of a perfectly dark room a piece of black paper suddenly pierced by the invisible rays, and the burning ring expanding on all sides from the centre of ignition.

(590) On November 15, 1864, a few experiments were

made on solar light. The heavens were not free from clouds, nor the London atmosphere from smoke, and at best only a portion of the action which a clear day would have given was obtained. Happening to possess this hollow lens, I filled it with the concentrated solution of iodine. Placed in the path of the solar rays, a faint red ring was imprinted on a sheet of white paper held behind the lens, the ring contracting to a faint red spot when the focus of the lens was reached. It was immediately found that this ring was produced by the light which had penetrated the thin rim of the liquid lens. Pasting a zone of black paper round the rim, the ring was entirely cut off and no visible trace of solar light crossed the lens. At the focus, whatever light passed would be intensified nine hundredfold; still even here no light was visible.

(591) Not so, however, with the sun's obscure rays; the focus was burning hot. A piece of black paper placed there was instantly pierced and set on fire; and by shifting the paper, aperture after aperture was formed in quick succession. Gunpowder was also exploded.

(592) From the setting of paper on fire and the fusion of non-refractory metals to the rendering of refractory bodies incandescent by the invisible rays, the step was immediate and inevitable. And here the enquiry derived a stimulus from the fact, that on theoretic grounds some eminent minds doubted whether the attainment of incandescence by invisible rays was possible. A moment's reflection will make plain to you that the success of the experiment involved *a change of period* on the part of the calorific waves. For if without the aid of combustion, waves of too slow a recurrence to excite the sense of vision were to render a refractory body luminous, it could only be by compelling the molecules of that body to vibrate more rapidly than the waves which fell upon them. Whether

this change of period could be effected had been long considered doubtful.

(593) A few preliminary experiments with platinum foil, which resulted in failure, raised the question whether, even with the *total radiation*, bright and dark, of the electric light, it would be possible to obtain incandescence without combustion. Abandoning the use of lenses altogether, a thin leaf of platinum was caused to approach the ignited coal-points. It was observed by myself from behind, while my assistant stood beside the lamp, and, looking through a dark glass, watched the distance between the platinum-foil and the electric light. At half an inch from the carbon-points the metal became red-hot. The problem now before me was to obtain, at a greater distance, a focus of rays which should possess a heating-power equal to that of the direct rays at a distance of half an inch.

(594) In the first attempt the direct rays were utilised as much as possible. A piece of platinum-foil was placed at a distance of an inch from the carbon-points, there receiving the direct radiation. The rays emitted *backwards* from the points were at the same time converged by this small mirror upon the foil, and were found more than sufficient to compensate for the diminution of intensity due to withdrawal of the foil to the distance of an inch. By the same method, incandescence was subsequently obtained when the foil was removed two, and even three, inches from the carbon-points.

(595) This enabled me to introduce between the focus and the source of rays a cell containing the solution of iodine. The dark rays transmitted were found of sufficient power to *inflame* paper, or to raise platinum-foil to incandescence.

(596) The experiments, however, were not unattended with danger. The bisulphide of carbon is extremely inflammable; and on the 2nd of November 1864, while

employing a very powerful battery and intensely-heated carbon-points, the substance took fire, and instantly enveloped the electric lamp and all its appurtenances in flame. Happily the precaution had been taken of placing the entire apparatus in a flat vessel containing water, into which the flaming mass was summarily turned. The bisulphide of carbon being heavier than the water, sank to the bottom, so that the flames were speedily extinguished. Similar accidents occurred twice subsequently.

(597) Such occurrences caused me to seek earnestly for a substitute for the bisulphide. Pure chloroform, though not so diathermic, transmits the invisible rays pretty copiously, and it freely dissolves iodine. In layers of the thickness employed, however, the solution was not sufficiently opaque; and its absorptive power enfeebled the effects. The same remark applies to the iodides of methyle and ethyle, to benzole, acetic ether, and other substances. They all dissolve iodine, but they weaken the results by their action on the dark rays.

(598) Special cells were then constructed for the element bromine and for chloride of sulphur. Neither of these substances is inflammable; but they are both intensely corrosive, and their action upon the lungs and eyes is so irritating as to render their employment impracticable. With both liquids, however, powerful effects were obtained; still their diathermancy did not come up to that of the dissolved iodine. Bichloride of carbon would be invaluable if its solvent power were equal to that of the bisulphide. It is not at all inflammable, and its own diathermancy appears equal to that of the bisulphide. But in reasonable thicknesses the iodine which it can dissolve is not sufficient to render the solution perfectly opaque. The solution forms a purple colour of exquisite beauty; and though unsuited to strict crucial experiments on dark rays,

this filter may be employed with excellent effect in class experiments.

(599) Thus foiled in my attempts to obtain a solvent equally good as, and less dangerous than, the bisulphide of carbon, I sought to reduce the danger of employing it to a minimum. A tin camera was constructed, within which were placed both the lamp and its converging mirror. Through an aperture in front, 2¾ inches wide, the cone of reflected rays issued, forming a focus outside the camera. Underneath this aperture was riveted a stage, on which the solution of iodine rested, thus closing the aperture and cutting off all the light. At first nothing intervened between the cell and the carbon-points; but the peril of thus exposing the bisulphide caused me to make the following improvements. A perfectly transparent plate of rock-salt, secured in a proper cap, was employed to close the aperture; and by it all direct communication between the solution and the incandescent carbons was cut off. The aperture was then surrounded by an annular space, about 2½ inches wide and a quarter of an inch deep, through which cold water was caused to circulate. The cell containing the solution was moreover surrounded by a jacket, and the current of water, having completed its course round the aperture, passed round the cell. Thus the apparatus was kept cold. The neck of the cell was stopped by a closely-fitting cork; through this passed a piece of glass tubing, which, when the cell was placed upon its stage, ended at a considerable height above the focus. Experiments on combustion might therefore be carried on at the focus without fear of igniting the vapour which, even under the improved conditions, might escape from the bisulphide of carbon.

(600) The arrangement will be at once understood by reference to figs. 100 and 101, which show the camera, lamp, and filter both from the side and from the front. $x\,y$ is the mirror, from which the reflected cone of rays passes, first,

through the rock-salt window, and afterwards through the iodine filter *m n*. The rays converge to the focus *k*, where they would form an invisible image of the lower carbon-point; the image of the upper would be thrown below *k*. *Both images spring vividly forth when a leaf of platinised platinum is exposed at the focus.* At *s s*, figs. 100 and 101, is shown, in section, and in plan, the annular space in which the cold water circulates.

Fig. 100.

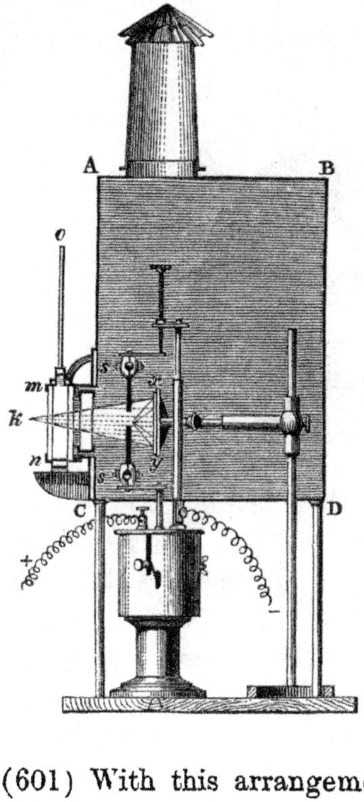

Fig. 101.

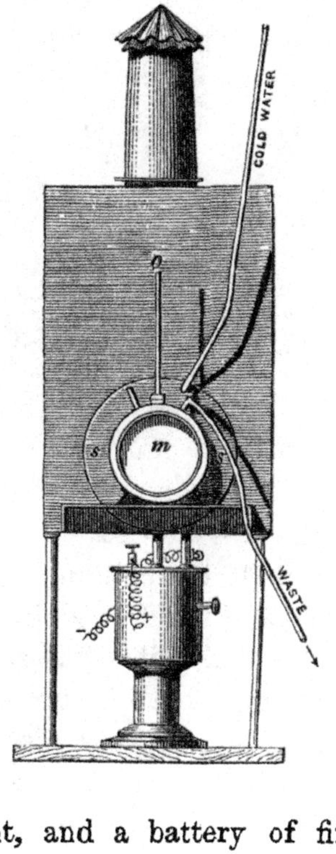

(601) With this arrangement, and a battery of fifty cells, the following results were obtained:—

A piece of silver leaf, fastened to a wire ring, and tarnished by exposure to the fumes of sulphide of ammonium, being held in the dark focus, the film was raised to vivid redness.

(602) Copper-leaf tarnished in a similar manner, when placed at the focus, was raised to redness.

(603) A piece of platinised platinum-foil was supported in an exhausted receiver, the vessel being so placed that the focus fell upon the platinum. The heat of the focus was instantly converted into light, a clearly-defined and inverted image of the points being stamped upon the metal. Fig. 102 represents the thermograph of the carbons.

(604) Blackened paper was now substituted for the platinum in the exhausted receiver. Placed at the focus of invisible rays, the paper was instantly pierced, a cloud of

FIG. 102.

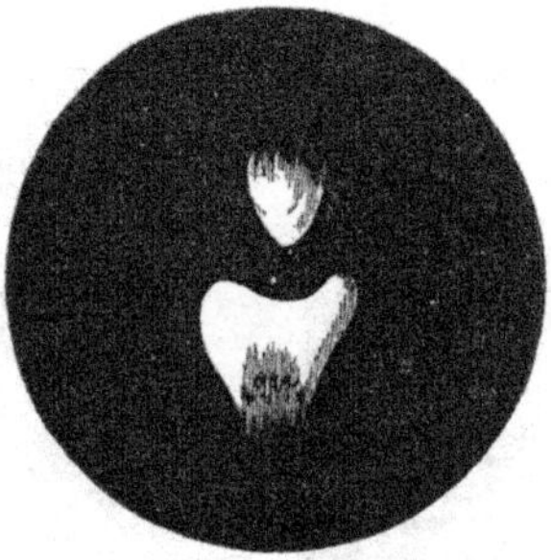

smoke was poured through the opening, and fell like a cascade to the bottom of the receiver. The paper seemed to burn without incandescence. Here also a thermograph of the coal-points was stamped out. When black paper is placed at the focus, where the thermal image is well defined, it is always pierced in two points, answering to the images of the two carbons. The superior heat of the positive carbon is shown by the fact that its image first pierces the paper; it burns out a large space, and shows its peculiar crater-like top, while the negative carbon usually pierces a small hole.

(605) Paper reddened by the iodide of mercury had its colour discharged at the places on which the invisible

image of the coal-points fell upon it, though not with the expected promptness.

(606) Discs of paper reduced to carbon by different processes were raised to brilliant incandescence, both in the air and in the exhausted receiver.

(607) In these earlier experiments an apparatus was employed which had been constructed for other purposes. The mirror, for example, was one detached from a Duboscq's camera; it was first silvered at the back, but afterwards silvered in front. The cell employed for the iodine solution was also that which usually accompanies Duboscq's lamp, being intended by its maker for a solution of alum. Its sides are of good white glass, the width from side to side being 1·2 inch.

(608) A point of considerable theoretic importance was involved in these experiments. In his excellent researches on fluorescence, Professor Stokes had invariably found the refrangibility of the incident light to be *lowered*. This rule was so constant as almost to enforce the conviction that it was a law of nature. But if the rays which in the foregoing experiments raised platinum and gold and silver to a red heat were wholly ultra-red, the rendering visible of the metallic films would be an instance of *raised* refrangibility.

And here I thought it desirable to make sure that no trace of visible radiation passed through the solution, and also that the invisible radiation was exclusively ultra-red.

(609) This latter condition might seem to be unnecessary, because the calorific action of the ultra-violet rays is so exceedingly feeble (in fact is immeasurably small) that, even supposing them to reach the platinum, their heating power would be an utterly vanishing quantity. Still the exclusion of *all* rays of high refrangibility was

necessary to the complete solution of the problem. Hence, though the iodine employed in the foregoing experiments was sufficient to cut off the light of the sun at noon, I wished to submit its opacity to a severer test. The following experiments were accordingly executed.

(610) The rays from the electric lamp being duly converged by the mirror, the iodine-cell was placed in the path of the convergent beam, its light being thereby to all appearance totally intercepted. With a piece of platinum-foil the focus was found and marked, and a cell containing a solution of alum was then placed between the focus and the iodine-cell. The alum solution diminished materially the invisible radiation, but it was without sensible influence upon visible rays.

(611) All stray light issuing from the crevices in the lamp being cut off, and the daylight also being excluded from the room, the eye was caused slowly to approach the focus. On reaching it, a singular appearance presented itself. The incandescent coke-points of the lamp were seen black, projected on a deep-red ground. Their motions could be followed, and when brought into contact, a white space was seen at the extremities of the points, appearing to separate them. The points were seen erect. By careful observation the whole of the points could be observed, and even the holders which supported them. The black appearance of the incandescent portion of the points was here only *relative*; they appeared dark because they intercepted more of the light reflected from the mirror behind than they could make good by their direct emission.

(612) The solution of iodine, 1·2 inch in thickness, proving thus unequal to the test applied to it, I had two other cells constructed—the one with transparent rock-salt sides, the other with glass ones. The width of th. former was 2 inches, that of the latter nearly 2½ inches

Filled with the solution of iodine, these cells were placed in succession in front of the camera, and the concentrated beam was sent through them. Determining the focus as before, and afterwards introducing the alum-cell, the eye on being brought up to the focus received no impression of light.

(612*a*) The alum-cell was then abandoned, and the undefended eye was caused to approach the focus: the heat was intolerable, but it seemed to affect the eyelids and not the eye itself. An aperture somewhat larger than the pupil being made in a metal screen, the eye was placed behind it, and brought slowly and cautiously up to the focus. The concentrated beam entered the pupil; but no impression of light was produced, nor was the retina sensibly affected by the heat. The eye was then withdrawn, and a plate of platinised platinum was placed in the position occupied by the retina a moment before. It instantly rose to vivid redness.* The failure to obtain, with the most sensitive media, the slightest evidence of fluorescence at the obscure focus, proved the invisible rays to be exclusively ultra-red. It will be subsequently shown that a considerable portion of these rays actually reached the retina.

(613) When intense effects are sought after, we collect as many of the invisible rays as possible, and concentrate them on the smallest possible space. The nearer the mirror is to the source of rays, the more of these rays will it intercept and reflect, and the nearer the focus is to the same source, the smaller will the image be. To secure proximity both of focus and mirror, the latter must be of short focal length. If a mirror of long focal length be employed, its distance from the source of rays must be considerable to bring the focus near the source, but when placed thus at a distance, a great number of rays escape the

* I do not recommend the repetition of these experiments.

mirror altogether. If, on the other hand, the mirror be too deep, spherical aberration comes into play; and though a vast quantity of rays may be collected, their convergence at the focus is imperfect. To determine the best form of mirror, three of them were constructed: the first 4·1 inches in diameter, and of 1·4 inch focal length; the second 7·9 inches in diameter, and of 3 inches focal length; the third 9 inches in diameter, with a focal length of 6 inches. Fractures caused by imperfect annealing repeatedly occurred; but at length I was so fortunate as to obtain the three mirrors, each without a flaw. The most convenient distance of the focus from the source, was found to be about 5 inches; and the position of the mirror ought to be arranged accordingly. This distance permits of the introduction of an iodine-cell of sufficient width, while the heat at the focus is exceedingly powerful.

(614) And now with this improved apparatus I will run through the principal experiments on invisible heat-rays. These dense volumes of smoke which rise from a blackened block of wood when it is placed in the dark focus are very striking: matches are at once ignited, and gunpowder instantly exploded at the focus. This dry paper held there bursts into flame. These chips of wood are also inflamed: the dry wood of a hat-box is very suitable for this experiment. When a sheet of brown paper is placed a little beyond the focus, it is first brought to vivid incandescence over a large space; the paper then yields, and the combustion propagates itself as a burning ring round the centre of ignition. Charcoal is made an ember at the focus, and these discs of charred paper glow with extreme vividness. When blackened zinc-foil is placed at the focus it bursts into flame; and by slowly moving the foil about, its ignition may be kept up till the whole of it is consumed. This magnesium wire, flattened at the end and blackened, also bursts into vivid combus-

tion when held at the focus. A cigar of course is instantly lighted there. The bodies experimented on may be enclosed in glass receivers; the concentrated rays will still burn them after having crossed the glass. This glass jar, for example, contains oxygen; and in the oxygen by means of a suitable holder is plunged a bit of charcoal bark. When the dark rays are concentrated upon the charcoal it instantly throws out showers of scintillations.

(615) In all these cases the body exposed to the action of the invisible rays was more or less combustible. It was first heated and then exposed to the attack of oxygen. The vividness observed was in part due to combustion, and does not furnish a conclusive proof that the refrangibility of the incident rays was elevated. This, however, is effected by exposing non-combustible bodies at the focus, or by enclosing combustible ones in a space devoid of oxygen. Both in air and *in vacuo* platinised platinum-foil has been repeatedly raised to a white heat. The same result has been obtained with a sheet of charcoal or coke suspended *in vacuo*. Now the waves from which this light was extracted had neither the visible nor the ultra-violet rays commingled with them; they were exclusively ultra-red. The action, therefore, of the atoms of platinum, copper, silver, and carbon upon these rays transmutes them from heat-rays into light-rays. They impinge upon these atoms at a certain rate; they return from them at a quicker rate, the invisible being thus rendered visible.

(616) *On looking at the white-hot platinum through a prism of bisulphide of carbon, a rich and complete spectrum was obtained. All the colours, from red to violet, shone vividly.*

(617) To express this transmutation of heat-rays into others of higher refrangibility, I propose the term *calorescence.* It harmonises well with the term 'fluorescence' introduced by Professor Stokes, and is also suggestive of the

character of the effects to which it is applied. The phrase 'transmutation of rays,' introduced by Professor Challis, covers both classes of effects.*

(618) I have sought to *fuse* platinum with the invisible rays of the electric light, but hitherto without success. In some experiments a large model of Foucault's lamp was employed, with a battery of 100 cells. In other experiments two batteries were employed, one of 100 cells and one of 70, making use of two lamps, two mirrors, and two filters, and converging the heat of both lamps in opposite directions upon the same point. When a leaf of platinum was placed at the common focus, the converged beams struck it at opposite sides, and raised it to dazzling whiteness. I am persuaded that the metal could be fused, if the platinum-black upon its surface could be retained. But this was immediately dissipated by the intense heat, and, the reflecting-power of the metal coming into play, the absorption was so much lowered that fusion was not effected. By coating the platinum with lampblack it has been brought to the verge of fusion, the incipient yielding of the mass being perfectly apparent after it had cooled. Here, however, as in the case of platinised platinum, the absorbing substance disappears too quickly. Copper and aluminium, however, when thus treated, are speedily burnt up.

(619) The isolation of the luminiferous ether from the air is strikingly illustrated by these experiments. The air at the focus may be of a freezing temperature, while the ether possesses an amount of the heat competent, if absorbed, to impart to that air the temperature of flame. An air-thermometer is unaffected where platinum is raised to a white heat.

(620) Arrangements have already been described with a view of avoiding the danger incidental to the use of so

* Philosophical Magazine, 1865. Vol. xxix. p. 336.

inflammable a substance as the bisulphide of carbon. I have thought of accomplishing this end by simpler means, and thus facilitating the repetition of the experiments. The arrangement now before you, fig. 103, may be adopted with safety.

A B C D is an outline of the camera.

x y the silvered mirror within it.

c the carbon-points of the electric light.

FIG. 103.

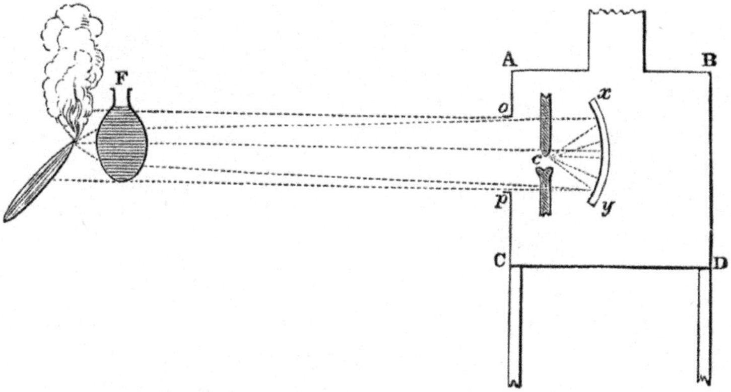

o p the aperture in front of the camera, through which issues the beam reflected by the mirror *x y*.

(621) Let the distance of the mirror from the carbon-points be such as to render the reflected beam slightly convergent. Fill a glass flask with the solution of iodine, and place the flask in the path of the reflected beam at a safe distance from the lamp. The flask acts as a lens and filter at the same time, the bright rays are intercepted, and the dark ones are powerfully converged. At F such a flask is represented; and at the focus formed a little beyond it combustion and calorescence may be produced. Flasks with diameters from 1½ to 3 inches are well suited for the experiments.

(622) By the arrangement here described, platinum

has been raised to redness at a distance of 22 feet from the source of the rays.

(623) *The best mirror, however, scatters the rays more* or less; and by this scattering, the beam at a great distance from the lamp becomes much enfeebled. The effect is therefore intensified when the beam is caused to pass through a tube polished within, which prevents the lateral

FIG. 104.

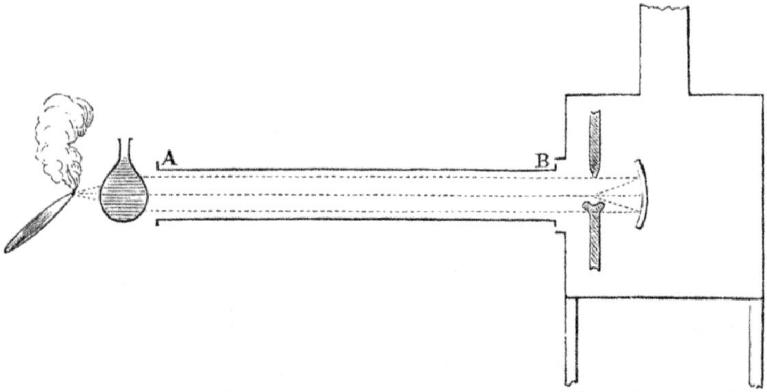

waste of radiant heat. Such a tube, placed in front of the camera, is represented at A B, fig. 104. The flask may be held against its end by the hand, or it may be permanently fixed there. With a battery of fifty cells, platinum may be raised to a white heat at the focus of the flask.

(624) Again let a lens of glass or rock-salt (L, fig. 105), 2·5 inches wide, and having a focal length of 3 inches, be placed in the path of the reflected beam. The rays are converged; and at their point of convergence all the effects of caloresence and combustion may be obtained, the luminous rays being cut off by a cell *m n*,* with plane glass sides and containing the opaque solution.

(625) Finally, the arrangement shown in fig. 106 may be adopted. The beam reflected by the mirror within the camera is received and converged by a second mirror $x' y'$. At

* The cell *m n* may be placed at a distance from the carbon-points; if a reflecting tube be used it is all the more effective.

the point of convergence, which may be several feet from the camera, all the effects hitherto described may be obtained. The light of the beam may be cut off at any convenient point of its course; but in ordinary cases the experiment is best made by employing the bichloride instead of the bisulphide of carbon, and placing the cell (*m n*) containing the opaque solution close to the camera. The moment

FIG. 105.

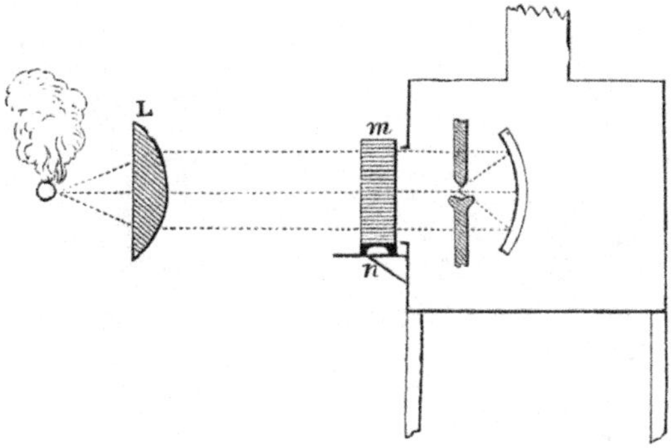

the coal-points are ignited, explosion, combustion, or calorescence, as the case may be, occurs at the focus.

(626) Thus far I have dealt exclusively with the invisible radiation of the electric light; but all solid bodies raised to incandescence emit these invisible calorific rays. The denser the incandescent body, moreover, the more powerful is its obscure radiation. We possess at the Royal Institution very dense cylinders of lime for the production of the Drummond light; and when a copious oxyhydrogen-flame is projected against one of them it shines with an intense yellowish light, while the obscure radiation is exceedingly powerful. Filtering the latter from the total emission by the solution of iodine, all the effects of combustion and calorescence which have been just described

may be obtained at the focus of the invisible rays. The light obtained by projecting the oxyhydrogen-flame upon compressed magnesia, after the manner of Signor Carlevaris, is whiter than that emitted by our lime; but the substance being light and spongy, its obscure radiation is surpassed by that of our more solid cylinders.

(627) The invisible rays of the sun have also been transmuted. A concave mirror, 3 feet in diameter, was mounted

FIG. 106.

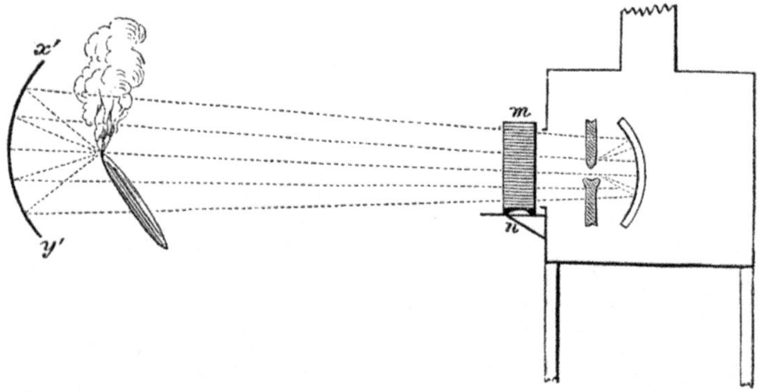

on the roof of the Royal School of Mines in Jermyn Street. The focus was formed in a darkened chamber, in which the platinised platinum-foil was exposed. Cutting off the visible rays by the solution of iodine, feeble but distinct incandescence was there produced by the invisible rays.

(628) To obtain a clearer sky, this mirror was transferred to the garden of my friend Mr. Lubbock (now Sir John Lubbock, Bart.), near Chislehurst. A blackened tin tube (A B, fig. 107) with square cross section and open at one end, was furnished at the other with a plane mirror (*x y*), forming an angle of 45° with the axis of the tube. A lateral aperture (*x o*), about 2 inches square, was cut out in front of the mirror. Over this aperture was placed a leaf of platinised platinum. Turning the leaf towards the concave mirror, the concentrated sunbeams were

permitted to fall upon it. In the full glare of daylight it is quite impossible to see whether the platinum was incandescent or not; but placing the eye at B, the glow of the platinum could be seen by reflection from the plane mirror. Incandescence was thus obtained at the focus of the large mirror, X, Y, after the removal of the visible rays by the iodine solution, *m n*.*

FIG. 107.

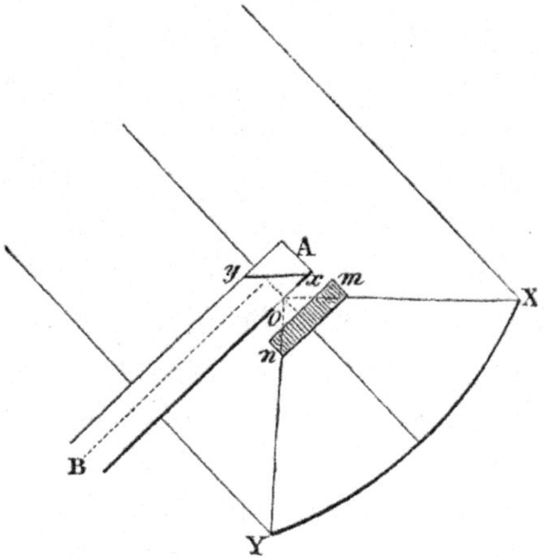

(629) The effects obtained with the total solar radiation were extraordinary. Large spaces of the platinum-leaf, and even thick foil, when exposed at the focus, disappeared as if vaporised. The handle of a pitchfork, similarly exposed, was soon burnt quite across. Paper placed at the focus burst into flame with almost explosive suddenness. The high ratio which the visible radiation of the sun bears to the invisible was strikingly manifested in these experiments. With a *total* radiation vastly inferior, the invisible rays of the electric light, or of the lime-light, raise platinum to whiteness, while, when the visible constituents of the concentrated sunbeam were intercepted, the most that

* Experiments on the sun had been previously, but unsuccessfully, attempted by others.

could be obtained from the dark rays was a bright red-heat. The heat of the luminous rays, moreover, is so great as to render it exceedingly difficult to experiment with the solution of iodine. It boiled up incessantly, exposure for two or three seconds being sufficient to raise it to ebullition. This high ratio of the luminous to the non-luminous radiation, is doubtless to be ascribed in part to the absorption of a large portion of the latter by the aqueous vapour of the air. From it, however, may also be inferred the enormous temperature of the sun.

(630) Converging the sun's rays with a hollow lens filled with the solution of iodine, incandescence was obtained at the invisible focus of the lens on the roof of the Royal Institution.

(631) Knowing the permeability of good glass to the solar rays, I requested Mr. Mayall to permit me to make a few experiments with his fine photographic lens at Brighton. Though exceedingly busy at the time, he in the kindest manner abandoned to my late assistant, Mr. Barrett, the use of his apparatus for the three best hours of a bright summer's day. A red heat was obtained at the focus of the lens after the complete withdrawal of the luminous portion of the radiation.

(632) Black paper has been very frequently employed in the foregoing experiments, the action of the invisible rays upon it being most energetic. This suggests that the absorption of the dark rays is not independent of colour. A red powder is red because of the entrance and absorption of the luminous rays of higher refrangibility than the red, and the ejection of the unabsorbed red light by reflection at the limiting surfaces of the particles of the red body. This feebleness of absorption of the red rays extends in many cases to the obscure rays beyond the red; and the consequence is that red paper when exposed at the focus

of invisible rays is often scarcely charred, while black paper bursts in a moment into flame. The following table exhibits the condition of paper of various kinds when exposed at the dark focus of an electric light of moderate intensity:—

Paper.	Condition.
Glazed orange-coloured paper .	Barely charred.
„ red- „ .	Scarcely tinged; less than the orange.
„ green- „ .	Pierced with a small burning ring.
„ blue- „ .	The same as the last.
„ black- „ .	Pierced; and immediately set ablaze.
„ white- „ .	Charred; not pierced.
Thin foreign-post	Barely charred; less than the white.
Foolscap	Still less charred; about the same as the orange.
Thin white blotting paper . .	Scarcely tinged.
„ whitey-brown „ . .	The same; a good deal of heat seems to get through these two last papers.
Ordinary brown „ . .	Pierced immediately, a beautiful burning ring expanding on all sides.
Thick brown	Pierced, not so good as the last.
Thick white sand-paper . . .	Pierced with a burning ring.
Brown emery „ . . .	The same as the last.
Dead-black „ . . .	Pierced, and immediately set ablaze.

(633) We have here an almost total absence of absorption on the part of the red paper. Even white absorbs more, and is consequently more easily charred. Rubbing the red iodide of mercury over paper, and exposing the reddened surface at the focus, a thermograph of the coal-points is obtained, which shows itself by the discharge of the colour at the place on which the invisible image falls. Expecting that this change of colour would be immediate, I was at first surprised at the time necessary to produce it.

(634) And here we find ourselves in a position to properly qualify and explain a popular experiment which has been fruitful in erroneous inferences. The celebrated Dr. Franklin placed cloths of various colours upon snow and allowed the sun to shine upon them. They absorbed

the solar rays in different degrees, became differently heated, and sank therefore to different depths in the snow beneath them. His conclusion was that dark colours were the best absorbers, and light colours the worst; and to this hour we appear to have been content to accept Franklin's generalisation without qualification. Did the emission from luminous sources consist exclusively of visible rays, we might fairly infer from the colour of a substance its capacity to absorb the heat of such sources. But we now know that the emission from luminous sources is by no means all visible. In terrestrial sources by far the greater part, and in the case even of the sun a very great part, of the emission consists of invisible rays, regarding which colour teaches us nothing.

(635) It remained therefore to examine whether the results of Franklin were the expression of a law of nature. Two cards were taken of the same size and texture; over one of them was shaken the white powder of alum, and over the other the dark powder of iodine. Placed before a glowing fire and permitted to assume the maximum temperature due to their position, it was found that the card bearing the alum became extremely hot, while that bearing the iodine remained cool. No thermometer was necessary to demonstrate this difference. Placing the back of the iodine card against the forehead or cheek, no inconvenience was experienced; while the back of the alum card similarly placed proved intolerably hot.

(636) This result was corroborated by the following experiments:—One bulb of a differential thermometer was covered with iodine, and the other with alum powder. A red-hot spatula being placed midway between both, the liquid column associated with the alum-covered bulb was immediately forced down, and maintained in an inferior position. Two delicate mercurial thermometers had their bulbs coated, the one with iodine, the other with alum.

On exposing them at the same distance to the radiation from a gas-flame, the mercury of the alum-covered thermometer rose nearly twice as high as that of its neighbour. Two sheets of tin were coated, the one with alum, and the other with iodine powder. The sheets were placed parallel to each other, and about 10 inches asunder; at the back of each was soldered a little bar of bismuth, which with the tin plate to which it was attached constituted a thermo-electric couple. The two plates were connected together by a wire, and the free ends of the bismuth bars were connected with a galvanometer. Placing a red-hot ball midway between both, the calorific rays fell with the same intensity on the two sheets of tin, but the galvanometer immediately declared that the sheet which bore the alum was the most highly heated.

(637) In some of the foregoing cases the iodine was simply shaken through a muslin sieve; in other cases it was mixed with bisulphide of carbon and applied with a camel's-hair brush. When dried afterwards it was almost as black as soot; but as an absorber of radiant heat it was no match for the perfectly white powder of alum.

(638) This difficulty of warming iodine by radiant heat is evidently due to the diathermic property which it manifests so strikingly when dissolved in bisulphide of carbon. The heat enters the powder, is reflected at the limiting surfaces of the particles, but it does not lodge itself among the atoms of the iodine. When shaken in sufficient quantity on a plate of rock-salt and placed in the path of a calorific beam, iodine intercepts the heat. But its action is mainly that of a white powder to light; it is impervious, not through absorption, but through repeated internal reflection. Ordinary roll sulphur, even in thin cakes, allows no radiant heat to pass through it; but its opacity is also due to internal reflection. The temperature of ignition of sulphur is about 244° C.; but on placing a small

piece of the substance at the obscure focus of the electric lamp, where the heat was sufficient to raise platinum-foil in a moment to whiteness, it required exposure for a considerable time to fuse and ignite the sulphur. Though impervious to the heat, it was not through absorption. Sugar is a much less inflammable substance than sulphur, but it is a far better absorber; exposed at the focus, it is speedily fused and burnt up. The heat moreover which is competent to inflame powdered sugar is scarcely competent to warm table-salt of the same white appearance.

(639) A fragment of almost black amorphous phosphorus was exposed at the dark focus of the electric lamp, but refused to be ignited. A still more remarkable result was obtained with ordinary phosphorus. A small fragment of this exceedingly inflammable substance could be exposed for twenty seconds without ignition at a focus where platinum was almost instantaneously raised to a white heat. The fusing-point of phosphorus is about 44° C., that of sugar is 160°; still at the focus of the electric lamp the sugar fuses before the phosphorus. All this is due to the diathermancy of the phosphorus: a thin disc of the substance placed between two plates of rock-salt permits of a copious transmission. This substance therefore takes its place with other elementary bodies as regards deportment towards radiant heat.

(640) The more diathermic a body is, the less it is warmed by radiant heat. No perfectly transparent body could be warmed by purely luminous heat. The surface of a vessel covered with a thick fur of hoar frost was exposed to the beam of the electric lamp condensed by a powerful mirror, the beam having been previously sent through a cell containing water. The sifted beam was powerless to remove the frost, though it was competent to set wood on fire. We may largely apply this result. It is not, for example, the luminous rays, but the dark rays of the sun

which sweep the snows of winter from the slopes of the Alps. Every glacier-stream that rushes through the Alpine valleys is almost wholly the product of invisible radiation. It is also the invisible solar rays which lift the glaciers from the sea-level to the summits of the mountains; for the luminous rays penetrate the tropical ocean to great depths, while the non-luminous ones are absorbed close to the surface, and become the main agents in evaporation.

(641) We will end this subject by fulfilling a promise formerly made (§ 612*a*). The method by which Melloni determined the ratio of the visible to the invisible rays emitted by any luminous source has been described to you (§ 370). It was explained to you, that assuming a solution of alum to transmit all the visible rays, which is sensibly the case, and to absorb all the invisible rays, the difference between the transmission through alum and rock-salt gives the action of the obscure rays. But is this assumption regarding the absorptive power of alum correct? Is a solution of this substance, of the thickness at which it has hitherto been examined, really competent to absorb all heat-rays of a lower refrangibility than those which produce light?

(642) The solution of iodine, with which you are now so intimately acquainted, was placed in front of an electric lamp, the luminous rays being thereby intercepted. Behind the rock-salt cell containing the opaque solution was placed a glass cell, empty in the first instance. The deflection produced by the obscure rays which passed through both produced a deflection of

80°.

The glass cell was now filled with a concentrated solution of alum; the deflection produced by the obscure rays passing through both solutions was

50°.

Calculating from the values of these deflections, it is found *that of the obscure heat emergent from the solution of iodine* 20 *per cent. was transmitted by the alum.**

(643) The question, whether the invisible rays emitted by luminous sources reach the retina of the eye, we have hitherto left in abeyance. But there cannot be a doubt that the invisible rays which have shown themselves competent to traverse such a thickness of the most powerful diathermic liquid yet discovered are also able to pass through the humours of the eye. Dr. Franz has indeed proved this to be the case for the dark solar rays. The very careful and interesting experiments of M. Janssen,† prove moreover that the humours of the eye absorb an amount of radiant heat exactly equal to that absorbed by a layer of water of the same thickness as the humours; and in our solution the power of alum is added to that of water. Direct experiments on the vitreous humour of an ox lead me to conclude that one-fifth of the obscure rays emitted by an intense electric light reaches the retina; and inasmuch as in every ten parts of that radiation nine are obscure, it follows that nearly two-thirds of the whole radiant energy, visible and invisible, which the electric light sends to the retina is incompetent to excite vision.

(644) Measured by a photometer the intensity of the electric *light* used by me was, in some cases, 1000 times that of the *light* of a good composite candle, and as the non-luminous *heat-rays* from the coal-points which reach the retina have, in round numbers, twice the energy of the

* In passing from one medium to another, light is always reflected; the same is true of radiant heat. And in the case of our empty glass cell, radiant heat was reflected from its two interior surfaces when it was empty. The introduction of the alum solution, no doubt altered the quantity of heat reflected; for the sake of simplicity, I have neglected taking this into account; my doing so would not materially affect the results here enunciated.

† *Annales de Chimie et de Physique*, tom. lx. p. 71.

luminous, it follows that at a common distance, say of a foot, the energy of the radiant *heat* which reaches the optic nerve, but is incompetent to provoke vision, is 2000 times that of the *light* of a candle. But on a tolerably clear night a candle-flame can be readily seen at the distance of a mile; and the intensity of the candle's light at the distance of a mile is less than one twenty-millionth of its intensity at the distance of a foot, hence the energy which renders the candle perfectly visible a mile off would have to be multiplied by $2000 \times 20,000,000$, or by forty thousand millions, to bring it up to the intensity of the radiation which the retina actually receives from the carbon points at a foot distance, without vision. Nothing, I think, could more forcibly illustrate the special relationship which subsists between the optic nerve and the oscillating periods of the molecules of luminous bodies. That nerve, like a musical string, responds to the periods with which it is in accordance, while it refuses to be excited by others of almost infinitely greater energy which are not in unison with its own.

(645) When we see a vivid light incompetent to affect our most delicate thermoscopic apparatus, the idea naturally presents itself that light and heat must be totally different things. The pure light emerging from a combination of water and green glass, even when rendered intense by concentration, has, according to Melloni, no sensible heating power. The light of the moon is also a case in point. Concentrated by a polyzonal lens more than a yard in diameter upon the face of his pile, it required all Melloni's acuteness to *nurse* the calorific action up to a measurable quantity. Such experiments, however, demonstrate, not that the two agents are dissimilar, but that the sense of vision can be excited by an amount of energy almost infinitely small.

(646) Here also we are able to offer a remark as to the

applicability of radiant heat to fog-signalling. The proposition, in the abstract, is a philosophical one; for were our fogs of a physical character similar to that of the iodine held in solution by the bisulphide of carbon, or to that of iodine or bromine vapours, it would be possible to transmit through them, from our signal lamps, powerful fluxes of radiant heat, even after the entire stoppage of the light. But our fogs are not of this character. They are unfortunately so constituted as to act very destructively upon the purely calorific rays; and this fact, taken in conjunction with the marvellous sensitiveness of the eye, leads to the conclusion that long before the *light* of our signals ceases to be visible, their radiant heat has lost the power of affecting, in any sensible degree, the most delicate thermoscopic apparatus that we could apply to their detection.

CHAPTER XIV.

DEW:—A CLEAR SKY AND CALM BUT DAMP ATMOSPHERE NECESSARY FOR ITS COPIOUS FORMATION—DEWED SUBSTANCES COLDER THAN UNDEWED ONES—DEWED SUBSTANCES BETTER RADIATORS THAN UNDEWED ONES—DEW IS THE CONDENSATION OF THE ATMOSPHERIC VAPOUR ON SUBSTANCES WHICH HAVE BEEN CHILLED BY RADIATION—LUNAR RADIATION—CONSTITUTION OF THE SUN—THE BRIGHT LINES IN THE SPECTRA OF THE METALS—AN INCANDESCENT VAPOUR ABSORBS THE RAYS WHICH IT CAN ITSELF EMIT—KIRCHHOFF'S GENERALISATION—FRAUNHOFER'S LINES—SOLAR CHEMISTRY—EMISSION OF THE SUN—HERSCHEL AND POUILLET'S EXPERIMENTS—MAYER'S METEORIC THEORY—THEORIES OF HELMHOLTZ AND THOMSON—EFFECT OF THE TIDES ON THE EARTH'S ROTATION—ENERGIES OF THE SOLAR SYSTEM—HELMHOLTZ, THOMSON, WATERTSON—RELATION OF THE SUN TO ANIMAL AND VEGETABLE LIFE—APPENDIX.

(647) WE have learned that our atmosphere is always more or less charged with aqueous vapour, the condensation of which forms our clouds, fogs, hail, rain, and snow. We have now to direct our attention to one particular case of condensation, of great interest and beauty—one, moreover, regarding which erroneous notions were for a long time entertained—the phenomenon of Dew. The aqueous vapour of our atmosphere is a powerful radiant, but it is diffused through air which usually exceeds its own mass more than one hundred times. Not only, then, its own heat, but the heat of the large quantity of air which surrounds it, must be discharged by the vapour, before it can sink to its point of condensation. The retardation of chilling due to this cause enables good solid radiators, at the earth's surface, to outstrip the vapour in their speed of refrigeration; and hence upon these bodies

aqueous vapour may be condensed to liquid, or even congealed to hoar-frost, while at a few feet above the surface it maintains its gaseous state. This is actually the case in the beautiful phenomenon which we have now to examine.

(648) We are indebted to a London physician for a true theory of dew. In 1818 Dr. Wells published his admirable essay on this subject. He made his experiments in a garden in Surrey, at a distance of three miles from Blackfriars Bridge. To collect the dew, he used little bundles of wool, which, when dry, weighed 10 grains each; and having exposed them during a clear night, the amount of dew deposited on them was determined by the augmentation of their weight. He soon found that whatever interfered with the view of the sky from his piece of wool, interfered also with the deposition of dew. He supported a board on four props; *on* the board he laid one of his wool parcels, and *under* it a second similar one; during a clear calm night, the former gained 14 grains in weight, while the latter gained only 4. He bent a sheet of pasteboard like the roof of a house, and placed underneath it a bundle of wool on the grass: by a single night's exposure the wool gained 2 grains in weight, while a similar piece of wool exposed on the grass, but quite unshaded by the roof, collected 16 grains of moisture.

(649) Is it steam from the earth, or is it fine rain from the heavens, that produces this deposition of dew? Both of these notions have been advocated. That it does not arise from the earth is, however, proved by the fact, that more moisture was collected *on* the propped board than under it. That it is not a fine rain is proved by the fact, that the most copious deposition occurs on the clearest nights.

(650) Dr. Wells next exposed thermometers, as he had done his wool-bundles, and found that *at those places*

where the dew fell most copiously, the temperature sank lowest. On the propped board already referred to, he found the temperature 9° Fahr. lower than under it; beneath the pasteboard roof the thermometer was 10° warmer than on the open grass. He also found, that when he laid his thermometer upon a grass plot, on a clear night, it sank sometimes 14° lower than a similar thermometer suspended in free air, at a height of 4 feet above the grass. A bit of cotton, placed beside the former, gained 20 grains; a similar bit, beside the latter, only 11 grains in weight. *The lowering of the temperature and the deposition of the dew went hand in hand.* Not only did artificial screens interfere with the lowering of the temperature and the formation of the dew, but a cloud-screen acted in the same manner. He once observed his thermometer, which, as it lay upon the grass, showed a temperature 12° Fahr. lower than the air a few feet above the grass, rise, on the passage of some clouds, until it was only 2° colder than the air. In fact, as the clouds crossed his zenith, or disappeared from it, the temperature of his thermometer rose and fell.

(651) A series of such experiments, conceived and executed with admirable clearness and skill, enabled Dr. Wells to propound a Theory of Dew, which has stood the test of all subsequent criticism, and is now universally accepted.

(652) It is an effect of chilling by radiation. 'The upper parts of the grass radiate their heat into regions of empty space, which, consequently, send no heat back in return; its lower parts, from the smallness of their conducting power, transmit little of the earth's heat to the upper parts, which, at the same time, receiving only a small quantity from the atmosphere, and none from any other lateral body, must remain colder than the air, and condense into dew its watery vapour, if this be sufficiently

abundant in respect to the decreased temperature of the grass.' Why the vapour itself, being a powerful radiant, is not so quickly chilled as the grass, has been already explained, on the ground that the vapour has not only its own heat to discharge, but also that of the large mass of air by which it is surrounded.

(653) Dew, then, is the result of the condensation of atmospheric vapour, on substances which have been sufficiently cooled by radiation; and as bodies differ widely in their radiative powers, we may expect corresponding differences in the deposition of dew. This Wells proved to be the case. He often saw dew copiously deposited on grass and painted wood, when none could be observed on gravel walks adjacent. He found plates of metal, which he had exposed, quite dry, while adjacent bodies were covered with dew: *in all such cases the temperature of the metal was found to be higher than that of the dewed substances.* This is quite in accordance with our knowledge that metals are the worst radiators. On one occasion he placed a plate of metal upon grass, and upon the plate he laid a glass thermometer; the thermometer, after some time, exhibited dew, while the plate remained dry. This led him to suppose that the instrument, though lying on the plate, did not share its temperature. He placed a second thermometer, with a *gilt bulb*, beside the first; the naked glass thermometer—a good radiator—remained 9° Fahr. colder than its companion. To determine the true temperature of a body is a task of some difficulty: a glass thermometer, suspended in the air, will not give the temperature of the air; its own power as a radiant or an absorbent comes into play. On a clear day, when the sun shines, the thermometer will be warmer than the air; on a clear night, on the contrary, the thermometer will be colder than the air. We have seen that the passage of a cloud can raise the temperature of a thermometer 10° in a

few minutes. This augmentation, it is manifest, does not indicate a corresponding augmentation of the temperature of the air, but merely the interception and reflection, by the cloud, of the rays of heat emitted by the thermometer.

(654) Dr. Wells applied his principles to the explanation of many curious effects, and to the correction of many popular errors. Moon blindness he refers to the chill produced by radiation from the eyes, the shining of the moon being merely an accompaniment to the clearness of the atmosphere. The putrefying influence ascribed to the moonbeams is really due to the deposition of moisture, as a kind of dew, on the exposed animal substances. The nipping of tender plants by frost, when the air of the garden is some degrees above the freezing temperature, is also to be referred to chilling by radiation. A cobweb screen would be sufficient to preserve them from injury.*

(655) Wells was the first to explain the formation, artificially, of ice in Bengal, where the substance is never formed naturally. Shallow pits are dug, which are partially filled with straw, and on the straw flat pans containing water are exposed to the clear firmament. The water is a powerful radiant, and sends off its heat copiously into space. The heat thus lost cannot be supplied from the earth—this source being cut off by the non-conducting straw. Before sunrise a cake of ice is

* With reference to this point we have the following beautiful passage in the Essay of Wells:—'I had often, in the pride of half-knowledge, smiled at the means frequently employed by gardeners to protect tender plants from cold, as it appeared to me impossible that a thin mat, or any such flimsy substance, could prevent them from attaining the temperature of the atmosphere, by which alone I thought them liable to be injured. But when I had learned thatbodies on the surface of the earth become, during a still and serene night, colder than the atmosphere, by radiating their heat to the heavens, I perceived immediately a just reason for the practice which I had before deemed useless.'

formed in each vessel. This is the explanation of Wells, and it is, no doubt, the true one. I think, however, it needs supplementing. It appears, from the description, that the condition most suitable for the formation of ice, is not only a clear air, but a *dry* air. The nights, says Sir Robert Barker, most favourable for the production of ice, are those which are clearest and most serene, and *in which very little dew appears after midnight.* The italicised phrase is very significant. To produce the ice in abundance, the atmosphere must not only be clear, but it must be comparatively free from aqueous vapour. When the straw on which the pans were laid became wet, it was always changed for dry straw; and the reason Wells assigned for this was, that the straw, by being wetted, was rendered more compact and efficient as a conductor. This may have been the case, but it is also certain that the vapour rising from the wet straw, and overspreading the pans like a screen, would check the chill, and retard the congelation.

(656) With broken health Wells pursued and completed this beautiful investigation; and, on the brink of the grave, he composed his Essay. It is a model of wise enquiry and of lucid exposition. He made no haste, but he took no rest till he had mastered his subject, looking steadfastly into it until it became transparent to his gaze. Thus he solved his problem, and stated its solution in a fashion which renders his work imperishable.*

(657) Since his time various experimenters have occupied themselves with the question of nocturnal radiation; but, though valuable facts have been accumulated, if we except a supplement contributed by Melloni, nothing of importance has been added to the theory of Wells. Mr. Glaisher, M. Martins, and others, have

* The tract of Wells is preceded by a personal memoir written by himself. It has the solidity of an essay of Montaigne.

illustrated the subject. The following table contains some results obtained by Mr. Glaisher, by exposing thermometers at different heights above the surface of a grass field. The chilling observed when the thermometer was exposed on long grass, is represented by the number 1000; while the succeeding numbers represent the relative chilling of the thermometers placed in the positions indicated :—

			Radiation.
Long grass			1000
One inch above the points of the grass			671
Two inches	"	"	570
Three inches	"	"	477
Six inches	"	"	282
One foot	"	"	129
Two feet	"	"	86
Four feet	"	"	69
Six feet	"	"	52

(658) It may be asked why the thermometer, which is a good radiator, is not, when suspended in free air, just as much chilled as at the earth's surface. Wells has answered this question. It is because the thermometer, when chilled, cools the air in immediate contact with it; this air contracts, becomes heavy, and descends, thus allowing its place to be taken by warmer air. In this way the free thermometer is prevented from falling very low beneath the temperature of the air. Hence, also, the necessity of a still night for the copious formation of dew; for, when the wind blows, fresh air continually circulates amid the blades of grass, and prevents any considerable chilling by radiation.

(659) When a radiator is exposed to a clear sky it tends to keep a certain thermometric distance, if the term may be used, between its temperature and that of the surrounding air. This distance will depend upon the energy of the radiator, but it is to a great extent independent

of the temperature of the air. Thus, M. Pouillet has proved that in the month of April, when the temperature of the air was 3·6° C., swansdown fell by radiation to —3·5°; the whole chilling, therefore, was 7·1°. In the month of June, when the temperature of the air was 17·75° C., the temperature of the radiating swansdown was 10·54°; the chilling of the swansdown, by radiation, is here 7·21°, almost precisely the same as that which occurred in April. Thus, while the general temperature varies within wide limits, the *difference* of temperature between the radiating body and the surrounding air, remains sensibly constant.

(660) These facts enabled Melloni to make an important addition to the theory of dew. He found that a glass thermometer, placed on the ground, is never chilled more than 2° C., or 3·6° F., below an adjacent thermometer, *with silvered bulb*, which hardly radiates at all. These 2° C., or thereabouts, mark the thermometric distance above referred to, which the glass tends to preserve between it and the surrounding air. But Six, Wilson, Wells, Parry, Scoresby, Glaisher, and others, have found differences of more than 10° C., or 18° F., between a thermometer on grass, and a second thermometer hung a few feet above the grass. How is this to be accounted for? Very simply, according to Melloni, thus: The grass blades first chill themselves by radiation, 2° C. below the surrounding air; the air is then chilled by contact with the grass, and forms around it a cold aërial bath. But the tendency of the grass is to keep the above constant difference between its own temperature and that of the surrounding medium. It therefore sinks lower. The air sinks in its turn, being still further chilled by contact with the grass; the grass, however, seeks to re-establish the former difference; it is again followed by the air, and thus, by a series of actions and reactions, the entire stratum of air in contact

with the grass becomes lowered to a temperature far below that which corresponds to the actual radiative energy of the grass.

(661) So much for terrestrial radiation; that of the moon will not occupy us so long. Many futile attempts have been made to detect the warmth of the moon's beams. No doubt every luminous ray is also a heat ray; but the light-giving power is not even an approximate measure of the calorific energy of a beam. With a large polyzonal lens, Melloni converged an image of the moon upon his pile; but he found the cold of his lens far more than sufficient to mask the heat, if such there were, of the moon. He then screened off his lens from the heavens, placed his pile in the focus of the lens, waited until the needle came to zero, and then suddenly removing his screen, allowed the concentrated light to strike his pile. The slight air-draughts in the place of experiment were sufficient to disguise the effect. He then stopped the tube in front of his pile with glass screens, through which the light went freely to the blackened face of the pile, where it was converted into heat. *This heat could not get back through the glass screen*, and thus Melloni, following the example of Saussure, accumulated his effects, and obtained a galvanometric deflection of 3° or 4°. The deflection indicated warmth.

(662) By far the greater part of the heat emitted by the full moon must consist of obscure rays, and these are almost wholly absorbed by our atmospheric vapour. Even such obscure rays as might happen to reach the earth would be utterly cut off by such a lens as Melloni made use of. It might be worth while to make the experiment with a metallic reflector, instead of with a lens. I have myself tried a conical reflector of very large dimensions, but have hitherto been defeated by the unsteadiness of London air.

(663) We have now to turn our thoughts to the source from which terrestrial and lunar heat is almost wholly derived. This source is the sun; for if the earth has ever been a molten sphere, which is now cooling, the quantity of heat which reaches its surface from within, has long ceased to be sensible. First, then, let us enquire what is the constitution of this wondrous body, to which we owe both light and life.

(664) Let us approach the subject gradually, and prepare our minds, by previous discipline, for the treatment of so grand a problem. You already know how the spectrum of the electric light is formed. Such a spectrum is now upon the screen, two feet wide and eight long, with all its magnificent gradations of colour, one passing into the other, without solution of continuity. The light from which this spectrum is derived, is emitted from the solid incandescent carbon points within our electric lamp. All other white-hot solids give a similar spectrum. When I raise this platinum wire to whiteness by an electric current, and examine its light by a prism, I find the same gradation of colours, and no gap whatever between one colour and the other. But by intense heat—by the heat of the electric lamp, for example—I can volatilise the metal, and throw upon the screen, not the spectrum of the incandescent solid, but of its *incandescent vapour*. The spectrum is now changed; instead of being a continuous gradation of colours, it consists of a series of brilliant lines, separated from each other by spaces of darkness.

(665) The pieces of carbon are now arranged in the following manner:—the lower one is a cylinder, about half an inch in diameter, in the top of which is scooped a small hollow; into this hollow is put this piece of zinc, and the upper carbon-point is then brought down upon it. The current passes; and when the points are drawn apart, the image of the arc that unites them is projected on the

screen as a stream of purple light, eighteen inches long. That coloured stream is zinc vapour; it contains the atoms of the zinc discharged across from carbon to carbon. These atoms are now oscillating in certain definite periods, and the colour which we perceive is the composite impression of these oscillations.

(665*a*) Resolving, by a prism, the light of the arc into its component colours, we have no longer a continuous spectrum, but these splendid bands of red and blue light.

(666) I interrupt the current, remove the zinc, and put in its place a bit of copper. On forming the arc we obtain this stream of green light, which we can analyse as we did the purple light of the zinc. In the spectrum of the copper you have these bands of brilliant green, which were absent in the case of zinc. We may therefore infer, with certainty, that the atoms of copper, in the voltaic arc, vibrate in periods different from those of zinc. Let us now enquire whether these different periods create any confusion, when we operate upon a substance composed of zinc and copper,—the familiar substance brass. Its spectrum is now before you, and if you have retained the impression made by our two last experiments, you will recognise here a spectrum, formed by the superposition of the two separate spectra of zinc and copper. The alloy emits, without confusion, the rays peculiar to both the metals of which it is composed.

(667) Every metal emits its own system of bands, which are as characteristic of it as those other physical and chemical qualities which give it its individuality. By a method of experiment sufficiently refined, we can measure, accurately, the position of the bright lines of every known metal. Acquainted with such lines, we should, by the mere inspection of the spectrum of any single metal, be able at once to declare its name. And not only so, but in the case of a mixed spectrum we should be able to

declare the constituents of the mixture from which it emanated. From the exhibition of unknown lines, the existence of new metals has been inferred. Bunsen and Kirchhoff, for example, thus discovered Cœsium and Rubidium; and Mr. Crooks, by the same method, discovered Thallium, which gives us this single line of brilliant green.

(668) This law is true, not only of the metals themselves, but also of their compounds, if they be volatile. I place a bit of sodium on the lower cylinder, and cause the voltaic discharge to pass from it to the upper coal-point; the spectrum of the sodium gives us this single band of brilliant yellow. With greater delicacy of experiment, that band is divided into two, with a narrow dark interval between them: a still greater amount of precision would further subdivide the yellow space. Let us now remove the sodium from the lamp and put in its place a little common salt, or chloride of sodium. At this high temperature the salt is volatile, and produces the exact yellow band yielded by the metal. Thus, also, from the chloride of strontium, we obtain the bands of the metal strontium; and by means of the chlorides of calcium, magnesium, and lithium, are produced the spectra of these respective metals.

(669) Finally, this carbon cylinder is perforated with holes, into which is crammed a mixture of all the compounds just mentioned. Surely nothing more magnificent can be imagined than the spectrum of the mixture now upon the screen. Each substance gives out its own peculiar rays, which cut transversely the whole eight feet of the spectrum into parallel bars of richly coloured light, Having previously made yourselves acquainted with the lines emitted by all the metals, you would be able to unravel this spectrum, and to state the metals concerned in its production.

(670) The voltaic arc is here employed simply because its light is so intense as to be visible to a large audience

like the present; * but the same experiments might be made with a common blowpipe flame. The introduction of sodium, or chloride of sodium, turns the flame yellow; strontium turns it red; copper, green, &c. The flames thus coloured, when examined by a prism, show the exact bands which have been displayed before you.

(671) We have here, then, the *radiation* of definite groups of rays by incandescent vapours. Let us now turn our attention to the *absorption* of definite groups of rays by gaseous substances. A famous experiment of Sir David Brewster's, thrown into a form suited to the lecture-room, will illustrate this power of selection. Into this cylinder, the ends of which are stopped by plates of glass, is introduced a quantity of nitrous acid gas, the presence of which is now indicated by its rich brown colour. Projecting a brilliant spectrum on the screen, and placing the cylinder, containing the brown gas, in the path of the beam as it issues from the lamp, you see the continuous spectrum furrowed by numerous dark bands. The rays answering to these bands are intercepted by the nitric gas, while it permits the intervening bands of light to pass without hindrance.

(672) We now come to the great principle on which these phenomena depend, and which we have already, in part, illustrated. This principle, first announced by Professor Kirchhoff, is, that *a gas, or vapour, absorbs those precise rays which it can itself emit.* Atoms which swing at a certain rate intercept waves which swing at the same rate. The atoms which vibrate red light will stop red light; the atoms that vibrate yellow will stop yellow; those that vibrate green will stop green, and so of the rest. Absorption, you already know, is a transference of motion from the ether to the molecules immersed in it,

* The splendid blue band of Lithium was discovered by means of the electric lamp in the theatre of the Royal Institution.

and the absorption of any atom is exerted chiefly upon those waves which arrive in periods coinciding with its own rate of oscillation.

(673) Let us endeavour to prove this experimentally. We already know that a sodium flame, when analysed, gives a brilliant band of yellow. This flat vessel contains a mixture of alcohol and water; when the mixture is warmed we can ignite its vapour: it then gives a flame so feebly luminous as to be scarcely visible. Let us mix salt with the liquid, and again ignite it: the flame, which a moment ago was scarcely to be seen, is now a brilliant yellow. Projecting a continuous spectrum upon the screen, in the track of the beam, as it issues from the electric lamp, I place the yellow sodium flame. Observe the spectrum narrowly: you see in the yellow a flickering grey band, very faint, but sufficient to show that the flame has, at least in part, intercepted the yellow of the spectrum: it has partially absorbed the precise light which it can itself emit.

(674) But the effect can be made much plainer. Abandoning the salt flame, I place the intensely hot flame of a Bunsen's burner in front of the lamp, so that the beam, whose decomposition is to form our spectrum, shall pass through the flame. In this little spoon of platinum wire is placed a bit of the metal sodium, about the size of a pea. The sodium, when ignited, emits a powerful light, and it is necessary to cut off that light from the screen on which the spectrum is to fall. First forming the spectrum, I introduce the platinum net containing the sodium into the flame through which the beam from the lamp passes. The sodium instantly colours the flame intensely yellow, and you already see a shadow coming over the yellow of the spectrum. But the effect is not yet at its maximum. After a little time the sodium bursts into intense combustion, and at the same

moment you see the yellow dug utterly out of the spectrum, a bar of intense darkness taking its place. This violent combustion will endure for a few seconds. On withdrawing the flame, the yellow reappears upon the screen; on reintroducing it, the yellow band is again cut out. This may be done ten times in succession, and in the whole range of optics there is scarcely a more striking experiment. We have thus conclusively proved that the light which the sodium flame absorbs is the light which it can emit.*

(675) Let us be still more precise in our experiments. The yellow of the spectrum spreads over a certain interval and we have now to examine whether it is not the particular portion of the yellow emitted by the sodium, that is absorbed by its flame. I place a little brine on the ends of the coal-points; you now see the continuous spectrum, with the yellow band of the sodium brighter than the rest of the yellow. When the sodium flame is placed in front, that particular band, which now stands out from the spectrum, is cut away.

(676) You have already seen a spectrum, derived from a mixture of various substances, and composed of a succession of sharply defined and brilliant bars, separated from each other by intervals of darkness. Could the temperature of the mixture which produced that striped spectrum be so exalted by a flame as to render its vapours incandescent; on placing the flame, and vapours, in the path of a beam producing a continuous spectrum, we should cut out of the latter the precise rays emitted by the components of the mixture. We should thus, instead of furrowing the spectrum by a single dark band, as in the case of sodium,

* Before trying the combustion of the metal, I had tried the salt-flame in a trough ten feet long: the effect, however, is far less fine than that attainable by the combustion of the metal. The experiment was first made during my preparations for a lecture on the 'Physical Basis of Solar Chemistry,' given in June, 1861.

furrow it by a series of dark bands, equal in number to the bright bands, produced by the mixture itself when the source of light.

(677) We now possess knowledge sufficient to raise us to the level of one of the most remarkable generalisations of our age. When the light of the sun is properly decomposed, the spectrum is seen furrowed by innumerable *dark* lines. A few of these were observed for the first time by Dr. Wollaston; but they were investigated with profound skill by Fraunhofer, and called, after him, Fraunhofer's lines. It has long been supposed that these dark bands were due to the absorption of the rays which correspond to them, by the atmosphere of the sun; but nobody knew how. Having once proved that an incandescent vapour absorbs the precise rays which it can itself emit, and knowing that the body of the sun is surrounded by an incandescent photosphere, the supposition at once flashes on the mind, that this photosphere may cut off those rays of the central incandescent orb, which the photosphere itself can emit. We are thus led to a theory of the constitution of the sun, which renders a complete account of the lines of Fraunhofer.

(678) The sun, according to Kirchhoff, consists of a central orb, molten or solid, of exceeding brightness, which emits all kinds of rays, and would therefore give a continuous spectrum. The radiation from the nucleus, however, has to pass through the photosphere, which wraps the sun like a flame, and this vaporous envelope cuts off those particular rays of the nucleus which it can itself emit—the lines of Fraunhofer marking the position of the failing rays. Could we abolish the central orb, and obtain the spectrum of the gaseous envelope, we should obtain a striped spectrum, each bright band of which would coincide with one of Fraunhofer's dark lines. These lines, therefore, are spaces of *relative*, not of absolute

darkness; upon them the rays of the absorbent photosphere fall; but these, not being sufficiently intense to make good the light intercepted, the spaces which they illuminate are dark, in comparison to the general brilliancy of the spectrum.

(679) It has long been supposed that sun and planets have had a common origin, and that hence the same substances are more or less common to them all. Can we detect the presence of any of our terrestrial substances in the sun? We have learned that the bright bands of a metal are characteristic of the metal; that we can, without seeing the metal, declare its name from the inspection of its bands. The bands are, so to speak, the *voice* of the metal declaring its presence. Hence, if any of our terrestrial metals be contained in the sun's atmosphere, the dark lines which they produce ought to coincide exactly with the bright lines emitted by the vapour of the metal itself. About sixty bright lines have been determined as belonging to the single metal iron. When the light from the incandescent vapour of iron, obtained by passing electric sparks between two iron wires, is allowed to pass through one half of a fine slit, and the light of the sun through the other half, the spectra from both sources of light may be placed side by side. When this is done, it is found that for every bright line of the iron spectrum there is a dark line of the solar spectrum. Reduced to actual calculation, this means that the chances are more than 1,000,000,000,000,000,000 to 1, that iron is in the atmosphere of the sun. Comparing the spectra of other metals in the same manner, Professor Kirchhoff, to whose genius we owe this splendid generalisation, finds iron, calcium, magnesium, sodium, chromium, and other metals, in the solar atmosphere; but as yet he has been unable to detect gold, silver, mercury, aluminium, tin, lead, arsenic, or antimony.

(680) We can imitate, in a way more precise than that hitherto employed, the solar constitution here supposed. In the electric lamp is placed a cylinder of carbon about half an inch thick; and round about the upper edge a ring of sodium, the central portion of the cylinder being left clear. I bring down the upper coal-point upon the middle of the cylinder, thus producing the ordinary electric light. Its proximity to the sodium, is sufficient to volatilise the latter, and thus the little central sun is surrounded with an atmosphere of sodium vapour, as the real sun is surrounded by its photosphere. You see that the yellow band is absent in the spectrum of this light.*

(681) The quantity of heat emitted by the sun has been measured by Sir John Herschel at the Cape of Good Hope, and by M. Pouillet in Paris. The agreement between the measurements is very remarkable. Sir John Herschel finds the direct heating effect of a vertical sun, at the sea level, to be competent to melt 0·00754 of an inch of ice per minute; while, according to M. Pouillet, the quantity is 0·00703 of an inch. The mean of the determinations cannot be far from the truth; this gives 0·00728 of an inch of ice per minute, or nearly half an inch per hour. Before you (fig. 108) I have placed an instrument, similar in form to that used by M. Pouillet, and called by him a pyrheliometer. The particular instrument which you now see is composed of a shallow cylinder of steel *a a*, filled with mercury. Into the cylinder is introduced the thermometer, *d*, the stem of which is protected by a piece of brass tubing. The flat end of the cylinder is to be turned towards the sun, and the surface thus presented is coated with lampblack. By means of a collar and screw,

* At this time the experiment of placing a bit of sodium, like the bit of zinc and copper already referred to, on the top of the lower cylinder of the lamp, was repeatedly made, the dark band being produced. The form described above was given to the experiment, simply to render its analogy with the effect of the solar atmosphere more apparent.

c c, the instrument may be attached to a stake driven into the ground, or into the snow, if the observations are made at considerable heights. It is necessary that the surface which receives the sun's rays should be perpendicular to them, and this is secured by attaching to the brass tube which shields the stem of the thermometer, a disc, *e e*, of precisely the same diameter as the steel cylinder. When the shadow of the cylinder accurately covers the disc, we are sure that the rays fall, as perpendiculars, on the upturned surface of the cylinder.

Fig. 108.

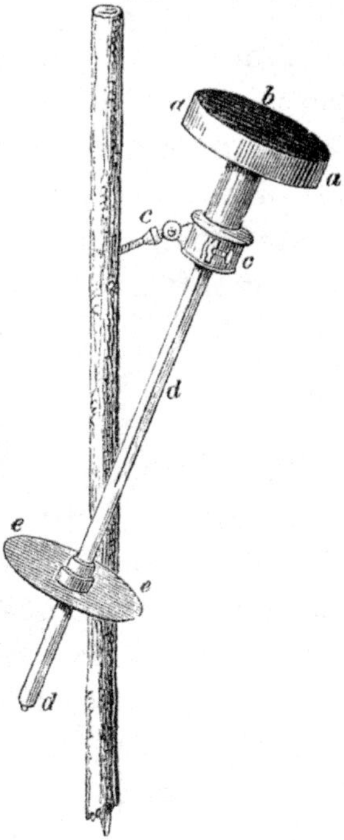

(682) The observations are made in the following manner:—First, the instrument is permitted, not to receive the sun's rays, but, to radiate its own heat for five minutes against an unclouded part of the firmament; the decrease of the temperature of the mercury consequent on this radiation being noted. Next, the instrument is turned towards the sun, so that the solar rays fall perpendicularly upon it for five minutes—the augmentation of temperature is noted. Finally, the instrument is turned again towards the firmament, away from the sun, and allowed to radiate for another five minutes, the sinking of the thermometer being noted as before. You might, perhaps, suppose that exposure to the sun alone would be sufficient; but we must not forget, that during the whole time of exposure to the sun's action, the blackened surface of the cylinder is also radiating into space; it is not therefore a case of pure gain: the heat received from the

sun is, in part, thus wasted, even while the experiment is going on; and to find the quantity thus lost, the first and last experiments are needed. In order to obtain the whole heating power of the sun, we must add the quantity lost during the time of exposure, and this quantity is the mean of the first and last observations. Supposing the letter R to represent the augmentation of temperature by five minutes' exposure to the sun, and that t and t' represent the reductions of temperature observed before and after, then the whole force of the sun, which we may call T, would be thus expressed:

$$T = R + \frac{t + t'}{2}$$

(683) The area of the surface on which the sun's rays here fall is known; the quantity of mercury within the cylinder is also known; hence, we can express the effect of the sun's heat upon a given area, by stating that it is competent, in five minutes, to raise so much mercury or so much water, so many degrees in temperature. Water, indeed, instead of mercury, was used in M. Pouillet's pyrheliometer.

(684) The observations were made at different hours of the day, and consequently, through different thicknesses of the earth's atmosphere; augmenting from the minimum thickness at noon, up to the maximum at 6 P.M., which was the time of the latest observation. It was found that the solar energy diminished, according to a certain law, as the thickness of the air crossed by the sunbeams increased; and from this law M. Pouillet was enabled to infer that the absorption, if the rays were directed downwards to his instrument from the zenith, would be 25 per cent. of the whole radiation. Doubtless, this absorption would be chiefly exerted upon the longer undulations emitted by the sun; the aqueous vapour of our air, not the air itself, being the principal agent. Taking into account

the whole terrestrial hemisphere turned towards the sun, the amount intercepted by the atmospheric envelope is four-tenths of the entire radiation. Thus, were the atmosphere removed, the illuminated hemisphere of the earth would receive nearly twice the amount of heat from the sun that now reaches it. The total amount of solar heat received by the earth in a year, if distributed uniformly over the earth's surface, would be sufficient to liquefy a layer of ice 100 feet thick, and covering the whole earth. It would also heat an ocean of fresh water 66 miles deep, from the temperature of melting ice to the temperature of ebullition.

(685) Knowing thus the annual receipt of the earth, we can calculate the entire quantity of heat emitted by the sun in a year. Conceive a hollow sphere to surround the sun, its centre being the sun's centre, and its surface at the distance of the earth from the sun. The section of the earth cut by this surface is to the whole area of the hollow sphere, as 1 : 2,300,000,000 ; hence, the quantity of sola heat intercepted by the earth is only $\frac{1}{2300000000}$ of the total radiation.

(686) The heat emitted by the sun, if used to melt a stratum of ice applied to the sun's surface, would liquefy the ice at the rate of 2400 feet an hour. It would boil per hour, 700,000 millions of cubic miles of ice-cold water. Expressed in another form, the heat given out by the sun, per hour, is equal to that which would be generated by the combustion of a layer of solid coal, ten feet thick, entirely surrounding the sun; hence the heat emitted in a year is equal to that which would be produced by the combustion of a layer of coal seventeen miles in thickness.

(687) This, then, is the sun's expenditure which has been going on for ages, without our being able, in historic times, to detect the loss. When the tolling of a bell is heard at a distance the sonorous vibrations are quickly

wasted, and renewed strokes are necessary to maintain the sound. Like the bell,

Die Sonne tönt nach alter Weise.

But how is its tone sustained? How is the perennial loss made good? We are apt to overlook the wonderful in the common. Possibly to many of us—and even to some of the most enlightened among us—the sun appears as a fire, differing from our terrestrial fires only in the magnitude and intensity of its combustion. But what is the burning matter which can thus maintain itself? All that we know of cosmical phenomena declares our brotherhood with the sun—affirms that the same constituents enter into the composition of his mass as those already known to chemistry. But no earthly substance with which we are acquainted—no substance which the fall of meteors has landed on the earth—would be at all competent to maintain the sun's combustion. The chemical energy of such substances would be too weak, and their dissipation too speedy. Were the sun a block of burning coal, and were it supplied with oxygen sufficient for the observed emission, it would be utterly consumed in 5000 years. On the other hand, to imagine it a body originally endowed with a store of heat—a hot globe now cooling—necessitates the ascription to it of qualities wholly different from those possessed by terrestrial matter. If we knew the specific heat of the sun, we could calculate its rate of cooling. Assuming the specific heat to be the same as that of water—the terrestrial substance which possesses the highest specific heat—at its present rate of emission, the entire mass of the sun would cool down 15,000° Fahr. in 5000 years. In short, if the sun be formed of matter like our own, some means must exist of restoring to it its wasted power.

(688) The facts are so extraordinary, that the soberest hypothesis regarding them must appear wild. The sun

we know rotates upon his axis once in about twenty-five days; and the notion has been entertained that the friction of the periphery of this wheel against something in surrounding space produces the light and heat. But what forms the brake, and by what agency is it held, while it rubs against the sun? Granting, moreover, the existence of the brake, we can calculate the total amount of heat which the sun could generate by such friction. We know his mass, we know his time of rotation; we know the mechanical equivalent of heat; and from these data we can deduce, with certainty, that the force of rotation, if entirely converted into heat, would cover less than two centuries of emission.* There is nothing hypothetical in this calculation.

(689) I have already alluded to another theory, which, however bold it may at first sight appear, deserves our serious attention—the Meteoric Theory of the Sun. Kepler's celebrated statement that 'there are more comets in the heavens than fish in the ocean,' implies that a small portion only of the total number of comets belonging to our system are seen from the earth. But besides comets, and planets, and moons, a numerous class of bodies belong to our system which, from their smallness, might be regarded as cosmical atoms. Like the planets and the comets these smaller asteroids obey the law of gravity, and revolve in elliptic orbits round the sun. It is they which, when they come within the earth's atmosphere, and are fired by friction, appear to us as meteors and falling stars.

(690) On a bright night, twenty minutes rarely pass at any part of the earth's surface, without the appearance of at least one meteor. Twice a year (on the 12th of August and 14th of November) they appear in enormous numbers. During nine hours in Boston, when they were described as

* Mayer, Dynamik des Himmels, p. 10.

falling as thick as snowflakes, 240,000 meteors were observed. The number falling in a year might, perhaps, be estimated at hundreds or thousands of millions, and even these would constitute but a small portion of the total crowd of asteroids that circulate round the sun. From the phenomena of light and heat, and by direct observations on Encke's comet, we learn that the universe is filled by a resisting medium, through the friction of which all the masses of our system are drawn gradually towards the sun. And though the larger planets show, in historic times, no diminution of their periods of revolution, it may be otherwise with the smaller bodies. In the time required for the mean distance of the earth to alter a single yard, a small asteroid may have approached thousands of miles nearer to the sun.

(691) Following up these reflections, we should be led to the conclusion, that while an immeasurable stream of ponderable meteoric matter moves unceasingly towards the sun, it must augment in density as it approaches its centre of convergence. And here the conjecture naturally rises, whether that vast nebulous mass, the Zodiacal Light, which embraces the sun, may not be a crowd of meteors. It is at least proved that this luminous phenomenon arises from matter which circulates in obedience to planetary laws; hence, the entire mass of the zodiacal light must be constantly approaching, and incessantly raining its substance down upon the sun.

(692) It is easy to calculate both the maximum and the minimum velocity, imparted by the sun's attraction to an asteroid circulating round him. The maximum is generated when the body approaches the sun from an infinite distance; the *entire pull* of the sun being then exerted upon it. The minimum is that velocity which would barely enable the body to revolve round the sun close to his surface. The final velocity of the former, just before striking

the sun, would be 390 miles a second, that of the latter 276 miles a second. The asteroid, on striking the sun, with the former velocity, would develope more than 9000 times the heat generated by the combustion of an equal asteroid of solid coal; while the shock, in the latter case, would generate heat equal to that of the combustion of upwards of 4000 such asteroids. It matters not, therefore, whether the substances falling into the sun be combustible or not; their being combustible would not add sensibly to the tremendous heat produced by their mechanical collision.

(693) Here, then, we have an agency competent to restore his lost energy to the sun, and to maintain a temperature at his surface which transcends all terrestrial combustion. In the fall of asteroids we find the means of producing the solar light and heat. It may be contended that this showering down of matter necessitates the growth of the sun; it does so; but the quantity necessary to maintain the observed calorific emission for 4000 years, would defeat the scrutiny of our best instruments. If the earth struck the sun, it would utterly vanish from perception; but the heat developed by its shock would cover the expenditure of a century.

(694) To the earth itself we might apply considerations similar to those which we have applied to the sun. From the present form of the earth, we infer that it was once in a fluid condition. The combination of the theory of gravitation and the mechanical theory of heat, suggests to us the possible origin of the earth's former fluidity. It enables us to regard the molten condition of a planet as resulting from the mechanical shock of cosmical masses, and it thus reduces to the same cause the internal heat of the earth and the radiant heat of the sun.

(695) Without doubt, the whole surface of the sun displays an unbroken ocean of molten matter. On this

ocean rests an atmosphere of glowing gas—a flame atmosphere, or photosphere. But gaseous substances emit, even when their temperature is very high, only a feeble light. Hence, it is probable that the dazzling white light of the sun comes to us, through the atmosphere, from the denser matter underneath.*

(696) There is one other consideration connected with the permanence of our present terrestrial conditions, which is well worthy of our attention. Standing upon one of the London bridges, we observe the current of the Thames reversed, and the water poured upwards twice a day. The water thus moved rubs against the river's bed and sides, and heat is the consequence of this friction. The heat thus generated is, in part, radiated into space, and there lost, as far as the earth is concerned. What is it that supplies this incessant loss? The earth's rotation. Let us look a little more closely into this matter. Imagine the moon fixed, and the earth turning like a wheel from west to east in its diurnal rotation. A mountain on the earth's surface, on approaching the moon's meridian, is, as it were, laid hold of by the moon; it forms a kind of handle, by which the earth is pulled more quickly round. But when the meridian is passed, the pull of the moon on the mountain will be in the opposite direction; it now tends to diminish the velocity of rotation as much as it previously augmented it; and thus the action of all fixed bodies on the earth's surface is neutralised.

(697) But suppose the mountain to lie *always* to the east of the moon's meridian, the pull would then be always exerted against the earth's rotation, the velocity of which would be diminished in a degree corresponding to the strength of the pull. *The tidal wave occupies this*

* I am quoting here from Mayer, but this is the exact view now entertained by Kirchhoff. We see the solid or molten mass of the sun *through* his photosphere.

position—it lies always to the east of the moon's meridian; the waters of the ocean are, in part, dragged as a brake along the surface of the earth, and as a brake they must diminish the velocity of the earth's rotation. The diminution, though inevitable, is, however, too small to make itself felt within the period over which observations on the subject extend. Supposing, then, that we turn a mill by the action of the tide, and produce heat by the friction of the millstones; that heat has an origin totally different from the heat produced by another pair of millstones, which are turned by a mountain stream. The former is produced at the expense of the earth's rotation; the latter at the expense of the sun's heat, which lifted the millstream to its source.*

(698) Such is an outline of the Meteoric Theory of the Sun, as extracted from Mayer's 'Essay on Celestial Dynamics.' I have held closely to his statements, and in most cases simply translated his words. But the sketch conveys no adequate idea of the firmness and consistency with which he has applied his principles. He deals with true causes; and the only question that can affect his theory refers to the quantity of action ascribed by him to these causes. I do not pledge myself to this theory, nor do I ask you to accept it as demonstrated; still, it would be a great mistake to regard it as chimerical. It is a noble speculation; and depend upon it, the true theory, if this, or some form of it, be not the true one, will not appear less wild or less astounding.†

* Dynamik des Himmels, p. 38, &c.

† While preparing these sheets finally for press, I had occasion to look once more into the writings of Mayer, and the effect was a revival of the interest with which I first read them. Dr. Mayer was a practising physician in the little German town of Heilbronn, and in 1840, he made the observation that the venous blood of a feverish patient in the tropics was redder than in more northern latitudes. Starting from this fact while engaged in the duties of a laborious profession, and apparently without a single kindred spirit to support and animate him, he raised his mind to the level indicated

(699) Mayer published his Essay in 1848; five years afterwards, Mr. Waterston sketched, independently, a similar theory at the Hull Meeting of the British Association. The Transactions of the Royal Society of Edinburgh for 1854 contain an extremely beautiful memoir, by Professor Sir William Thomson, in which Mr. Waterston's sketch is fully developed. He considers that the meteors, which are to furnish stores of energy for our future sunlight, lie principally within the earth's orbit, and that we see them there, as the Zodiacal Light, 'an illuminated shower, or rather tornado, of stones.'

(700) Sir William Thomson adduces the following forcible considerations to show the inadequacy of chemical combination to produce the sun's heat. 'Let us consider,' he says, 'how much chemical action would be required to produce the same effects. . . . . Taking the former estimate, 2781 thermal units Centigrade (each 1390 foot pounds, § 38) or 3,869,000 foot pounds, which is equivalent to 7000 horse-power, as the rate per second of emission of energy from every square foot of the sun's surface, we find that more than 0·42 of a pound of coal per second, 1500 lbs. per hour, would be required to produce heat at the same rate. Now if all the fires of the whole Baltic fleet (this was written in 1854) were heaped up and kept in full combustion over one or two square yards of surface, and if the surface of a globe all round had every square yard so occupied, where could a sufficient supply of air come from to sustain the combustion? Yet such is the condition we must suppose the sun to be in, according to the hypothesis now under

by the references made to his works, throughout this book. In 1842 he published his first memoir 'On the Forces of Inorganic Nature;' in 1845, his 'Organic Motion' was published; in 1848, his 'Celestial Dynamics' appeared; and in 1851, he published his 'Remarks on the Mechanical Equivalent of Heat.' After this his overtasked brain gave way, and a cloud settled on the intellect which had accomplished so much. The shade, however, was but temporary, and Dr. Mayer is now restored.

consideration. . . . . If the products of combustion were gaseous, they would, in rising, check the necessary supplies of fresh air; if they were solid and liquid (as they might be if the fuel were metallic) they would interfere with the supply of elements from below. In either, or in both ways, the fire would be choked, and I think it may be safely affirmed that no such fire could keep alight for more than a few minutes, by any conceivable adaptation of air and fuel. If the sun be a burning mass it must be more analogous to burning gunpowder than to a fire burning in air; and it is quite conceivable that a solid mass, containing within itself all the elements required for combustion, provided the products of combustion are permanently gaseous, could burn off at its surface all round, and actually emit heat as copiously as the sun. Thus, an enormous globe of gun-cotton might, if at first cold, and once set on fire round its surface, get to a permanent rate of burning, in which any internal part would become heated sufficiently to ignite, only when nearly approached by the burning surface. It is highly probable indeed that such a body might for a time be as large as the sun and give out luminous heat as copiously, to be freely radiated into space, without suffering more absorption from its atmosphere of transparent gaseous products than the light of the sun actually does experience from the dense atmosphere through which it passes. Let us therefore consider at what rate such a body, giving out heat so copiously, would burn away, the heat of combustion could probably not be so much as 4000 thermal units per pound of matter burned, the greatest thermal equivalent of chemical action yet ascertained falling considerably short of this. But 2781 thermal units (as found above) are emitted per second from each square foot of the sun; hence there would be a loss of about 0·7 of a pound of matter per square foot per second. . . . . or a layer half a foot thick in a minute, or 55 miles thick in a year. At the same rate continued, a mass as large as the sun is at present would burn away in 8000 years. If the sun has been burning at that rate in past time he must have been of double diameter, of quadruple heating power, and of eight-fold mass only 8000 years ago. We may therefore quite safely conclude that the sun does not get its heat by chemical action. . . . and we must therefore look to the meteoric theory for fuel.'

(701) The eminent physicist I have just quoted subsequently modified his view of the origin and maintenance of solar heat. He showed in 1854 that the conclusion of physical is against the idea of the meteoric matter being extra-planetary. He inferred that if this were the case the year would be so shortened by the augmentation of the sun's mass that in reckoning back 2000 of our present years we should findourselves one-eighth of a year in error. Hence he concluded that the meteors which supply the sun with heat had existed long previously within the earth's orbit.

But the researches of Le Verrier on the motion of the planet Mercury, though they indicate the existence of such circulating matter round the sun, show it to be small in quantity. Hence Sir William Thomson in 1862, arrived at the conclusion that if any appreciable portions of the sun's heat be due to the present raining down of meteoric matter, the matter must circulate round the sun close to his surface. But if such matter existed, it is difficult to imagine how bodies so attenuated as comets could escape from the sun without any sensible loss of energy after having passed at a distance from his surface less than one-eighth of his radius. Sir William Thomson therefore concludes, that though the sun was formed by the collision of small masses, this collision being demonstrably able to supply us with twenty million years of solar heat at the present rate of emission, the sun's expenditure, though thus *originated*, is not *main-*

tained by the mechanical collision of gravitating masses, the low rate of cooling and the consequent constancy of the emission being due in great part to the high specific heat of the matter of the sun.

(702) From the first memoir of Sir William Thomson I extract the following interesting data, showing the amount of heat equivalent to the rotation of the sun and the orbital revolutions of the planets, or the amounts of heat which would be generated if a brake were applied at the surface of the sun, so as to stop the motion of rotation, and if the planets were stopped in their orbits; also the heat obtainable from gravitation, or that which would be developed by each of the planets falling into the sun. The quantity of heat is expressed by the time during which it would cover the solar emission.

	Heat of Gravitation, equal to Solar emission for a period of	Heat of Revolution, equal to Solar emission for a period of
Sun		116 years 6 days
Mercury	6 years 214 days	15 "
Venus	83 " 227 "	99 "
Earth	94 " 303 "	81 "
Mars	12 " 252 "	7 "
Jupiter	32240 " "	14 " 144 "
Saturn	9650 " "	2 " 127 "
Uranus	1610 " "	71 "
Neptune	1890 " "	

(703) Thus, if the planet Mercury were to strike the sun, the quantity of heat generated would cover the solar emission for nearly seven years; while the shock of Jupiter would cover the loss of 32,240 years. Our earth would furnish a supply for 95 years. The heat of rotation of the sun and planets, taken together, would cover the solar emission for 134 years; while the total heat of gravitation (that produced by the planets falling into the sun) would cover the emission for 45,589 years.

(704) Whatever be the ultimate fate of the theory here sketched, it is a great thing to be able to state the con-

ditions which certainly would produce a sun,—to be able to discern in the force of gravity, acting upon dark matter, the source from which the starry heavens *may* have been derived. For, whether the sun be produced and his emission maintained by the collision of cosmical masses,—whether the internal heat of the earth be the residue of that developed by the impact of cold dark asteroids, or not, there cannot be a doubt as to the competence of the cause assigned to produce the effects ascribed to it. Solar light and solar heat lie latent in the force which pulls an apple to the ground. 'Created simply as a difference of position of attracting masses, the potential energy of gravitation was the original form of all the energy in the universe. As surely as the weights of a clock run down to their lowest position, from which they can never rise again, unless fresh energy is communicated to them from some source not yet exhausted, so surely must planet after planet creep in, age by age, towards the sun. When each comes within a few hundred thousand miles of his surface, if he is still incandescent, it must be melted and driven into vapour by radiant heat. Nor, if he be crusted over and become dark and cool externally, can the doomed planet escape its fiery end. If it does not become incandescent, like a shooting star, by friction in its passage through his atmosphere, its first graze on his surface must produce a stupendous flash of light and heat. It may be at once, or it may be after two or three bounds, like a cannon-shot ricochetting on a surface of earth or water, the whole mass must be crushed, melted, and evaporated by a crash, generating in a moment some thousands of times as much heat as a coal of the same size would produce by burning.'*

(705) Helmholtz, an eminent German physiologist,

* Thomson and Tait in 'Good Words,' Oct. 1862, p. 606.

physicist, and mathematician, takes a somewhat different view of the origin and maintenance of solar light and heat. He starts from the nebular hypothesis of Laplace, and assuming the nebulous matter, in the first instance, to have been of extreme tenuity, he determines the amount of heat generated by its condensation to the present solar system. Supposing the specific heat of the condensing mass to be the same as that of water, then the heat of condensation would be sufficient to raise the temperature 28,000,000° Centigrade. By far the greater part of this heat was wasted, ages ago, in space. The most intense terrestrial combustion that we can command is that of oxygen and hydrogen, and the temperature of the pure oxyhydrogen flame is 8061° C. The temperature of a hydrogen flame burning in air, is 3259° C.; while that of the lime light, which shines with such sunlight brilliancy, is estimated at 2000° C. What conception, then, can we form of a temperature more than thirteen thousand times that of the Drummond light? If our system were composed of pure coal, and burnt up, the heat produced by its combustion would only amount to $\frac{1}{3500}$th of that generated by the condensation of the nebulous matter, to form our solar system. Helmholtz supposes this condensation to continue; that a virtual falling down of the superficial portions of the sun towards the centre still takes place, a continual development of heat being the result. However this may be, he shows by calculation that the shrinking of the sun's diameter by $\frac{1}{10000}$th of its present length, would generate an amount of heat competent to cover the solar emission for 2000 years; while the condensation of the sun from its present mean density to that of the earth, would have its equivalent in an amount of heat competent to cover the present solar emission for 17,000,000 years.

(706) 'But,' continues Helmholtz, 'though the store of

our planetary system is so immense that it has not been sensibly diminished by the incessant emission which has gone on during the period of man's history, and though the time which must elapse before a sensible change in the condition of our planetary system can occur, is totally beyond our comprehension, the inexorable laws of mechanics show that this store, which can only suffer loss, and not gain, must finally be exhausted. Shall we terrify ourselves by this thought? We are in the habit of measuring the greatness of the universe, and the wisdom displayed in it, by the duration and the profit which it promises to our own race; but the past history of the earth shows the insignificance of the interval during which man has had his dwelling here. What the museums of Europe show us of the remains of Egypt and Assyria we gaze upon with silent wonder, in despair of being able to carry back our thoughts to a period so remote. Still, the human race must have existed and multiplied for ages before the Pyramids could have been erected. We estimate the duration of human history at 6000 years; but, vast as this time may appear to us, what is it in comparison with the period during which the earth bore successive series of rank plants and mighty animals, but no men? Periods during which, in our own neighbourhood (Königsberg), the amber-tree bloomed, and dropped its costly gum on the earth and in the sea; when in Europe and North America groves of tropical palms flourished, in which gigantic lizards, and, after them, elephants, whose mighty remains are still buried in the earth, found a home. Different geologists, proceeding from different premises, have sought to estimate the length of the above period, and they set it down from one to nine millions of years. The time during which the earth has generated organic beings is again small, compared with the ages during which the world was a mass of molten rocks.

The experiments of Bischof upon basalt show that our globe would require 350 millions of years to cool down from 2000° to 200° Centigrade. And with regard to the period during which the first nebulous masses condensed, to form our planetary system, conjecture must entirely cease. The history of man, therefore, is but a minute ripple in the infinite ocean of time. For a much longer period than that during which he has already occupied this world, the existence of a state of inorganic nature, favourable to man's continuance here, seems to be secured, so that for ourselves, and for long generations after us, we have nothing to fear. But the same forces of air and water, and of the volcanic interior, which produced former geologic revolutions, burying one series of living forms after another, still act upon the earth's crust. They, rather than those distant cosmical changes of which we have spoken, will put an end to the human race; and, perhaps, compel us to make way for new and more complete forms of life, as the lizard and the mammoth have given way to us and our contemporaries.'*

(707) The relationship of our planet and the powers active there, to the sun, demands special attention. Five and thirty years ago, the following remarkable passage, bearing upon this subject, was written by Sir John Herschel.† 'The sun's rays are the ultimate source of almost every motion which takes place on the surface of the earth. By its heat are produced all winds, and those disturbances in the electric equilibrium of the atmosphere which give rise to the phenomena of lightning, and probably also to terrestrial magnetism and the Aurora. By their vivifying action vegetables are enabled to draw support from inorganic matter, and become in their turn the support of animals and man, and the source of those

* Wechselwirkung der Naturkräfte, Phil. Mag., Ser. IV. vol. ix. p. 515.

† Outlines of Astronomy, 1833.

great deposits of dynamical efficiency which are laid up for human use in our coal strata. By them the waters of the sea are made to circulate in vapour through the air, and irrigate the land, producing springs and rivers. By them are produced all disturbances of the chemical equilibrium of the elements of nature, which by a series of compositions and decompositions give rise to new products and originate a transfer of materials. Even the slow degradation of the solid constituents of the surface, in which its chief geological change consists, is almost entirely due, on the one hand, to the abrasion of wind or rain and the alternation of heat and frost; on the other, to the continual beating of sea waves agitated by winds, the results of solar radiation. Tidal action (itself partly due to the sun's agency) exercises here a comparatively slight influence. The effect of oceanic currents (mainly originating in that influence), though slight in abrasion, is powerful in diffusing and transporting the matter abraded; and when we consider the immense transfer of matter so produced, the increase of pressure over large spaces in the bed of the ocean, and diminution over corresponding portions of the land, we are not at a loss to perceive how the elastic force of subterranean fires, thus repressed on the one hand and released on the other, may break forth in points where the resistance is barely adequate to their retention, and thus bring the phenomena of even volcanic activity under the general law of solar influence.'

(708) This fine passage requires but the breath of recent investigation to convert it into an exposition of the law of the conservation of energy, as applied to both the organic and inorganic world. Late discoveries have taught us that winds and rivers have their definite thermal values, and that, in order to produce their motion, an equivalent amount of solar heat has been consumed. While they exist as winds and rivers, the heat expended in producing

them has ceased to exist, being converted into mechanical motion; but when that motion is arrested, the heat which produced it is restored. A river, in descending from an elevation of 7720 feet, generates an amount of heat competent to augment its own temperature 10° Fahr., and this amount of heat was abstracted from the sun, in order to lift the matter of the river to the elevation from which it falls. As long as the river continues on the heights, whether in the solid form as a glacier, or in the liquid form as a lake, the heat expended by the sun in lifting it has disappeared from the universe. It has been consumed in the act of lifting. But at the moment that the river starts upon its downward course, and encounters the resistance of its bed, the heat expended in its elevation begins to be restored. The mental eye, indeed, can follow the emission from its source, through the æther as vibratory motion, to the ocean, where it ceases to be vibration, and takes the potential form among the molecules of aqueous vapour; to the mountain-top, where the heat absorbed in vaporisation is given out in condensation, while that expended by the sun in *lifting* the water to its present elevation is still unrestored. This we find paid back to the last unit by the friction along the river's bed; at the bottom of the cascades where the plunge of the torrent is suddenly arrested; in the warmth of the machinery turned by the river; in the spark from the millstone; beneath the crusher of the miner; in the Alpine saw-mill; in the milk-churn of the châlet; in the supports of the cradle in which the mountaineer, by water power, rocks his baby to sleep. All the forms of mechanical motion here indicated are simply the parcelling out of an amount of calorific motion derived originally from the sun; and at each point at which the mechanical motion is destroyed, or diminished, it is the sun's heat which is restored.

(709) We have thus far dealt with the sensible motions and energies which the sun produces and confers; but there are other motions and other energies, whose relations are not so obvious. Trees and vegetables grow upon the earth, and when burned they give rise to heat, from which immense quantities of mechanical energy are derived. What is the source of this energy? Sir John Herschel answered this question in a general way; while Dr. Mayer and Professor Helmholtz fixed its exact relation to the more general question of conservation. Let me try to put their answers into plain words. You see this iron rust, produced by the falling together of the atoms of iron and oxygen; but though you cannot see this transparent carbonic acid gas, it is formed by the union of carbon and oxygen. These atoms thus united resemble a weight resting on the earth; their mutual attraction is satisfied. But as I can wind up the weight, and prepare it for another fall, even so these atoms can be wound up, separated from each other, and thus enabled to repeat the process of combination.

(710) In the building of plants, carbonic acid is the material from which the carbon of the plant is derived, while water is the substance from which it obtains its hydrogen. The solar beam winds up the weight; it is the agent which severs the atoms, setting the oxygen free, and allowing the carbon and the hydrogen to aggregate in woody fibre. If the sun's rays fall upon a surface of sand, the sand is heated, and finally radiates away as much heat as it receives; but let the same beams fall upon a forest; then the quantity of heat given back is less than that received, for a portion of the sunbeams is invested in the building of the trees. We have already seen how heat is consumed in forcing asunder the atoms of bodies; and how it reappears, when the attraction of the separated atoms comes again into play.* The precise considerations

* Chapter V.

which we then applied to heat, we have now to apply to light, for it is at the expense of the solar light that the chemical decomposition takes place. Without the sun, the reduction of the carbonic acid and water cannot be effected; and, in this act, an amount of solar energy is consumed, exactly equivalent to the molecular work done.

(711) Combustion is the reversal of this process of reduction, and all the energy invested in a plant reappears as heat, when the plant is burned. I ignite this bit of cotton, it bursts into flame; the oxygen again unites with its carbon, and an amount of heat is given out, equal to that originally sacrificed by the sun to form the bit of cotton. So also as regards the 'deposits of dynamical efficiency' laid up in our coal strata; they are simply the sun's rays in a potential form. We dig from our pits, annually, eighty-four millions of tons of coal, the mechanical equivalent of which is of almost fabulous vastness. The combustion of a single pound of coal in one minute is equal to the work of three hundred horses for the same time. It would require one hundred and eight millions of horses, working day and night with unimpaired strength for a year, to perform an amount of work equivalent to the energy which the sun of the Carboniferous epoch invested in one year's produce of our coalpits.

(712) The farther we pursue this subject, the more its interest and its wonder grow upon us. You have learned how a sun may be produced by the mere exercise of gravitating force; that by the collision of cold dark planetary masses the light and heat of our central orb, and also of the fixed stars, may be obtained. But here we find the physical powers, derived or derivable from the action of gravity upon dead matter, introducing themselves at the very root of the question of vitality. We find in solar light and heat the very mainspring of vegetable life.

(713) Nor can we halt at the vegetable world, for it, mediately or immediately, is the source of all animal life. Some animals feed directly on plants, others feed upon their herbivorous fellow-creatures; but all in the long run derive life and energy from the vegetable world; all, therefore, as Helmholtz has remarked, may trace their lineage to the sun. In the animal body the carbon and hydrogen of the vegetable are again brought into contact with the oxygen from which they had been divorced, and which is now supplied by the lungs. Reunion takes place, and animal heat is the result. Save as regards intensity, there is no difference between the combustion that thus goes on within us, and that of an ordinary fire. The products of combustion are in both cases the same, namely, carbonic acid and water. Looking then at the physics of the question, we see that the formation of a vegetable is a process of winding up, while the formation of an animal is a process of running down. This is the rhythm of Nature as applied to animal and vegetable life.

(714) But is there nothing in the human body to liberate it from that chain of necessity which the law of conservation coils around inorganic nature? Look at two men upon a mountain side, with equal health and physical strength; the one will sink and fail, while the other with determined energy, scales the summit. Has not volition, in this case, a creative power? Physically considered, the law that rules the operations of a steam-engine rules the operations of the climber. For every pound raised by the former, an equivalent quantity of its heat disappears; and for every step the climber ascends, an amount of heat, equivalent jointly to his own weight and the height to which it is raised, is lost to his body. The strong will can draw largely upon the physical energy furnished by the food; but it can *create* nothing. The function of the will is to *apply* and *direct*, not to create.

(715) I have just said that, as a climber ascends a mountain, heat disappears from his body; the same statement applies to animals performing work. It would appear to follow from this, that the body ought to grow colder, in the act of climbing or of working, whereas universal experience proves it to grow warmer. The solution of this seeming contradiction is found in the fact, that when the muscles are exerted, augmented respiration and increased chemical action set in. The bellows which urge oxygen into the fire within are more briskly blown, and thus, though heat actually disappears as we climb, the loss is more than covered by the increased activity of the chemical processes.

(716) By means of a modification of the thermo-electric pile, MM. Becquerel and Breschet proved that heat is developed in a muscle when it contracts. MM. Billroth and Fick have also found that in the case of persons who die from tetanus, the temperature of the muscles is sometimes nearly eleven degrees Fahrenheit in excess of the normal temperature. M. Helmholtz has shown that the muscles of dead frogs in contracting produce heat; and an extremely important result as regards the influence of contraction has been obtained by Professor Ludwig of Vienna and his pupils. Arterial blood, you know, is charged with oxygen: when this blood passes through a muscle in an ordinary uncontracted state, it is changed into venous blood which still retains about $7\frac{1}{2}$ per cent. of oxygen. But if the arterial blood pass through a *contracted* muscle, it is almost wholly deprived of its oxygen, the quantity remaining amounting, in some cases, to only $1\frac{3}{10}$ per cent. As a result of the augmented combustion within the muscles when in a state of activity, we have an increased amount of carbonic acid expired from the lungs. Dr. Edward Smith has shown that the quantity of this gas

expired during periods of great exertion may be five times that expired in a state of repose.

(717) Now when we augment the temperature of the body by labour, *a portion* only of the excess of molecular motion generated is applied to the performance of the work. Suppose a certain amount of food to be oxidized in the body of a man in a state of repose, the quantity of heat produced in the process is exactly that which we should obtain from the direct combustion of the food in an ordinary fire. But suppose the oxidation of the food to take place while the man is performing work, then the heat generated in the body falls short of that which could be obtained from direct combustion. An amount of heat is missing, equivalent to the work done. Supposing the work to consist in the development of heat by friction, then the amount of heat thus generated outside of the man's body would be exactly that which was wanting within his body, to make the heat there generated equal to that produced by direct combustion.

(718) It is, of course, easy to determine the amount of heat consumed by a mountaineer, in lifting his own body to any elevation. When lightly clad, I weigh about 145 lbs.; what is the amount of heat consumed, in my case, in climbing from the sea-level to the top of Mont Blanc? The height of the mountain is 15,774 feet; and for every pound of my body raised to a height of 772 feet, a quantity of heat is consumed, sufficient to raise the temperature of a pound of water 1° Fahr. Consequently, on climbing to a height of 15,774, or about $20\frac{1}{2}$ times 772 feet, an amount of heat is consumed sufficient to raise the temperature of 145 lbs. of water $20\frac{1}{2}$° Fahr. If, on the other hand, I could perform a glissade from the top of the mountain to the sea-level, the quantity of heat generated during the descent would be precisely equal to that consumed in the ascent. Your attention has been more

than once directed to the energy of molecular forces, and here the subject appears once more. Measured by one's feelings, the amount of exertion necessary to reach the top of Mont Blanc is very great. Still the energy which performs this feat would be derived from the combustion of about two ounces of carbon. In the case of an excellent steam-engine, about one-tenth of the heat employed is converted into work; the remaining nine-tenths being wasted in the air, the condenser, &c. In the case of an active mountaineer, as much as one-fifth of the heat due to the oxidation of his food may be converted into work; hence, as a working machine, the animal body is much more perfect than the steam-engine.

(719) We see, however, that the engine and the animal derive, or may derive, these powers from the selfsame source. We can work an engine by the direct combustion of the substances which we employ as food; and if our stomachs were so constituted as to digest coal, we should, as Hemholtz has remarked,* be able to derive our energy from this substance. The grand point permanent throughout all these considerations is, that *nothing is created.* We can make no movement which is not accounted for by the contemporaneous extinction of some other movement. And how complicated soever the motions of animals may be, whatever may be the change which the molecules of our food undergo within our bodies, the whole energy of animal life consists in the falling of the atoms of carbon and hydrogen and nitrogen from the high level which they occupy in the food, to the low level which they occupy when they quit the body. But what has enabled the carbon and the hydrogen to fall? What first raised them to the level which rendered the fall possible? We have already learned that it is the sun. It is at his cost that animal heat is produced, and animal motion accomplished.

* Phil. Mag. 1856, vol. ix. p. 510.

Not only then is the sun chilled, that we may have our fires, but he is likewise chilled that we may have our powers of locomotion.

(720) The subject is of such vast importance, and is so sure to tinge the whole future course of philosophic thought, that I will dwell upon it a little longer, and endeavour, by reference to analogical processes, to give you a clearer idea of the part played by the sun in vital actions. We can raise water by mechanical action to a high level; and that water, in descending by its own gravity, may be made to assume a variety of forms, and to perform various kinds of mechanical work. It may be made to fall in cascades, rise in fountains, twirl in eddies, or flow along a uniform bed. It may, moreover, be employed to turn wheels, lift hammers, grind corn, or drive piles. Now there is no power *created* by the water during its descent. All the energy which it exhibits is merely the parcelling out and distribution of the original energy which raised it up on high. Thus also as regards the complex motions of a clock or a watch; they are entirely derived from the energy of the hand which winds it up. Thus also the singing of the little Swiss bird in the International Exhibition; the quivering of its artificial organs, the vibrations of the air which strike the ear as melody, the flutter of its little wings, and all other motions of the pretty automaton, were simply derived from the force by which it was wound up. It gives out nothing that it has not received. In this precise sense, you will perceive, is the energy of man and animals the parcelling out and distribution of an energy originally exerted by the sun. In the vegetable, as we have remarked, the act of elevation, or of winding up, is performed; and it is during the descent, in the animal, of the carbon, hydrogen, and nitrogen, to the level from which they started, that the powers of life appear.

(721) But the question is not yet exhausted. The water which we used in our first illustration produces all the motion displayed in its descent, but the *form* of the motion depends on the character of the machinery interposed in the path of the water. And thus the primary action of the sun's rays is qualified by the atoms and molecules among which their power is distributed. Molecular forces determine the *form* which the solar energy will assume. In the one case this energy is so conditioned by its atomic machinery as to result in the formation of a cabbage; in another case it is so conditioned as to result in the formation of an oak. So also as regards the reunion of the carbon and the oxygen—the *form* of their reunion is determined by the molecular machinery through which the combining force acts. In one case the action may result in the formation of a man, while in another it may result in the formation of a grasshopper.

(722) The matter of our bodies is that of inorganic nature. There is no substance in the animal tissues which is not primarily derived from the rocks, the water, and the air. Are the forces of organic matter, then, different in kind from those of inorganic? All the philosophy of the present day tends to negative the question; and to show that it is the directing and compounding, in the organic world, of forces belonging equally to the inorganic, that constitutes the mystery and the miracle of vitality.

(723) In discussing the material combinations which result in the formation of the body and the brain of man, it is impossible to avoid taking side glances at the phenomena of consciousness and thought. Science has asked daring questions, and will, no doubt, continue to ask such. Problems will assuredly present themselves to men of a future age, which, if enunciated now, would appear to most people as the direct offspring of insanity. Still, though the progress and development of science may seem to

be unlimited, there is a region beyond her reach—a line with which she does not even tend to osculate. Given the masses and distances of the planets, we can infer the perturbations consequent on their mutual attractions. Given the nature of a disturbance in water, air, or ether, we can infer from the properties of the medium how its particles will be affected. In all this we deal with physical laws, and the mind runs freely along the line which connects the phenomena from beginning to end. But when we endeavour to pass, by a similar process, from the region of physics to that of thought, we meet a problem not only beyond our present powers, but transcending any conceivable expansion of the powers we now possess. We may think over the subject again and again, but it eludes all intellectual presentation. The origin of the material universe is equally inscrutable. Thus, having exhausted science, and reached its very rim, the real mystery of existence still looms around us. And thus it will ever loom—ever beyond the bourne of man's intellect—giving the poets of successive ages just occasion to declare that

> We are such stuff
> As dreams are made of, and our little life
> Is rounded by a sleep.

(724) Still, presented rightly to the mind, the discoveries and generalisations of modern science constitute a poem more sublime than has ever yet addressed the human imagination. The natural philosopher of to-day may dwell amid conceptions which beggar those of Milton. Look at the integrated energies of our world,—the stored power of our coal-fields; our winds and rivers; our fleets, armies, and guns. What are they? They are all generated by a portion of the sun's energy, which does not amount to $\frac{1}{2300000000}$ of the whole. This is the entire fraction of the sun's force intercepted by the

earth, and we convert but a small fraction of this fraction into mechanical energy. Multiplying all our powers by millions of millions, we do not reach the sun's expenditure. And still, notwithstanding this enormous drain, in the lapse of human history we are unable to detect a diminution of his store. Measured by our largest terrestrial standards, such a reservoir of power is infinite; but it is our privilege to rise above these standards, and to regard the sun himself as a speck in infinite extension—a mere drop in the universal sea. We analyse the space in which he is immersed, and which is the vehicle of his power. We pass to other systems and other suns, each pouring forth energy like our own, but still without infringement of the law, which reveals immutability in the midst of change, which recognises incessant transference or conversion, but neither final gain nor loss. This law generalises the aphorism of Solomon, that there is nothing new under the sun, by teaching us to detect everywhere, under its infinite variety of appearances, the same primeval force. The energy of Nature is a constant quantity, and the utmost man can do in the pursuit of physical truth, or in the applications of physical knowledge, is to shift the constituents of the never-varying total, sacrificing one if he would produce another. The law of conservation rigidly excludes both creation and annihilation. Waves may change to ripples, and ripples to waves—magnitude may be substituted for number, and number for magnitude—asteroids may aggregate to suns, suns may invest their energy in floræ and faunæ, and floræ and faunæ may melt in air—the flux of power is eternally the same. It rolls in music through the ages, while the manifestations of physical life as well as the display of physical phenomena are but the modulations of its rhythm.

CHAPTER XV.

ACTION OF ETHER WAVES OF SHORT PERIOD UPON GASEOUS MATTER—CLOUDS FORMED BY ACTINIC DECOMPOSITION—COLOUR PRODUCED BY SMALL PARTICLES—POLARISATION OF LIGHT BY NEBULOUS MATTER—THE BLUE OF THE SKY AND THE POLARISATION OF ITS LIGHT.

(725) YOU have had laid before you an abstract of the principal researches which have occupied my attention for the last ten years. In these investigations, my chief aim was to render the longer waves of the prismatic spectrum interpreters and expositors of molecular condition, and we wound up our 13th Chapter by filtering the waves of visible period from those of invisible period, and then broke up the larger heat-waves so as to enable them to produce all the phenomena of light. Unlike the beautiful researches of Melloni and Knoblauch, the investigations here referred to made radiant heat a means to an end. I endeavoured to place before my mind such images of molecules and their constituent atoms as modern science renders probable, and such images of the luminiferous ether and its motions as the undulatory theory of light enables us to form, and to found upon these conceptions experimental enquiries which should give us a more sure and certain hold of molecular constitution.

(726) One result among many now known to you of these researches is the sudden change of relations between the ether of space and ordinary matter which accompanies the act of combination. Preserving the quantity and ultimate quality of the matter traversed by the ethereal

waves constant, vast changes in the amount of wave-motion intercepted may be produced by the act of chemical union. If nitrogen and oxygen, for example, be mixed mechanically together in the proportion, by weight, of seven to four, radiant heat will pass through the mixture as through a vacuum. At all events, the quantity of heat intercepted is multiplied a thousandfold the moment the oxygen and nitrogen combine to form laughing-gas. So, in like manner, if nitrogen and hydrogen be mixed mechanically in the proportion of fourteen to three, the amount of radiant heat which they absorb in this condition is multiplied by thousands—it may be by millions—the moment they unite chemically to form ammonia. No single experiment shows the air we breathe to be a mechanical mixture, and not a chemical compound, with the same conclusiveness as that which proves it to be as practically pervious as a vacuum to the rays of heat.

(727) But the molecules which, like those of ammonia and laughing-gas, can intercept the waves of ether, must be shaken by those waves—possibly shaken asunder. That ordinary thermometric heat can produce chemical changes is one of the commonest facts. Radiant heat also, if sufficiently intense, and if absorbed with sufficient avidity, could produce all the effects of ordinary thermometric heat. The dark rays, for example, which can make platinum white-hot, could also, if absorbed, produce the chemical effects of white-hot platinum. They could, for example, decompose water, as they can now in a moment boil water. But the decomposition in this case would be effected through the virtual conversion of the radiant heat into thermometric heat. There would be nothing in the act characteristic of radiation, or demanding it as an essential element in the decomposition.

(728) The chemical actions for which the radiant form

seems essential are mainly produced by the least energetic rays of the spectrum. Thus the photographer has his heat-focus in advance of his chemical focus; which latter, though potent for his special purpose, possesses almost infinitely less mechanical energy than its neighbour. The mechanical energy depends upon the amplitude or range of vibration of the individual particles that constitute a wave of ether. Now, the heat waves have enormously greater amplitudes than the chemical waves; hence decomposition is less a matter of amplitude than of period of vibration. The quicker motions of the shorter and weaker waves are so related to the periods of vibration possible to the atoms that, like the timed impulses of a boy in a swing, they accumulate so as finally to jerk the atoms asunder; thus effecting what is called chemical decomposition.

(729) It is this jerking asunder of the constituent atoms of molecules that we have to examine during the coming hour. Our previous investigations dealt with the action of the long waves; this with the action of the short waves upon gaseous matter. Vapours of various kinds were sent into a glass experimental tube a yard in length, and about three inches in diameter. As a general rule, the vapours were perfectly transparent; the tube, when they were present, appearing as empty as when they were absent. In two or three cases, however, a faint cloudiness showed itself within the tube. This caused me a momentary anxiety, for I did not know how far, in describing my previous experiments, actions might have been ascribed to pure cloudless vapour which were really due to those newly-observed nebulæ. Intermittent discomfort, however, is the normal feeling of the investigator; for it drives him to closer scrutiny, to greater accuracy, and often, as a consequence, to new discovery. It was soon found that the nebulæ revealed by the beam were all

generated by the beam, and the observation opened a new door into regions inaccessible to sense, but which embrace so much of the intellectual life of the physical investigator.

(730) What *are* those vapours of which we have been speaking? They are aggregates of *molecules*, or small masses of matter, and every molecule is itself an aggregate of smaller parts called *atoms*. A molecule of aqueous vapour, for example, consists of two atoms of hydrogen and one of oxygen. A molecule of ammonia consists of three atoms of hydrogen and one of nitrogen, and so of other substances. Thus the molecules, themselves inconceivably small, are made up of distinct parts still smaller. When, therefore, a compound vapour is spoken of, the corresponding mental image is an aggregate of molecules separated from each other, though exceedingly near, each of these being composed of a group of atoms still nearer to each other. So much for the *matter* which enters into our conception of a vapour.* To this must now be added the idea of *motion*. The molecules have motions of their own *as wholes*; their constituent atoms have also motions of their own, which are executed independently of those of the molecules; just as the various movements of the earth's surface are executed independently of the orbital revolution of our planet.

(731) The vapour molecules are kept asunder by forces which, virtually or actually, are forces of repulsion. Between these elastic forces and the atmospheric pressure under which the vapour exists, equilibrium is established as soon as the proper distances between the molecules have been assumed. If, after this, the molecules be

* Newton seemed to consider that the molecules might be rendered visible by microscopes; but of atoms he appears to have entertained a different opinion. He finely remarks:—'It seems impossible to see the more secret and noble works of nature within the corpuscles, by reason of their transparency.' (Herschel, 'On Light,' art. 1145.)

urged nearer to each other by a momentary force, they recoil as soon as the force is expended. If by the exercise of a similar force they be separated more widely, when the force ceases to act they again approach each other. The case is different as regards the constituent atoms.

(732) And here let me remark that we are now upon the very outmost verge of molecular physics; and that I am attempting to familiarise your minds with conceptions which have not yet obtained universal currency even among chemists; which many chemists, moreover, might deem untenable. But, tenable or untenable, it is of the highest scientific importance to discuss them. Let us, then, look mentally at our atoms grouped together to form a molecule. Every atom is held apart from its neighbours by a force of repulsion; why, then, do not the mutually repellent members of this group part company? The molecules *do* separate from each other when the external pressure is lessened or removed, but the atoms do not. The reason of this stability is that *two* forces, the one attractive and the other repulsive, are in operation between every two atoms; and the position of every atom —its distance from its fellows—is determined by the equilibration of these two forces. If the atoms come too near, repulsion predominates and drives them apart; if too distant, attraction predominates and draws them together. The point at which attraction and repulsion are equal to each other is the atom's *position of equilibrium.* If not absolutely cold—and there is no such thing as absolute coldness in our corner of nature—the atoms are always in a state of vibration, their vibrations being executed to and fro *across their positions of equilibrium.*

(733) Into a vapour thus constituted, we have now to pour a beam of light. But what, in the first instance, *is* a beam of light? You know it is a train of innumerable waves, excited in, and propagated through, an almost

infinitely attenuated and elastic medium, which fills all space, and which we name the Ether. You know that these waves of light are not all of the same size; that some of them are much longer and higher than others; that the short waves and the long ones move with the same rapidity through space, just as short and long waves of sound travel with the same rapidity through air, and that hence the shorter waves must follow each other in quicker succession than the longer ones. You know that the different rapidities with which the waves of light impinge upon the retina, or optic nerve, give rise in consciousness to differences of *colour*; that there are, however, numberless waves emitted by the sun and other luminous bodies which reach the retina, but which are incompetent to excite the sensation of light; that if the lengths of the waves exceed a certain limit, or if they fall short of a certain other limit, they cannot generate vision. And it is to be particularly borne in mind that the capacity to produce *light* does not depend so much on the *strength* of the waves, as on their *periods of recurrence.*

(734) The elements of all the conceptions with which we shall have subsequently to deal are now in your possession. And you will observe that though we are speaking of things which lie entirely beyond the range of the senses, the conceptions are as truly *mechanical* as they would be if we were dealing with ordinary masses of matter, and with waves of sensible magnitude. I do not think that any really scientific mind at the present day will be disposed to draw a substantial distinction between chemical and mechanical phenomena. They differ from each other as regards the magnitude of the masses involved; but in this sense the phenomena of astronomy differ, also, from those of ordinary mechanics. The main bent of the natural philosophy of a future age will probably be to

chasten into order, by subjecting it to mechanical laws, the existing chaos of chemical phenomena.

(735) Whether we see rightly or wrongly—whether our notions be real or imaginary—it is of the utmost importance in science to aim at perfect clearness in the description of all that comes, or seems to come, within the range of the intellect. For if we are right, clearness of utterance forwards the cause of right; while if we are wrong, it ensures the speedy correction of error. In this spirit, and with the determination at all events to speak plainly, let us deal with our conceptions of ether waves and molecules. Supposing a wave, or a train of waves, to impinge upon a molecule so as to urge all its parts with the same motion, the molecule would move bodily as a whole, but because they are animated by a *common motion* there would be no tendency to its constituent atoms to separate from each other. *Differential motions* among the atoms themselves would be necessary to effect a separation, and if such motions be not introduced by the shock of the waves, there is no mechanical ground for the decomposition of the molecule.

(736) It is, however, difficult to conceive the shock of a wave, or a train of waves, so distributed among the atoms as to cause no strain amongst them. For atoms are of different weights, probably of different sizes; at all events it is almost certain that the ratio of the mass of the atom to the surface it presents to the action of the ether waves is different in different cases. If this be so, and I think the probabilities are immensely in favour of its being so, then every wave which passes over a molecule tends to decompose it—tends to carry away from their weightier and more sluggish companions those atoms which, in relation to their mass, present the largest resisting surfaces to the motion of the waves. The case may be illustrated by reference to a man standing on the deck of a ship. As

long as both of them share equally the motions of the wind or of the sea, there is no tendency to separation. In chemical language, they are in a state of combination. But a wave passing over it finds the ship less rapid in yielding to its motion than the man; the man is consequently carried away, and we have what may be roughly regarded as decomposition.

(737) Thus the conception of the decomposition of compound molecules by the waves of ether comes to us recommended by *à priori* probability. But a closer examination of the question compels us to supplement, if not materially to qualify, this conception. It is a most remarkable fact, that the waves which have thus far been most effectual in shaking asunder the atoms of compound molecules are those of least mechanical power. *Billows*, to use a strong comparison, are incompetent to produce effects which are readily produced by *ripples*. It is, for example, the violet and ultra-violet rays of the sun that are most effectual in producing these chemical decompositions; and, compared with the red and ultra-red solar rays, the energy of these 'chemical rays' is infinitesimal. This energy would probably in some cases have been multiplied by millions to bring it up to that of the ultra red-rays; and still the latter are powerless where the smaller waves are potent. We here observe a remarkable similarity between the behaviour of chemical molecules and that of the human retina.

(738) Whence, then, the power of these smaller waves to unlock the bonds of chemical union? If it be not a result of their strength, it must be, as in the case of vision, a result of their periods of recurrence. But how are we to figure this action? I should say thus: the shock of a single wave produces no more than an infinitesimal effect upon an atom or a molecule. To produce a larger effect, the motion must *accumulate*, and for wave-

impulses to accumulate, they must arrive in periods identical with the periods of vibration of the atoms on which they impinge. In this case each successive wave finds the atom in a position which enables that wave to add its shock to the sum of the shocks of its predecessors. The effect is mechanically the same as that due to the timed impulses of a boy upon a swing. The single tick of a clock has no appreciable effect upon the unvibrating and equally long pendulum of a distant clock; but a succession of ticks, each of which adds, at the proper moment, its infinitesimal push to the sum of the pushes preceding it, will, as a matter of fact, set the second clock going. So likewise a single puff of air against the prong of a heavy tuning-fork produces no sensible motion, and, consequently, no audible sound; but a succession of puffs, which follow each other in periods identical with the tuning-fork's period of vibration, will render the fork sonorous. I think the chemical action of light is to be regarded in this way. Fact and reason point to the conclusion that it is the heaping up of motion on the atoms, in consequence of their synchronism with the shorter waves, that causes them to part company. This I take to be the mechanical cause of these decompositions which are effected by the waves of ether.

(739) And now let us return to that faint cloudiness, already mentioned, from which, as from a germ, these considerations and speculations have sprung. It has been long known that light effected the decomposition of a certain number of bodies. The transparent iodide of ethyl, or of methyl, for example, becomes brown and opaque on exposure to light, through the discharge of its iodine. The art of photography is founded on the chemical actions of light; so that it is well known that the effects for which the foregoing theoretic considerations would have prepared us, are not only probable, but actual.

(740) But the method now to be employed, and which consists simply in offering the *vapours* of volatile substances to the action of light, enables us not only to present such experiments in a beautiful form, but also to give a vast extension to the operations of light, or rather of radiant force, as a chemical agent. It also enables us to imitate in our laboratories actions which have been hitherto performed only in the laboratory of nature. The substances now to be examined are so constituted that when their molecules are broken by the waves of light, the newly-formed bodies are comparatively *involatile*. To keep them in the gaseous form these products of decomposition require a greater heat than is required by the vapours from which they are derived; and hence, if the space in which these new bodies are liberated be of the proper temperature, they will not remain in the vaporous condition, but will be precipitated as clouds upon the beam to the action of which they owe their existence.

(741) I hold in my hand a little flask, F (fig. 109), stopped by a cork, pierced in two places. Through one orifice passes a narrow glass tube, *a*, which terminates immediately under the cork; through the other orifice passes a similar tube, *b*, descending to the bottom of the little flask, which is filled to a height of about an inch with a transparent liquid. The name of this liquid is *nitrite of amyl*, in every molecule of which we have 5 atoms of carbon, 11 of hydrogen, 1 of nitrogen, and 2 of oxygen. Upon this group of atoms the waves of our electric light will be immediately let loose. Our experimental tube is here connected with the small flask F, a stop-cock, however, intervening between them, by means of which the passage between the flask and the experimental tube can be opened or closed at pleasure. The other tube, passing through the cork of the flask and descending into the liquid, is connected with a U-shaped vessel, filled with fragments of clean glass,

covered with sulphuric acid. In front of the U-shaped vessel is a narrow tube stuffed with cotton-wool. At the other end of the experimental tube is our electric lamp; and here, finally, is an air-pump, by means of which the tube has been exhausted. We are now ready for experiment.

FIG. 109.

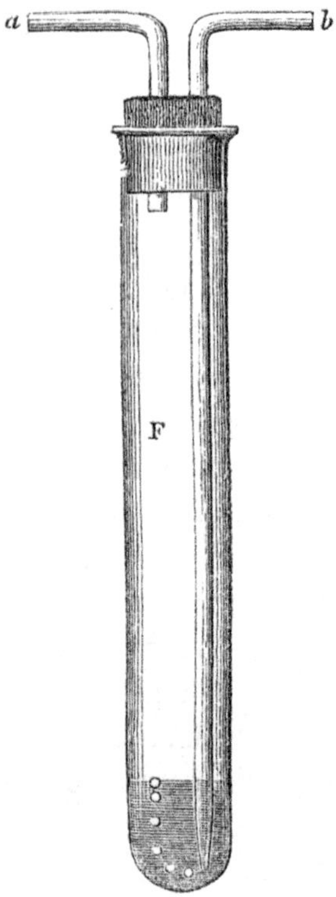

(742) Opening the cock cautiously, the air of the room passes, in the first place, through the cotton wool, which holds back the numberless organic germs and dust-particles floating in the atmosphere. The air, thus cleansed, passes into the U-shaped vessel, where it is dried by the sulphuric acid. It then descends through the narrow tube to the bottom of the little flask, and escapes there through a small orifice into the liquid. Through this it bubbles, loading itself to some extent with the nitrite of amyl vapour, and then the air and vapour enter the experimental tube together.

(743) We will now permit the electric beam to play upon this invisible vapour. The lens of the lamp is so situated as to render the beam convergent, the focus being formed near the middle of the tube. You will notice that the space remains dark for a moment after the turning on of the beam; but the chemical action will be so rapid that attention is requisite to mark this interval of darkness. I ignite the lamp; the tube for a moment seems empty; but a luminous white cloud immediately fills the beam. It

has, in fact, shaken asunder the molecules of the nitrite of amyl, and brought down upon itself a shower of particles which cause it to flash forth in your presence like a solid luminous spear. This experiment, moreover, illustrates the fact, that however intense a beam of light may be, it remains invisible unless it has something to shine upon. *Space*, though traversed by the rays from all suns and all stars, is itself unseen. Not even the ether which fills space, and whose motions are the light of the universe, is itself visible.

(744) You notice that the end of the experimental tube most distant from the lamp is free from cloud. Now the nitrite of amyl vapour is there also, but it is unaffected by the powerful beam passing through it. Let us concentrate the transmitted beam by receiving it on a concave silver mirror, and cause it to return into the tube. It is still powerless. Though a cone of light of extraordinary intensity now traverses the vapour, no precipitation occurs, no trace of cloud is formed. Why? Because the very small portion of the beam competent to decompose the vapour is quite exhausted by its work in the frontal portions of the tube. The great body of the light which remains, after this sifting out of the few effectual rays, has no power over the molecules of nitrite of amyl. We have here strikingly illustrated, what has been already stated regarding the influence of *period*, as contrasted with that of *strength*. For the portion of the beam which is here ineffectual has probably more than a million times the absolute energy of the effectual portion. It is energy specially related to the atoms that we here need, which specially related energy being possessed by the feeble waves, invests them with their extraordinary power. When the experimental tube is reversed so as to bring the undecomposed vapour under the action

of the *unsifted* beam, you have instantly this fine luminous cloud precipitated.

(745) The light of the sun also effects the decomposition of the nitrite of amyl vapour. In the track of a sunbeam I placed a large plano-convex lens, which formed a fine convergent cone in the dust of the room behind it. On thrusting one end of the tube into the light behind the lens, precipitation within the cone was copious and immediate. Here also the vapour at the distant end of the tube was shielded by that in front; but on reversing the tube, a second and similar cloud-cone was precipitated.

(746) And here I would ask you to make familiar to your minds the idea that no chemical action can be produced by a ray that does not involve the destruction of the ray. But abandoning the term ray as loose and indefinite, let us fix our thoughts upon the *waves* of light. We have to render clear to our minds that those waves which produce chemical action do so by delivering up their own motion to the molecules which they decompose. We have here forestalled to some extent a question of great importance in molecular physics, which, however, is worthy of being further dwelt upon ; it is this : When the waves of ether are intercepted by a compound vapour, is the motion of the waves transferred to the molecules of the vapour, or to the atoms of the molecules ? We have thus far leaned to the conclusion that the motion is communicated to the atoms; for if not to these individually, why should they be shaken asunder ? The question, however, is capable of, and is worthy of, another test, the bearing and significance of which you will immediately appreciate.

(747) As already explained, the molecules are held in their positions of equilibrium by their mutual repulsion on the one side, and by an external pressure on the other. Like a stretched string, their rate of vibration, if they vibrate at all, must depend upon the elastic force existing between

them. If this force be changed, the rate of vibration must change along with it; and after the change the molecules could no longer absorb the waves which they absorbed prior to the change. Now the elastic force between molecule and molecule is utterly altered when a vapour passes to the liquid state. Hence, if the liquid absorb waves of the same period as its vapour, it is a proof that the absorption is not effected by the molecules. Let us be perfectly clear on this important point. Waves are absorbed whose vibrations synchronise with those of the molecules or atoms on which they impinge—a principle which is sometimes expressed by saying that bodies radiate and absorb the same rays. This great law, as you know, is the foundation of spectrum-analysis; it enabled Kirchhoff to explain the lines of Frauenhofer, and to determine the chemical composition of the atmosphere of the sun. If then, after the passage of a vapour to the liquid state, the same waves be absorbed as were absorbed prior to the passage, it is a proof that the molecules, which must have utterly changed *their* periods, cannot be the seat of the absorption; and we are driven to conclude that it is to the *atoms*, whose rates of vibration are unchanged by the change of aggregation, that the wave-motion is transferred. If experiment should prove this identity of action on the part of a vapour and its liquid, it would establish in a new and striking manner the conclusion to which we have previously leaned.

(748) We will now resort to the experimental test. In front of the experimental tube, which contains a quantity of the nitrite of amyl vapour, is placed a glass cell a quarter of an inch in thickness, filled with the liquid nitrite of amyl. I send the electric beam first through the liquid and then through its vapour. The luminous power of this beam is very great, but it can make no impression upon the vapour. The liquid has robbed it completely of its effective waves. On the removal of the liquid, chemical action immediately

commences, and in a moment we have the apparently empty tube filled with this bright cloud, precipitated by one portion of the beam, and illuminated by another. I reintroduce the liquid: the chemical action instantly ceases. I again remove it, and the action commences once more. Thus we uncover in part the secrets of this world of molecules and atoms.

(749) Instead of employing air as the vehicle by which the vapour is carried into the experimental tube, we may employ oxygen, hydrogen, or nitrogen, and besides the nitrite of amyl a great number of other substances might be employed, which, like the nitrite, have been hitherto not known to be chemically susceptible to light. One point in addition I wish to illustrate, chiefly because the effect is similar in kind to one of great importance in nature. In our atmosphere you know floats carbonic acid gas, which furnishes food to the vegetable world. But this food could not be consumed by plants and vegetables without the intervention of the sun's rays. And yet, as far as we know, these rays are powerless upon the free carbonic acid of our atmosphere. The sun can only decompose the gas when it is drunk in by the leaves of plants. In the leaves it is in close proximity with substances ready to take advantage of the loosening of its molecules by the waves of light. Incipient disunion being thus introduced, the carbon of the gas is seized upon by the leaf and appropriated, while the oxygen is discharged into the atmosphere.

(750) The experimental tube now before you contains a different vapour from that which we have hitherto employed. It is called the nitrite of butyl.* On sending the electric beam through the tube the chemical action is scarcely sensible. I add to the vapour a quantity of air

* I have to thank Mr. Ernest Chapman for a portion of this precious substance.

which has been permitted to bubble through hydrochloric acid. When the beam is now turned on, so rapid is the action and so dense the cloud precipitated, that you could hardly by an effort of attention observe the dark interval which preceded the precipitation. This enormous augmentation of the action is due to the presence of the hydrochloric acid. Like the chlorophyl and carbonic acid in the leaves of plants, the two substances interact under the influence of the waves of the electric light.

(751) The nitrite of amyl furnishes a similar example. The decomposition of this substance by light is very energetic when alone, but the energy and brilliancy of the action are greatly augmented by the presence of hydrochloric acid. Air which has bubbled through the liquid nitrite has been admitted into this experimental tube till the mercury gauge of the pump has sunk eight inches. Eight additional inches of air which had bubbled through liquid hydrochloric acid were then admitted; and now we will permit the powerful beam of the electric lamp to act upon the mixture. A cloud of extraordinary density and brilliancy is immediately precipitated on the beam; which seems to pierce like a share the shining nebula, tossing in heaps the precipitated particles right and left as it advances among them.

(752) I connect the nozzle of a bellows by a bit of india-rubber tubing with a glass tube passing through a cork into a vessel containing the nitrite vapour, a sharp tap on the bellows sends a puff of the vapour through a second tube passing through the same cork. The puff of vapour in ordinary diffuse light is invisible; it in fact remains vapour. But when projected into a concentrated sun-beam, or into the beam from the electric lamp, on crossing the limit of light and darkness the vapour is instantly precipitated as cloud, and forms a shining white ring. This ring has the same mechanical cause as the smoke-rings

puffed from the mouth of a cannon; but it is latent until revealed by actinic precipitation.

(753) It is possible to impart to these clouds any required degree of tenuity, for it is in our power to limit at pleasure the amount of vapour in our experimental tube. When the quantity is duly limited, the precipitated particles are at first inconceivably small, defying the highest microscopic power to bring them within the range of vision. Probably their diameters are then not greater than the millionth of an inch. They grow gradually, and as they augment in size, throw from them a continually increasing quantity of wave-motion, until, finally, the cloud which they form becomes so luminous as to fill this theatre with light. During the growth of the particles the most splendid iridescences are often exhibited. Such I have sometimes seen with delight and wonder in the atmosphere of the Alps, but never anything so gorgeous as those which our laboratory experiments reveal. It is not, however, with the iridescences, however beautiful they may be, that we have now to occupy our thoughts, but with other effects which bear upon the two great standing enigmas of meteorology—the colour of the sky and the polarisation of its light.

(753*a*) First, then, with regard to the sky; how is it produced, and can we not reproduce it? Its colour has not the same origin as that of ordinary colouring matter, in which certain portions of the white solar light are extinguished, the colour of the matter being that of the portion of light which remains. A violet is blue because its molecular texture enables it to quench the yellow and red constituents of white light, and to allow the blue free transmission. A geranium is red because its molecular texture is such as quenches all rays except the red. Such colours are called colours of absorption; but the hue of the sky is not of this character. The blue light of the sky is *reflected* light, and were there nothing in our atmosphere competent to reflect the solar rays we should see no blue firmament, but should look into the darkness of infinite space. The reflection of the blue is effected by perfectly colourless particles. Smallness of size alone is requisite to ensure the selection and reflection of this colour. Of all the visual waves emitted by the sun, the shortest and smallest are those which correspond to the colour blue. On such waves small particles have more power than upon large ones, hence the predominance of blue colour in all light reflected from exceedingly small particles. The crimson glow of the evening and the morning seen so finely in the Alps is due, on the other hand, to *transmitted* light; that is to say, to light which in its passage through great atmospheric distances has its blue constituents sifted out of it by repeated reflection.

(754) It is possible, as stated, by duly regulating the quantity of vapour, to make our precipitated particles grow from an infinitesimal and altogether ultra-microscopic size to masses of sensible magnitude; and by means of these particles, in a certain stage of their growth, we can produce a blue which shall rival, if it does not transcend, that of the deepest and purest Italian sky. Let this point be in the first place established. Associated with our experimental tube is a barometer, the mercurial column of which now indicates that the tube is exhausted. Into the tube I introduce a quantity of the mixed air and nitrite of butyl vapour sufficient to depress the mercurial column one-twentieth of an inch; that is to say, the air and vapour together exert a pressure of one six-hundredth of an atmosphere. I now add a quantity of air and hydrochloric acid sufficient to depress the mercury half-an-inch further, and into this compound and highly attenuated atmosphere I discharge the beam of the electric light.

The effect is slow; but gradually within the tube arises this splendid azure, which strengthens for a time, reaches a maximum of depth and purity, and then, as the particles grow larger, passes into whitish blue. This experiment is representative, and it illustrates a general principle. Various other colourless substances of the most diverse properties, optical and chemical, might be employed for this experiment. The *incipient cloud* in every case would exhibit this superb blue; thus proving to demonstration that particles of infinitesimal size, without any colour of their own, and irrespective of those optical properties exhibited by the substances in a massive state, are competent to produce the colour of the sky.

(755) But there is another subject connected with our firmament, of a more subtle and recondite character than even its colour. I mean that 'mysterious and beautiful phenomenon,'* the polarisation of the light of the sky. Brewster, Arago, Babinet, Herschel, Wheatstone, Rubeson and others, have made us masters of the phenomenon, but its cause remains a mystery still. The polarity of a magnet consists in its *two-endedness*, both ends, or poles, acting in opposite ways. Polar forces, as most of you know, are those in which the duality of attraction and repulsion is manifested. And a kind of *two-sidedness*—noticed by Huygens, commented on by Newton, and discovered by a French philosopher, named Malus, in a beam of light which had been reflected from one of the windows of the Luxembourg Palace in Paris—receives the name of *polarisation.* We must now, however, attach a distinctness to the idea of a polarised beam, which its discoverers were not able to affix to it. For in their day men's thoughts were not sufficiently ripe, nor optical theory sufficiently advanced, to seize upon or express the physical meaning of polarisation. When a gun is fired,

* Herschel's 'Meteorology,' art. 233.

the explosion is propagated as a wave through the air. The shells of air, if I may use the term, surrounding the centre of concussion, are successively thrown into motion, each shell yielding up its motion to that in advance of it, and returning to its position of equilibrium. Thus, while the *wave* travels through long distances, each individual particle of air concerned in its transmission performs merely a small excursion to and fro.* In the case of sound, the vibration of the air-particles are executed *in* the direction in which the sound travels. They are therefore called *longitudinal* vibrations. In the case of light, on the contrary, the vibrations are *transversal*; that is say, the individual particles of ether move to and fro *across* the direction in which the light is propagated. In this respect waves of light resemble ordinary water-waves, more than waves of sound. In the case of an *ordinary* beam of light, the vibrations of the ether particles are executed *in every direction* perpendicular to it; but let the beam impinge obliquely, upon a plane glass surface, as in the case of Malus, the portion reflected will no longer have its particles vibrating in all directions round it. By the act of reflection, *if it occur at the proper angle*, the vibrations are all confined to a single plane, and light thus circumstanced is called *plane polarised light.*

(756) A beam of light passing through ordinary glass executes its vibrations within the substance exactly as it would do in air, or in ether-filled space. Not so when it passes through many transparent crystals. For these also have their two-sidedness, the arrangement of their particles being such as to tolerate vibrations only in certain definite directions. There is the well-known crystal tourmaline, which shows a marked hostility to all vibrations executed at right angles to the axis of the

* 'Lectures on Sound,' p. 3. (Longmans.)

crystal. It speedily extinguishes such vibrations, while those executed parallel to the axis are freely propagated. The consequence is, that a beam of light, after it has passed through any thickness of this crystal, emerges from it polarised. So also as regards the beautiful crystal known as Iceland spar, or as double-refracting spar. In one direction, but in one only, this crystal shows the neutrality of glass; in all other directions it splits the beam of light passing through it into two distinct halves, both of which are perfectly polarised, their vibrations being executed in two planes, at right angles to each other.

(757) It is possible by a suitable contrivance to get rid of one of the two polarised beams into which Iceland spar divides an ordinary beam of light. This was done so ingeniously and effectively by a man named Nicol, that the spar, cut in his fashion, is now universally known as Nicol's prism. Such a prism can polarise a beam of light, and if the beam, before it impinges on the prism be already polarised, in one position of the prism it is stopped, while in another position it is transmitted. The same is true of radiant heat. Our way is now, to some extent, cleared. Looking at various points of the blue firmament through a Nicol's prism, and turning the prism round its axis, we soon notice variations of the brightness of the sky. In certain positions of the spar, and from certain points of the firmament, the light appears to be freely transmitted; while, looking at the same points, it is only necessary to turn the prism round its axis through an angle of ninety degrees to materially diminish the intensity of the light. On close scrutiny it is found that the difference produced by the rotation of the prism is greatest when the sky is regarded in a direction *at right angles to that of the solar rays.*

(758) Experiments of this kind prove that the blue

light sent to us by the firmament is polarised, and that the direction of most perfect polarisation is perpendicular to the solar rays. Were the heavenly azure like the ordinary light of the sun, the turning of the prism would have no effect upon it; it would be transmitted equally during the entire rotation of the prism. The light of the sky is in great part quenched, because it is in great part polarised.

(759) When a luminous beam impinges at the proper angle on a plane glass surface it is polarised by reflection. It is polarised, in part, by all oblique reflections; but at one particular angle, the reflected light is perfectly polarised. An exceedingly beautiful and simple law, discovered by Sir David Brewster, enables us readily to find *the polarising angle* of any substance whose refractive index is known. This law was discovered experimentally by Brewster; but the Wave Theory of light renders a complete reason for the law. A geometrical image of it is thus given. When a beam of light impinges obliquely upon a plate of glass it is in part reflected and in part refracted. At one particular incidence the reflected and the refracted portions of the beam are at right angles to each other. The angle of incidence is *then* the polarising angle. It varies with the refractive index of the substance; being for water $52\frac{1}{2}$, for glass $57\frac{1}{2}$, and for diamond 68 degrees.

(760) And now we are prepared to comprehend the difficulties which have beset the question before us. It has been already stated that in order to obtain the most perfect polarisation of the firmamental light, the sky must be regarded in a direction at right angles to the solar beams. This is sometimes expressed by saying that the place of maximum polarisation is at an angular distance of 90° from the sun. This angle, enclosed as it is between the direct and reflected rays, comprises both the

angles of incidence and reflection. Hence the angle of incidence, which corresponds to the maximum polarisation of the sky, is half of 90°, or 45°. This is the atmospheric polarising angle, and the question is, what known substance possesses an index of refraction to correspond with this polarising angle? If we knew this substance, we might be tempted to conclude that particles of it, scattered in the atmosphere, produce the polarisation of the sky. 'Were the angle of maximum polarisation,' says Sir John Herschel, '76° (instead of 90°), we should look to *water*, or ice, as the reflecting body, however inconceivable the existence in a cloudless atmosphere, and a hot summer day, of unevaporated particles of water.' But a polarising angle of 45° corresponds to a refractive index of 1; this means that there is no refraction at all, in which case we ought to have no reflection. To satisfy the law of Brewster, as Sir John Herschel remarks, 'the reflection would have to be made *in* air *upon* air!' 'The more the subject is considered,' adds the celebrated philosopher last named, 'the more it will be found beset with difficulties, and its explanation, when arrived at, will probably be found to carry with it that of the blue colour of the sky itself.'

(761) If you doubt the wisdom, acknowledge, at all events, the faith in your capacity which has caused me to bring a subject so entangled before you. I believe, however, that even the intellect which draws its strength and its associations from a totally different source, may have its interest excited in subjects like the present, dark and difficult though they be. It is not to be expected that you will all grasp the details of this discussion; but I think that everybody present will see the extremely important part hitherto played by the law of Brewster in speculations as to the colour and polarisation of the sky. I shall now seek to demonstrate in your presence, *firstly*, and in

confirmation of our former experiments, that sky-blue may be produced by exceedingly minute particles of any kind of matter; *secondly*, that polarisation identical with that of the sky is produced by such particles; and *thirdly*, that matter in this fine state of division, where its particles are probably small in comparison with the height and span of a wave of light, releases itself completely from the law of Brewster; the direction of maximum polarisation being absolutely independent of the polarising angle as hitherto defined. Why this should be the case, the wave theory of light, to make itself complete, will have subsequently to explain.

(762) Into this experimental tube, in the manner already described, I introduce a new vapour, and add to it air, which has been permitted to bubble through dilute hydrochloric acid. And now I permit the electric beam to play upon the mixture. For some time nothing is seen. The chemical action is doubtless progressing, and condensation going on; but the condensing molecules have not yet coalesced to particles sufficiently large to scatter sensibly the waves of light. As before stated—and the statement rests upon an experimental basis—the particles here generated are at first so small that their diameters do not probably exceed a millionth of an inch; while to form each of these *particles* whole crowds of *molecules* are probably aggregated. Helped by such considerations our intellectual vision plunges more profoundly into atomic nature, and shows us, among other things, how far we are from the realisation of Newton's hope that the molecules might one day be seen by microscopes. While I am speaking, you observe this delicate blue colour forming and strengthening within the experimental tube. No sky-blue could exceed it in richness and purity; but the particles which produce this colour lie wholly beyond our microscopic range. A uniform colour is

here developed, which has as little breach of continuity—which yields as little evidence of the individual particles concerned in its production, as that yielded by a body whose colour is due to true molecular absorption. This blue is at first as deep and dark as the sky seen from the highest Alpine peaks, and for the same reason. But it grows gradually brighter, still maintaining its blueness, until at length a whitish tinge mingles with the pure azure; announcing that the particles are now no longer of that infinitesimal size which reflects the shortest waves alone.*

(763) The liquid here employed is the iodide of allyl,† but I might choose any one of a dozen substances here before me to produce the effect. You have seen what may be done with the nitrite of butyl. With nitrite of amyl, bisulphide of carbon, benzol, benzoic ether, &c. the same blue colour may be produced. In all cases, where matter slowly passes from the molecular to the massive state the transition is marked by the production of the blue. More than this:—you have seen me looking at the blue colour (I hardly like to call it a blue 'cloud,' its texture and properties are so different from ordinary clouds) through this bit of spar. This is a Nicol's prism, and I could wish one of them to be placed in the hands of each of you. Well, this blue that I have been looking at turns out to be a bit of more perfect sky than the sky itself. Looking across the illuminating beam as we look across the solar rays at the sky we obtain not only partial polarisation, but *perfect* polarisation. In one position of the Nicol the blue light seems to pass unimpeded to the eye; in the other it is absolutely cut off, the experimental tube being reduced to optical *emptiness*. Behind the experimental

* Possibly a photographic impression might be taken long before the blue becomes visible, for the ultra-blue rays are first reflected.

† For which I have to thank the obliging kindness of Dr. Maxwell Simpson.

tube it is well to place a black surface, in order to prevent foreign light from troubling the eye. In one position of the Nicol this black surface is seen without softening or qualification; for the particles within the tube are themselves invisible, and the light which they reflect is quenched. If the light of the sky were polarised with the same perfection, on looking properly towards it through a Nicol we should also meet, not the mild radiance of the firmament, but the unillumed blackness of space.

(764) The construction of the Nicol is such that it permits to pass through it vibrations which are executed in a certain determinate direction, and these only. All vibrations executed at right angles to this direction are completely stopped; while components only of those executed obliquely to it are transmitted. It is easy, therefore, to see that from the position in which the Nicol must be held to transmit or to quench the light of our incipient cloud, we can infer the direction of the vibrations of that light. You will be able to picture those vibrations without difficulty. Suppose a line drawn from any point of the 'cloud' perpendicular to the illuminating beam. Along that line, the particles of ether which carry the light from the cloud to the eye, vibrate in a direction perpendicular both to the line and to the beam. And if any number of lines be drawn in the same way from the cloud, like the spokes of a wheel, the particles of ether along all of them oscillate in the same manner. Wherefore, if a *plane surface* be imagined cutting the incipient cloud at right angles to its length, the perfectly polarised vibrations discharged laterally will all be parallel to this surface. This is the plane of vibration of the polarised light. Or you may suppose a circle drawn round the experimental tube, and a series of strings attached to various points of this circle. If all the cords be stretched as perpendiculars to the experi-

mental tube, and caused to vibrate by a series of jerks imparted at right angles both to them and the tube, the motion of the particles of the strings will then represent those of the particles of ether. A distinct image of those vibrations is now, I hope, in your minds.

(765) Our incipient blue cloud is a virtual Nicol's prism, and between it and the real Nicol, we can produce all the effects obtainable between the polariser and analyser of a polariscope. When, for example, a thin plate of selenite, which is crystallised sulphate of lime, is placed between the Nicol and the incipient cloud, we obtain the splendid chromatic phenomena of polarised light. The colour of the gypsum-plate, as many of you know, depends upon its thickness. If this be uniform, the colour is uniform. If, on the contrary, the plate be wedge-shaped, thickening gradually and uniformly from edge to back, we have brilliant bands of colour produced parallel to the edge of the wedge. Perhaps the best form of plate for experiments of this character is that now in my hand, which was prepared for me some years ago by a man of genius in his way, the late Mr. Darker of Lambeth. It consists of a plate of selenite thin at the centre, and gradually thickening towards the circumference. Placing this film between the Nicol and the cloud, we obtain, instead of a series of parallel bands, a system of coloured rings. The colours are most vivid when the incipient cloud is looked at perpendicularly. Precisely the same phenomena are observed when we look at the blue firmament in a direction perpendicular to the solar rays.

(766) We have thus far illuminated our incipient cloud with ordinary light, and found the portion of this light reflected laterally from the cloud in all directions round it to be perfectly polarised. We will now examine the effects produced when the light which illuminates the cloud is itself polarised. In front of the electric lamp,

and between it and the experimental tube, is placed this fine Nicol's prism, which is sufficiently large to embrace and to polarise the entire beam. The prism is now placed so that the plane of vibration of the light emergent from it, and falling upon the cloud, is vertical. How does the cloud behave towards this light? This formless aggregate of infinitesimal particles, without definite structure, shows the two-sidedness of the light in the most striking manner. It is absolutely incompetent to reflect upwards or downwards, while it freely discharges the light horizontally, right and left. I turn the polarising Nicol so as to render the plane of vibration horizontal; the cloud now freely reflects the light vertically upwards and downwards, but it is absolutely incompetent to shed a ray horizontally to the right or left.

(767) Suppose the atmosphere of our planet to be surrounded by an envelope impervious to light, with an aperture on the sunward side, through which a solar beam could enter. Surrounded on all sides by air not directly illuminated, the track of the sunlight would resemble that of the electric beam in a dark space filled with our incipient cloud. The course of the sunbeam would be *blue*, and it would discharge laterally, in all directions round it, light in precisely the same polarised condition as that discharged from the incipient cloud. In fact, the azure revealed by the sunbeam would be the azure of such a cloud. And if, instead of permitting the ordinary light of the sun to enter the aperture, a Nicol's prism were placed there, which should polarise the sunlight on its entrance into our atmosphere, the particles producing the colour of the sky would act precisely like those of our incipient cloud. In two directions we should have the solar light reflected; in two others unreflected. In fact, out of such a solitary beam, traversing the unilluminated air, we should be able to extract every effect shown by our incipient cloud. In

the production of such clouds we virtually carry bits of the sky into our laboratories, and obtain with them all the effects obtainable in the open firmament of heaven.

(768) And here, had not a sufficient strain been already imposed upon your minds, I might enter upon the description of a series of extraordinary effects observed when the particles of our incipient clouds are allowed to augment in size, so as to approach the condition of true cloudy matter. The selenite ring-system, already referred to, is a most delicate reagent for the detection of polarised light. When we look *normally*, or perpendicularly, at an incipient cloud, the colours of the rings are most vividly developed, a diminution of the colour being immediately apparent when the incipient cloud is regarded *obliquely*. But let us continue to look through the Nicol and selenite normally at the cloud: the particles augment in size, the cloud becomes coarser and whiter, the strength of the selenite colours becoming gradually feebler. At length the cloud ceases to discharge polarised light along the normal, and the selenite colours entirely disappear. If *now* the cloud be regarded *obliquely* the colours are restored, very vividly, if not with their first vividness and clearness. Thus the cloud that has ceased to discharge polarised light at right angles to the illuminating beam, pours out such light copiously in oblique directions. The direction of maximum polarisation changes with the texture of the cloud.

(769) But this is not all; and to understand, even partially, what remains, a word must be said regarding the appearance of the colours of our plate of selenite. If, as before stated, the plate be of uniform thickness, its hue in white polarised light is uniform. Suppose, then, that by arranging the Nicol the colour of the plate is raised to its maximum brilliancy, and suppose the colour produced to be *green*; on turning the Nicol round its axis the green becomes fainter. When the angle of rotation amounts to 45

degrees the colour disappears; we then pass what may be called a neutral point, where the selenite behaves, not as a crystal, but as a bit of amorphous glass. Continuing the rotation, a colour reappears, but it is no longer green, but *red*. This attains its maximum at a distance of 45 degrees from the neutral point, or, in other words, at a distance of 90 degrees from the position which showed the green at its maximum. At a further distance of 45 degrees from the position of maximum red, the colour disappears a second time. We have there a second neutral point, beyond which the green comes again into view, attaining its maximum brilliancy at the end of a rotation of 180 degrees. By the rotation of the Nicol, therefore, through an angle of 90 degrees, we produce a colour *complementary* to that with which we started.

(770) As may be inferred from this result, the selenite ring system changes its character when the Nicol is turned. It is possible to have the centre of the circle dark, the surrounding rings being vividly coloured. The turning of the Nicol through an angle of 90 degrees renders the centre bright, while every point occupied by a certain colour in the first instance is occupied by the *complement* of that colour in the second. But what am I aiming at in these long preliminary statements? I want to be able to say, with full assurance of being understood by everybody present, that a cloud may so alter its texture as to produce upon a light an effect equivalent to the rotation of the Nicol through 90 degrees. By curious internal actions, not here to be described, the cloud in our experimental tube sometimes divides itself into sections of different textures. Some sections are coarser than others, while it often happens that some are iridescent to the naked eye, and others not. Looking normally at such a cloud through the selenite and Nicol, it often happens that in passing from section to section the whole character of the ring-

system is changed. You start with a section producing a *dark* centre and a corresponding system of rings; you pass to another section through a neutral point and find in that section the centre *bright,* and at the same radial distances find each of the first rings displaced by one of the complementary colour. Sometimes as many as four such reversions occur in the cloud of an experimental tube a yard long. Now, the changes here indicated mean that in passing from section to section of the cloud the plane of vibration of the polarised light turns suddenly through an angle of 90 degrees; this change being entirely due to the different texture of the two parts of the cloud.

(771) You will now be able to understand, as far as it is capable of being understood, a very beautiful effect which, under favourable circumstances, might be observed in our atmosphere. This experimental tube contains an inch of the iodide of allyl vapour, the remaining 29 inches necessary to fill the tube being air, which has bubbled through aqueous hydrochloric acid. Besides, therefore, the vapour of iodide of allyl, we have those of water and of acid within the tube. The light has been acting on the mixture for some time, a beautiful blue colour being produced. As before stated, the 'incipient cloud' is wholly different in texture and optical properties from an ordinary cloud; but it is possible to precipitate the aqueous vapour within this tube so as to cause it to form a cloud similar to those of our atmosphere. This new and real cloud will be precipitated in the midst of the azure of the incipient cloud. An exhausted vessel of about one-third of the capacity of the experimental tube is now connected with the tube, the passage uniting both being closed by a stop-cock. On opening this cock the mixed air and vapour will rush from the experimental tube into the empty vessel; and, in consequence of the chilling due to rarefaction, the vapour in the experimental tube will fall together as a

true cloud. You are now prepared for the experiment. I first look at this azure, so as to obtain a vivid ring-system with a dark centre. Turning on the cock, the air is rarefied and the cloud precipitated. What is the result? Instantly the centre of the system of coloured rings becomes bright, and the whole series of colours corresponding to definite radial distances, complementary. While I continue to look at the cloud, it gradually melts away as an atmospheric cloud might do in the azure of heaven. And *there* is our azure also remaining behind. The coarser cloud seems drawn aside like a veil, the blue reappears, the first ring-system, with its dark centre and correspondingly coloured circles, being restored.

(772) The vision of an object always implies a differential action on the retina of the observer. The object is distinguished from surrounding space by its excess or defect of light in relation to that space. By altering the illumination, either of the object itself or of its environment, we alter the appearance of the object. Take the case of clouds floating in the atmosphere with patches of blue between them. Anything that changes the illumination of either alters the appearance of both, that appearance depending, as stated, upon differential action. Now the light of the sky, being polarised, may, as you know, be in great part quenched by a Nicol's prism, while the light of a cloud, being unpolarised, cannot be thus extinguished. Hence the possibility of very remarkable variations, not only in the aspect of the firmament, which is really changed, but also in the aspect of the clouds which have that firmament as a background. When a reddish cloud at sunset chances to float in the region of maximum polarisation, the quenching of the sky behind it causes it to flash with a brighter crimson. Last Easter eve the Dartmoor sky, which had just been cleansed by a snow storm, wore a very wild appearance. Round the horizon

it was of steely brilliancy, while reddish cumuli and cirri floated southwards. When the sky was quenched behind them these floating masses seemed like dull embers suddenly blown upon, brightening into fire. In the Alps we have the most magnificent examples of crimson clouds and snows, so that the effects just referred to may be here studied under the best possible conditions. On the 23rd of August, 1869, the evening Alpen-glow was very fine, though it did not reach its maximum depth and splendour. Towards sunset I walked up the slopes to obtain a better view of the Weisshorn. The side of the peak seen from the Bel Alp, being turned from the sun, was tinted *mauve*; but I wished to see one of the rose-coloured buttresses of the mountain. Such was visible from a point a few hundred feet above the hotel. The Matterhorn also, though for the most part in shade, had a crimson projection, while a deep ruddy red lingered along its western shoulder. Four distinct peaks and buttresses of the Dom, in addition to its dominant head—all covered with pure snow—were reddened by the light of sunset. The shoulder of the Alphubel was similarly coloured, while the great mass of the Fletschorn was all aglow, and so was the snowy spine of the Monte Leone.

(773) Looking at the Weisshorn through the Nicol, the glow of its protuberance was strong or weak according to the position of the prism. The summit also underwent a change. In one position of the prism it exhibited a pale white against a dark background; in the rectangular position, it was a dark mauve against a light background. The red of the Matterhorn changed in a similar manner; but the whole mountain also passed through striking changes of definition. The air at the time was highly opalescent—filled in fact with a silvery haze, in which the Matterhorn almost disappeared. This could be wholly quenched by the Nicol, and then the mountain sprang

forth with astonishing solidity and detachment from the surrounding air. The changes of the Dom were still more wonderful. A vast amount of light could be removed from the sky behind it, for it occupied the position of maximum polarisation. When the sky was quenched, the four minor peaks and buttresses and the summit of the Dom, together with the shoulder of the Alphubel, glowed as if set suddenly on fire. This was immediately dimmed by turning the Nicol through an angle of 90°. It was not the stoppage of the light of the sky alone which produced this startling effect; the air between the Bel Alp and the Dom was, as I have said, highly opalescent, and the quenching of this intermediate glare augmented remarkably the distinctness of the mountain.

(774) On the morning of the 24th of August similar effects were finely shown. At 10 A.M. all three mountains, the Dom, the Matterhorn, and the Weisshorn, were powerfully affected by the Nicol. But in this instance also the line drawn to the Dom being accurately perpendicular to the direction of the solar shadows, and consequently very nearly perpendicular to the solar beams, the effects on this mountain were most striking. The grey summit of the Matterhorn at the same time could scarcely be distinguished from the opalescent haze around it; but when the Nicol quenched the haze, the summit became instantly isolated, and stood out in bold definition. It is to be remembered that in the production of these effects the only things changed are the sky behind and the luminous haze in front of the mountains; that these are changed because the light emitted from the sky and from the haze is *plane polarised light,* and that the light from the snows and from the mountains being sensibly unpolarised, is not directly affected by the Nicol. It will also be understood that it is not the interposition of the haze *as an opaque body* that renders the mountains indis-

tinct, but that it is the *light* of the haze which dims and bewilders the eye, and thus weakens the definition of objects seen through it.

(775) These results have a direct bearing upon what artists call 'aërial perspective.' As we look from the summit of the Aletschhorn, or from a lower elevation, at the serried crowd of peaks, especially if the mountains be darkly coloured—covered with pines, for example—every peak and ridge is separated from the mountains behind it by a thin blue haze which renders the relations of the mountains as to distance unmistakable. When this haze is regarded through the Nicol perpendicular to the sun's rays, it is in many cases wholly quenched, because the light which it emits in this direction is wholly polarised. When this happens, aërial perspective is abolished, and mountains very differently distant appear to rise in the same vertical plane. Close to the Bel Alp, for instance, is the gorge of the Massa, a river produced by the ablation of the Aletsch glacier, and beyond the gorge is a high ridge darkened by pines. This ridge may be projected upon the dark slopes at the opposite side of the Rhone valley, and between both we have the blue haze referred to, throwing the distant mountains far away. But at certain hours of the day this haze may be quenched, and then the Massa ridge and the mountains beyond the Rhone seem almost equally distant from the eye. The one appears, as it were, a vertical continuation of the other. The haze varies with the temperature and humidity of the atmosphere. At certain times and places it is almost as blue as the sky itself; but to see its colour, the attention must be withdrawn from the mountains and from the trees which cover them. In point of fact, the haze is a piece of more or less perfect sky; it is produced in the same manner, and is subject to the same laws, as the firmament itself. We live *in* the sky, not *under* it.

(776) These points were further elucidated by the deportment of the selenite plate. On some of the sunny days of August the haze in the valley of the Rhone, as looked at from the Bel Alp, was very remarkable. Towards evening the sky above the mountains opposite to my place of observation yielded a series of the most splendidly-coloured iris-rings; but on lowering the selenite until it had the darkness of the pines at the opposite side of the Rhone valley, instead of the darkness of space as a background, the colours were not much diminished in brilliancy. I should estimate the distance across the valley, as the crow flies, to the opposite mountains, at nine miles; so that a body of air nine miles thick can, under favourable circumstances, produce chromatic effects of polarisation almost as vivid as those produced by the sky itself.

(777) Again: the light of a landscape, as of most other things, consists of two parts; the one part comes purely from superficial reflection, and this light is always of the same colour as that which falls upon the landscape; the other part comes to us from a certain depth within the objects which compose the landscape, and it is this portion of the total light which gives these objects their distinctive colours. The white light of the sun enters all substances to a certain depth, and is partially ejected by internal reflection; each distinct substance absorbing and reflecting the light in accordance with the laws of its own molecular constitution. Thus the solar light is *sifted* by the landscape, which appears in such colours and variations of colour as, after the sifting process, reach the observer's eye. Thus the bright green of grass, or the darker colour proper to the pine, never comes to us alone, but is always mingled with an amount of really foreign light derived from superficial reflection. A certain hard brilliancy is conferred upon the woods and meadows by this superficially-reflected light. Under certain circumstances, it

may be quenched by a Nicol's prism, and we then obtain the true colour of the grass and foliage. Trees and meadows thus regarded exhibit a richness and softness of tint which they never show as long as the superficial light is permitted to mingle with the true interior emission. The needles of the pines show this effect very well, large-leaved trees still better; while a glimmering field of maize exhibits the most extraordinary variations when looked at through the rotating Nicol.

(778) Thoughts and questions like those here referred to took me last August to the top of the Aletschhorn. The effects described in the foregoing paragraphs were for the most part reproduced in the summit of the mountain. I scanned the whole of the sky with my Nicol. Both alone and in conjunction with the selenite it pronounced the perpendicular to the solar beams to be the direction of maximum polarisation. But at no portion of the firmament was the polarisation complete. The artificial sky produced in the experiments already recorded could, in this respect, be rendered more perfect than the natural one; while the gorgeous 'residual blue' which makes its appearance when the polarisation of the artificial sky ceases to be perfect, was strongly contrasted with the lack-lustre hue which, in the case of the firmament, outlived the extinction of the brilliance. With certain substances, however, artificially treated, this dull residue may also be obtained.

(779) All along the arc from the Matterhorn to Mont Blanc the light of the sky immediately above the mountains was powerfully acted upon by the Nicol. In some cases the variations of intensity were astonishing. A little practice enables the observer to shift the Nicol from one position to another so rapidly as to render the alternate extinction and restoration of the light immediate. When this was done along the arc to which I have referred, the alternations of light and darkness resembled the play of

sheet lightning behind the mountains. There was an element of awe connected with the suddenness with which the mighty masses, ranged along the line referred to, changed their aspect and definition under the operation of the prism.

(780) I have endeavoured to show you that the colour and polarisation of the sky could be reproduced artificially, and that the only condition necessary to their production was the smallness of the particles by which the light was scattered. The effects were proved to be totally independent of the optical character of the substances from which the particles were derived. The parallelism of the artificial and the natural phenomena is so perfect as to leave no doubt upon the mind that they are due to a common cause. And here a practical issue of immense import reveals itself. Supposing those particles which now throw down upon us the blue light of the firmament to be abolished, what would be the result? The sun's rays would pass through the atmosphere without lateral scattering—the earth would lose the light of the sky. To form an idea of the magnitude of this loss we must have a clear idea of the *quality* of the light under consideration. It is now known to you that the vegetable world is nourished by the rays of the sun; and as animal life is sustained by vegetables, that life also is supported in the long run by the solar rays. Now, these rays are as composite as the coins of the realm. As regards their power to produce the chemical actions necessary to vegetable life, they differ from each other in value as widely as gold does from copper. It is the gold of the solar beams that is showered down upon us from the sky. Professor Roscoe has shown that the light of the sky, which is mainly produced by the shorter waves, has a chemical value at Kew Observatory greater than that of the unclouded sun at a height

of 42° above the horizon.* This would be the measure of the loss to the vegetable world at Kew if the sky were abolished. Roscoe's experiments were made with chemical substances sensitive to solar light, and they appear open to the objection that the rays effective in the plant-world may not be those which were effective upon his salts. But taking everything into account, and assuming the correctness of the observations, I think the probability great that the value of sky-light as a feeder of the vegetable world, and through it of the animal, cannot be much less than Roscoe makes it to be.

* Proceedings of the Royal Institution, vol. iv. p. 657. The whole article here referred to is exceedingly interesting.

REMARKS

ON THE

CONVERTIBILITY OF NATURAL FORCES.

I ADD here a few remarks, which may be useful to some of my readers. They are extracted from a little book called 'Faraday as a Discoverer,' and form a portion of the chapter entitled 'Unity and Convertibility of Natural Forces; Theory of the Electric Current:'—

'The whole stock of *energy* or *working-power* in the world consists of *attractions*, *repulsions*, and *motions*. If the attractions and repulsions are so circumstanced as to be able to produce motion, they are sources of working-power, but not otherwise. Let us for the sake of simplicity confine our attention to the case of attraction. The attraction exerted between the earth and a body at a distance from the earth's surface is a source of working power; because the body can be moved by the attraction, and in falling to the earth can perform work. When it rests upon the earth's surface it is *not* a source of power or energy, because it can fall no further. But though it has ceased to be a source of *energy*, the attraction of gravity still acts as a *force*, which holds the earth and weight together.

'The same remarks apply to attracting atoms and molecules. As long as distance separates them, they can move across it in obedience to the attraction, and the motion thus produced may, by proper appliances, be caused to perform mechanical work. When, for example, two atoms of hydrogen unite with one of oxygen, to form water, the atoms are first drawn towards each other—they move, they clash, and then by virtue of their resiliency, they recoil and *quiver*. To this quivering motion we give the name of heat. Now this quivering motion is merely the redistribution of the motion produced by the chemical affinity; and this is the only sense in which chemical affinity can be said to be converted into heat. We must not imagine the chemical *attraction* destroyed, or converted into anything else. For the atoms when mutually clasped to form a molecule of water, are held together by the very attraction which first drew them towards each other. That which has really been expended is the *pull* exerted through the space by which the distance between the atoms has been diminished.

'If this be understood it will be at once seen that *gravity* may in this sense be said to be convertible into heat; that it is in reality no more an outstanding and inconvertible agent, as it is sometimes stated to be, than chemical affinity. By the exertion of a certain pull through a certain space a body is caused to clash with a certain definite velocity against the earth. Heat is thereby developed, and this is the only sense in which gravity can be said to be converted into heat. In no case is the *force* which produces the motion annihilated or changed into anything else. The mutual *attraction* of the earth and weight exists when they are in contact as when they were separate; but the ability of that attraction to employ itself in the production of motion does *not* exist.

'The transformation, in this case, is easily followed by the mind's eye. First, the weight as a whole is set in motion by the attraction of gravity. This motion of the mass is arrested by collision with the earth, being broken up into molecular tremors, to which we give the name of heat.

'And when we reverse the process, and employ those tremors of heat to raise a weight, as is done through the intermediation of an elastic fluid in the steam-engine, a certain definite portion of the molecular motion is destroyed in raising the weight. In this sense, and this sense only, can the heat be said to be converted into gravity, or more correctly, into potential energy of gravity. It is not that the destruction of the heat has created any *new* attraction, but simply that the old attraction has now a power conferred upon it, of exerting a certain definite pull in the interval between the starting-point of the falling weight and its collision with the earth.

'So also as regards magnetic attraction: when a sphere of iron placed at some distance from a magnet rushes towards the magnet, and has its motion stopped by collision, an effect mechanically the same as that produced by the attraction of gravity occurs.

The magnetic attraction generates the motion of the mass, and the stoppage of that motion produces heat. In this sense, and in this sense only, is there a transformation of magnetism, or more correctly, magnetic work into heat. And if by the mechanical action of heat brought to bear by means of a suitable machine, the sphere be torn from the magnet and again placed at a distance, a power of exerting a pull through that distance, and producing a new motion of the sphere, is thereby conferred upon the magnet; in this sense, and in this sense only, is the heat converted into magnetism, or rather magnetic potential energy.

'When, therefore, writers on the conservation of energy speak of tensions being "consumed" and "generated," they do not mean thereby that old attractions have been annihilated and new ones brought into existence, but that, in the one case, the power of the attraction to produce motion has been diminished by the shortening of the distance between the attracting bodies, and that in the other case the power of producing motion has been augmented by the increase of the distance. These remarks apply to all bodies, whether they be sensible masses or molecules.

'Of the inner quality that enables matter to attract matter we know nothing; and the law of conservation makes no statement regarding that quality. It takes the facts of attraction as they stand, and affirms only the constancy of *working-power*. That power may exist in the form of MOTION; or it may exist in the form of FORCE, *with distance to act through*. The former is dynamic energy, the latter is potential energy, the constancy of the sum of both being affirmed by the law of conservation. The *convertibility* of natural forces consists solely in transformations of dynamic into potential, and of potential into dynamic energy, which are incessantly going on. In no other sense has the convertibility of force, at present, any scientific meaning.

'By the contraction of a muscle a man lifts a weight from the earth. But the muscle can contract only through the oxidation of its own tissue or of the blood passing through it. Molecular motion is thus converted into mechanical motion. Supposing the muscle to contract without raising the weight, oxidation would also occur, but the whole of the heat produced by this oxidation would be liberated *in the muscle itself*. Not so when it performs external work; to do that work a certain definite portion of the heat of oxidation must be expended. It is so expended in pulling the weight away from the earth. If the weight be permitted to fall, the heat generated by its collision with the earth would exactly make up for that lacking in the muscle during the lifting of the weight. In the case here supposed, we have a conversion of molecular muscular action into potential energy of gravity; and a conversion of that potential energy into heat; the heat, however, appearing at a distance from its real origin in the muscle. The whole process consists of a transference of molecular motion from the muscle to the weight, and gravitating force is the mere go-between, by means of which the transference is effected.'

ON COMETARY THEORY.

As an exercise for the reader, and as a possible application of some of the results recorded in this work, I here publish a hypothesis as to the constitution of comets which has found some favour among eminent men. I purpose to develop the hypothesis at a future day; did I not believe it to contain some germ of the truth I should not publish it here.

The speculation was communicated to the Cambridge Philosophical Society on the 8th of March, 1869, in the following words:—

'In the course of my experiments on actinic action I have been often astonished at the body of light which a perfectly infinitesimal amount of matter, when diffused in the form of a cloud, can discharge from it by reflection. I have been repeatedly perplexed and led into error by the action of residues so minute as to be simply inconceivable. In order to get rid of these residues, my experimental tubes, after having been employed for any vapour, are flooded with alcohol, sponged out with soap and hot water, and finally flooded with pure water. Let me give you some idea of the quantities of matter that here come into play. The tube before you, which is 3 feet long and 3 inches wide, was so thoroughly cleansed that when filled with air, or with the vapour of aqueous hydrochloric acid, no amount of exposure to an intense light produced the least cloudiness. Having thus assured myself of the perfect purity of the tube, I took a small bit of bibulous paper, rolled it up into a pellet not the fourth part of the size of a small pea, and moistened it with a liquid possessing a higher boiling-point than that of water. I held the pellet in my fingers until it had become almost dry, then introduced it into a connecting-piece and allowed dry air to pass over it into this tube. The air charged with the modicum of vapour thus taken up was subjected to the action of light. A blue actinic cloud began to form immediately, and in five minutes the blue colour had extended quite through the experimental tube. For some minutes this cloud continued blue, and could be completely quenched by Nicol's prism, no trace of its light reaching the eye when the Nicol was in its proper position. But its particles augmented gradually in magnitude, and at the end of fifteen minutes a dense white cloud filled the tube. Considering the amount of vapour carried in by the air, the appearance of a cloud so massive and luminous seemed like the creation of a world out of nothing.

'But this is not all; the pellet of bibulous paper was removed, and the experimental tube was cleared out by sweeping a current of dry air through it. *This current passed also through the connecting-piece in which the pellet of bibulous paper had rested.* The air was at length cut off and the experimental tube exhausted. Fifteen inches of hydrochloric acid were then sent into the tube through the same connecting-piece. Now it is here to be noted (1) that the total quantity of liquid absorbed by the pellet in the first instance was exceedingly small, (2) that nearly the whole of this small quantity had been allowed to evaporate between my fingers before the pellet was placed in the connecting-piece, (3) that the pellet had been ejected and the tube in which it rested rendered for some minutes the conduit of a strong current of pure air. It was part of such a residue as could linger in the connecting-piece after this process, that was carried into the experimental tube by the hydrochloric acid and subjected there to the action of light.

'One minute after the ignition of the electric lamp a faint cloud showed itself; in two minutes it had filled all the anterior portion of the tube and stretched a considerable way down it; it developed itself afterwards into a very beautiful cloud-figure; and at the end of fifteen minutes the body of light discharged by the cloud, considering the amount of matter involved in its production, was simply astounding. But though thus luminous, the cloud was far too fine to dim in any appreciable degree objects placed behind it. The flame of a candle seemed no more affected by it than it would be by a vacuum. Placing a page of print so that it might be illuminated by the cloud itself, it could be read *through* the cloud without any sensible enfeeblement. Nothing could more perfectly illustrate that "spiritual texture" which Sir John Herschel ascribes to a comet than these actinic clouds. Indeed the experiments prove that matter of almost infinite tenuity is competent to shed forth light far more intense than that of the tails of comets. The weight of the matter which sent this body of light to the eye would probably have to be multiplied by millions to bring it up to the weight of the air in which it hung.

'And now will you bear with me for five minutes while I endeavour to apply these results to cometary theory? I am encouraged to do so by a remark of Bessel's, who said that had any theory preceded his observations on Halley's comet, by fixing his attention either upon its verification or its confutation, it would have enabled him to return from his observations with a greater store of knowledge than he had actually derived from them. If time permitted, I should like to lead you by an easy gradient up to the view that I wish to submit to you; but time does not permit of this, and therefore the speculation must suffer from the baldness arising from the absence of such preparation.

'You are doubtless aware of the tremendous difficulties which beset cometry theory. The comet examined by Newton in 1680 shot out a tail sixty millions of miles in length in two days. The comet of 1843, if I remember aright, shot out in a single day a tail which covered 100 degrees of the heavens. This enormous reach of cloudy matter is supposed to be generated in the head of the comet and driven backwards by some mysterious force of repulsion exerted by the sun. Bessel devised a kind of magnetic polarity and repulsion to account for it. "It is clear," says Sir John Herschel, "that *if we have to deal here with matter such as we conceive it, viz. possessing inertia, at all,* it must be under the dominion of forces incomparably more energetic than gravitation, and quite of a different nature." And in another place he states the difficulties of the subject in the following remarkable words:—

'"There is beyond question some profound secret and mystery of nature concerned in the phenomenon of their tails. Perhaps it is not too much to hope that future observation, borrowing every aid from rational speculation, grounded on the progress of physical science generally (especially those branches of it which relate to the æthereal or imponderable elements), may ere long enable us to penetrate this mystery, and to declare whether it is really *matter*, in the ordinary acceptation of the term, which is projected from their heads with such extravagant velocity, and if not impelled, at least *directed* in its course by a reference to the sun as its point of avoidance. In no respect is this question as to the materiality of the tail more forcibly pressed on us for consideration than in that of the enormous sweep which it makes round the sun *in perihelio*, in the manner of a straight and rigid rod, in defiance of the law of gravitation, nay, even of the received laws of motion, extending (as we have seen in the comets of 1680 and 1843) from near the sun's surface to the earth's orbit, yet whirled round unbroken—in the latter case through an angle of 180° in little more than two hours. It seems utterly incredible that in such a case it is one and the same material object which is thus brandished. [I would especially invite the reader's attention to these words in reference to the following theory.—J. T.] If there could be conceived such a thing as a *negative shadow*, a momentary impression made upon the luminiferous æther behind the comet, this would represent in some degree the conception such a phenomenon irresistibly calls up."

'I now ask for permission to lay before you a speculation which seems to do away with all these difficulties, and which, whether it represents a physical verity or not, ties together the phenomena exhibited by comets in a remarkably satisfactory way.

'1. The theory is, that a comet is composed of vapour decomposable by the solar light, the visible head and tail being an actinic clould resulting from such decomposition; the texture of actinic clouds is demonstrably that of a comet.

'2. The tail, according to this theory, is not projected matter, but matter precipitated on the solar beams traversing the cometary atmosphere. It can be proved by experiment that this precipitation may occur either with comparative slowness along the beam, or that it may be practically momentary throughout the entire length of the beam. The amazing rapidity of the development of the tail would be thus accounted for without invoking the incredible motion of translation hitherto assumed.

'3. As the comet wheels round its perihelion, the tail is not composed throughout of the same matter, but of new matter precipitated on the solar beams, which cross the cometary atmosphere in new directions. The enormous whirling of the tail is thus accounted for without invoking a motion of translation.

'4. The tail is always turned from the sun for this reason:—Two antagonistic powers are brought to bear upon the cometary vapour,—the one an *actinic* power tending to produce precipitation, the other a *calorific* power tending to effect vaporisation. Where the former prevails, we have the cometary cloud; where the latter prevails, we have the transparent cometary vapour. As a matter of fact, the sun emits the two agents here invoked. There is nothing whatever hypothetical in the assumption of their existence. That precipitation should occur behind the head of the comet, or in the space occupied by the head's shadow, it is only necessary to assume that the sun's calorific rays are absorbed more copiously by the head and nucleus than the actinic rays. This augments the relative superiority of the actinic rays behind the head and nucleus, and enables them to bring down the cloud which constitutes the comet's tail.

'5. The old tail, as it ceases to be screened by the nucleus, is dissipated by the solar heat; but its dissipation is not instantaneous. The tail leans towards that portion of space last quitted by the comet, a general fact of observation being thus accounted for.

'6. In the struggle for mastery of the two classes of rays a temporary advantage, owing to variations of density or some other cause, may be gained by the actinic rays even in parts of the cometary atmosphere which are unscreened by the nucleus. Occasional lateral streamers, and the apparent emission of feeble tails towards the sun would be thus accounted for.

'7. The shrinking of the head in the vicinity of the sun is caused by the beating against it of the calorific waves, which dissipate its attenuated fringe and cause its apparent contraction.

'Throughout this theory I have dealt exclusively with true causes, and no agency has been invoked which does not rest on the sure basis either of observation or experiment. It remains with you to say whether in venturing to enunciate it I have transgressed the limits of "rational speculation."

'If I have done so, surely I could not have come to a place more certain to ensure my speedy correction. If the theory be a mere figment of the mind, your Adams and your Stokes (both happily here present), to whom I submit the speculation with the view of having it instantly annihilated by astronomy and physics, if it merit no better fate, will, I doubt not, effectually discharge that duty, and thus save both you and me from error before it has had time to lay any serious hold on our imagination.'

I may now add (1870) that cometary envelopes and various other appearances may be accurately reproduced through the agency of cyclonic movements introduced by heat among actinic nebulæ. It is needless to say that this hypothesis also accounts for the polarisation of the light of the comet's tail.*

* There may be comets whose vapour is undecomposable by the sun, or which, if decomposed, is not precipitated. This view opens out the possibility of invisible comets wandering through space, perhaps sweeping over the earth and affecting its sanitary condition without our being otherwise conscious of their passage. As regards tenuity, I entertain a strong persuasion that out of a few ounces (the possible weight assigned by Sir John Herschel to certain comets) of iodide-of-allyl vapour, an actinic cloud of the magnitude and luminousness of Donati's comet might be manufactured.

POLARISATION OF HEAT.

In the Philosophical Magazine for 1845 the late Principal Forbes gave an account of the experiments by which he demonstrated the polarisation of non-luminous heat. He first operated with tourmalines, and afterwards, by a happy inspiration, devised piles of mica plates, which from their greater power of transmission enabled him more readily and conclusively to establish the fact of polarisation. The subject was subsequently followed up by Melloni and other philosophers. With great sagacity Melloni turned to account his own discovery, that the obscure rays of luminous sources were in part transmitted by black glass. Intercepting by a plate of this glass the light emitted by his oil lamp and operating upon the transmitted heat, he obtained effects exceeding in magnitude any that could be obtained by means of the radiation from obscure sources. The possession of a more perfect ray-filter and a more powerful source of heat enables us now to obtain, on a greatly augumented scale, the effects obtained by Forbes and Melloni.

Two large Nicol's prisms, such as those employed in my experiments on the polarisation of light by nebulous matter, were placed in front of an electric lamp, and so supported that either of them could be turned round its horizontal axis. The beam from the lamp, rendered slightly convergent by the camera-lens, was sent through both prisms. But between them was placed a cell containing iodine dissolved in bisulphide of carbon in quantity sufficient to quench the strongest solar light. Behind the prisms was placed a thermo-electric pile, furnished with two conical reflectors. The front face of the pile received heat from the electric lamp, the hinder face from a spiral of platinum wire, through which passed a suitably regulated electric current.

The apparatus was so arranged that when the principal sections of the Nicols were at right angles to each other, the needle of the galvanometer connected with the pile showed a deflection of 90° in favour of the posterior source of heat. One of the prisms was then turned so as to render the principal sections parallel. The needle immediately descended to zero, and passed on to 90° at the other side of it. Reversing, or continuing the motion, so as to render the principal sections again perpendicular to each other, the calorific sheaf was intercepted, the needle descended to zero and went up to its first position.

So copious indeed is the flow of polarised heat that a prompt rotation of the Nicol would cause the needle to spin several times round over its graduated dial.

These experiments were made with the delicate galvanometer employed in my researches upon radiant heat. But the action is strong enough to cause a coarse lecture-room galvanometer, with needles 6 inches long and paper indexes a square inch each in area, to move through an arc of nearly 180°.

Reflection, refraction, dispersion, polarisation, plane and circular, double refraction, the formation of invisible images, both by mirrors and lenses, may all be strikingly illustrated by the employment of the iodine filter and the electric light.

Take, for example, the following experiments:—The Nicols being crossed, the needle of the galvanometer pointed to 78° in favour of the heated platinum spiral behind the pile. A plate of mica was then placed across the dark beam with its principal section inclined at an angle of 45° to those of the Nicols. The needle instantly fell to zero, and went up to 90° on the other side.

And for circular polarisation:—The Nicols being crossed and the needle pointing to 80° in favour of the platinum spiral, a plate of rock-crystal cut perpendicular to the axis was placed across the dark beam. The needle fell to zero, and went to 90° on the other side.

The penetrative power of the heat here employed may be inferred from the fact that it traversed about 12 inches of Iceland spar, and about 1½ inch of the cell containing the solution of iodine.

CONCLUDING ADDITIONS.

My friend Mr. Ingleby has directed my attention to three articles published in the 'Monthly Magazine' for 1820, vol. ii. pp. 33, 129, and 505, by a writer who signs himself Common Sense. The first article is headed 'Electricity and Galvanism explained on the Mechanical Theory of Matter and Motion.' The second is 'On the Nature of Motion and the Laws and Phenomena of its Propagation.' The third is entitled 'New Views of the Economy of Animal Nature in accordance with the Theory of Matter and Motion.' These titles indicate the character of the writer's thoughts. With a good deal of unavoidable error, these articles display in many cases a power of penetration, and a truth of insight, altogether remarkable for the time. Take the following quotations as examples:—

'But in a certain variety of cases transfer of motion does not produce change of place; and this exception gives rise to a new set of phenomena. Thus, if two bodies moving in contrary directions impinge against one another in a line which joins the centres of their masses, the disposition to change the place in both is destroyed, and apparently their motion. The motion, however, in such case is not destroyed; but only changes its appearance, and is imparted to the atoms of the body, which by the collision are thrown into active vibrations, representing the previous motions of the bodies. *Aggregate motion* is thus converted into *atomic motions*, and these give rise to many complicated and curious phenomena, as in heat, light, and gas.'

The italics are here the author's own. Until Mayer and Joule appeared, more than twenty years subsequently, nothing comparable as regards precision and completeness to the foregoing statement, to my knowledge, found utterance. Indeed, some of the phrases employed by myself might fairly be regarded as having been copied from this anonymous correspondent of the 'Monthly Magazine.'

The second one of the articles above referred to is thus summed up:—

'1. That all force, all weight, and all power of bodies are derived from the motion, or motions, imparted to them or possessed by them. And that force, weight, and motion are convertible terms and physical synonyms.

'2. That every force, weight, and motion is generated locally by its own set of proximate causes or motions.

'3. That although motion constantly changes its subject and its mode of exhibition, yet no motion is either lost or created, but is in constant circulation and varied appropriation.

'4. That motions of aggregates are convertible into motions or vibrations of atoms, and *vice versâ*; the mutual conversion producing many classes of phenomena.

'5. That action and reaction, inertia, resistance and friction, are so many phenomena of parting with motion, and of receiving, and dividing it with a moving body.

'6. That the medium in which a body in atomic motion is situated, conveys away the atomic motion, till the excitement exceeds its powers of transmission; when heat, evaporation, gaseous production, light, and decomposition take place as varieties and accelerated degrees of atomic motion.

'7. That atomic motion is heat—and being parted with from the air in the act of respiration, creates animal heat and vital action.

'8. That all local motions on the earth are derived from the deflection of the earth's motions [he missed the part played by the sun's rays], and are finally returned to the earth.

'9. Motion in all these enquiries and determinations is to be considered as the *secondary cause* of the sublime agency of Eternal Omnipotence.'

In his article on the Economy of Animal Nature, he says:—

'Animals consist therefore of a basis of bones for strength—of a continuity of muscles for motion—of a medullary system of brain and nerves for sensation, comparison and retention—of respiratory organs for appropriating gaseous atomic motion—and of arteries and veins for circulating nutriment and excitation to the whole.

'A steamboat deriving its internal energies from an engine wrought by the alternate introduction and fixation of aqueous gas, and put into motion by the reaction of wheels against water or land, is exactly coarsely analogous in all its operations to a locomotive animal, which derives its internal energies from the fixation of atmospheric gas, and its locomotion from the reactions of the feet or body against the earth.' Thus do great questions simmer before they receive complete expression.

PRODUCTION OF FIRE BY SAVAGES.

Extract from 'Adventures among the Dyaks of Borneo,' by F. BOYLE. Published by HURST & BLACKETT, 1865, pp. 67, 68.

'Among some of the Dyak tribes there is a manner of striking fire much more extraordinary. The instrument used is a slender cube of lead which fits tightly in a case of bamboo. The top of the cube is hollowed into a cup, and when fire is required, this cup is filled with tinder, the leaden piston is held upright in the left hand, the bamboo case is thrust sharply down over it, as quickly withdrawn, and the tinder is found to be alight. The natives say that no metal but lead will produce the effect. I must observe that we never saw this singular method in use, though the officers of the Rajah seemed acquainted with it.'

MORNING CHILL PRODUCING SNOW IN A ROOM.

'A curious phenomenon might also be observed, at Erzeroom, upon the door of one of the subterranean stables being opened, when, although the day was clear and fine without, the warm air within immediately congealed with a little fall of snow; this might be seen in great perfection every morning on the first opening of the outer door, when the house was warm from its having been shut up all night.'

'The preceding sentence is contained in a work by the Honourable R. Curzon, entitled "Armenia: a Year at Erzeroom, and on the Frontiers of Russia, Turkey, and Persia," and is quoted in the "Athenæum," 8th April, 1854, p. 431, and 1st column, from which it is here transcribed.' [The writer of these lines had heard me give Dove's account of a fall of snow in a Russian ball-room when one of the windows was broken. Hence his letter.—J. T.]

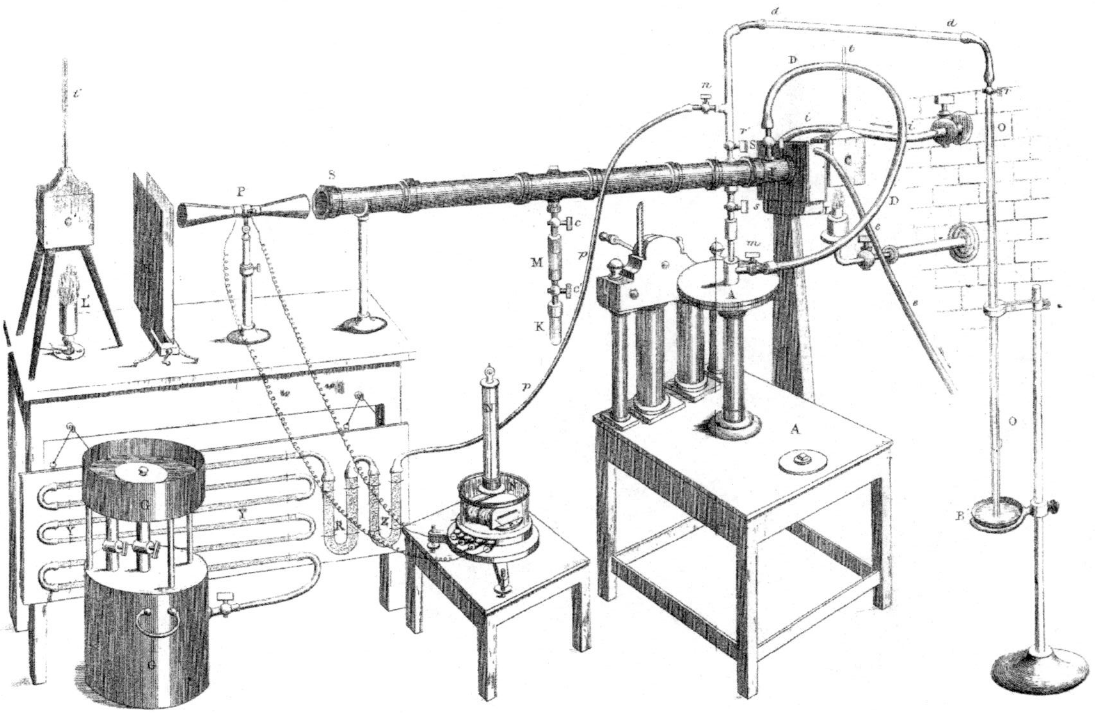

J. Basire sc.

INDEX.

The Numbers (except where p. is prefixed) refer to the Paragraphs, and not to the Pages.

ABSOLUTE zero of temperature, 96
Absorber, qualities necessary to form a good, 423
Absorption of heat by coats of whiting and tin, 344
— — — elective power possessed by bodies, 348
— — — takes place within a body, 355
— — — by different thicknesses of glass, 356; of selenite, 358
— — — by *solids*, Melloni's table, 351
— — — — *liquids*, ditto, 353
— — — — — experimental arrangement, 500
— — — — — at different thicknesses, table, 506
— — — — —vapours of those liquids, 509
— — — — *gases*, mode of experiment, 372, *et seq.*
— — — — — tables, 414, 418, 421
— — — — olefiant gas, 393, 402
— — — — proportional to density of gas in small quantities, 403
— — — the transference of motion, not annihilation, 412
— — — — *vapours*, tables, 433, 437, 439, 459, 522, 424
— — — — aqueous vapour, 470, *et seq.*, 515
— — — a molecular act, 511
— — — the physical cause of, 517
— — — from flames, by vapours, table, 526, 528
— and radiation of heat, reciprocity of, 343, 354, 415, 459
Absorption and radiation of heat by gases and vapours determined without external heat, 449
— — — — —, *dynamic*, table of gases, 455
— of the dark solar rays in relation to colour, 632
— — Franklin's experiments on this subject, 634
Acoustic experiments, Appendix to Chap. VIII., p. 254
Actinic decomposition, clouds formed by, 751
Actual energy defined. 154
Aërial perspective, 775
Aërolites, velocity of, 12
Æthrioscope, 493
Aggregation, change of state of, in bodies by heat, 176
Air, compressed, chilled by expansion, 15
— effect of stoppage of motion of, 16, 26
— compression of, containing bisulphide of carbon, 25
— expanded by heat, 67
— expansion of, under constant *pressure*, 69
— expansion of, under constant *volume*, 73
— heated, ascends, illustrations of 68, 205
— cooling effect of, 292
— passage of sound through, 298
— thermometer, uninfluenced by heat that has passed through air and glass, 362
— not warmed by passage of heat through, 364
— dry, transmission of heat by, 392
— — feeble dynamic radiation of,

The Numbers (except where p. is prefixed) refer to the Paragraphs, and not to the Pages.

Air, dry, powerful ditto, when varnished by vapours, 458
— difficulties in obtaining perfectly pure, 390, 471
— saturated with moisture, calorific absorption by, 474, 485
— humid, table of absorption by at different pressures, 482
— cause of slow nocturnal cooling of, 647
— distinction between clear and dry, 492
— from the lungs, its calorific absorption, 538
— — — — amount of carbonic acid in, determined, 540
Alcohol, expansion of, by heat, shown, 97
— evaporation of, produces cold, 186
Alps, formation and motion of glaciers on, 221
— the evening Alpen-glow, 772
Alum, powerful absorption and radiation of, 351, 354
— number of luminous and obscure rays transmitted by, 370
America, extreme cold of E. coast of, 223
Ammonia, powerful absorption of heat by, 420
Amplitude of an ether particle, 338*a*
Amyl, nitrate of, action of short waves of ether upon, 742, 751
Ancient glaciers, evidences of, 236, *et seq.*
Animal substances, table of conductive power of, 275, 277
Angular velocity of reflected ray explained, 323
Aqueous vapour, precipitated by rarefaction of air, 28
— — cause of precipitation of in England, 215
— — use of, in our climate, 215, 492, 544
— — precipitation of less, east of Ireland, 217
— — definition of, 467
— — amount of, in atmosphere, 469
— — action of, on radiant heat, 470, *et seq.*, 515, and Appendix to Chap. XI. p. 370
— — absorption of, in air obtained from various places, 480*a*
Aqueous vapour, objections to experiments on, answered, 472, 474, 484
— — cause of copious precipitation of, in tropics, 488
— — effect of removal of from English atmosphere, 492
— — absorbs same class of rays as water, 496, 544
Asbestos, cause of bad conduction of heat by, 285
Asia, cause of coldness of central parts of, 492
Asteroids, amount of heat developed by collision of, with sun, 692
Atmosphere, effect of its pressure on boiling point, 137, *et seq.*
— diminution of its pressure lowers boiling point, 138
— amount of aqueous vapour in, 469
— action of aqueous vapour in on radiant heat, 470, 480
— use of aqueous vapour in, 492
— absorption of solar heat by, 364
— the earth's radiation through, Appendix to Chap. XI. p. 370
its influence on temperature of the planets, 546
Atoms, collision of carbon and oxygen, 48
— when separated, heat consumed, 159
— enormous attractions of, 99, 158
— their relative weights, 161
— possess the same amount of heat, 161
— absorb and emit same rays, 672
— atoms entering into the formation of molecules, 730
— forces in operation between every two atoms, 732
Atomic oscillations of a body increased by heat, 58, 159
— motion, how propagated, 61
— forces, power of, 99, 158
— constitution, influence of, on absorption of heat, 422

BACON, extract from 2nd Book of Novum Organum, Appendix to Chap. II. p. 53
— his experiment on the compression of water, 158
Bark of trees, bad conductive power of, 272

The Numbers (except where p. is prefixed) refer to the Paragraphs, and not to the Pages.

Beeswax, contraction of, in cooling, 120
Bell struck by hammer, motion not lost, 23
Bismuth, expansion of, in cooling, 102
Bisulphide of carbon, vapour of, ignited by compression, 25
— — — transparency of, to heat, 348, 353, 433, 507
Blagden and Chantrey, their exposure of themselves in heated ovens, 256
Blood, heat of, why so constant in all climates, 256
Body, cause of its resisting high temperatures, 256
Boiler explosions, 134, 201
Boiling of water by friction, Rumford's experiments, 13, and Appendix to Chap. II. p. 54
— — — to what due, 135
— point of water raised by being freed of air, 132
— — true definition of, 137
— — lowered by ascending, 138
— — on summit of Mt. Blanc, Mte. Rosa, &c., 138
— — depends on external pressure, 139
Boracic ether, large absorption of heat by vapour of, 433
— — table of dynamic radiation of vapour of, 461
Boutigny, M., his experiments on the spheroidal state of liquids, 200
— — water first frozen in a red-hot crucible by, 202
Brass, expansion of, by heat, 103
Breath, the human, its absorption of heat at different pressures, 538
— — — a physical analysis of, 540
Breeze, land and sea, how produced, 212
British Isles, cause of dampness of, 215
Bromine, opacity of, to light, but transparency to heat proved, 426
Bullet, heat generated by destruction of its motion, 45
Bunsen, Prof., description of his burner, 53
— — his determination of the temperature of Geysers, 145
— — his Geyser-theory, 148
Bunsen's burner, radiation from flame of, through vapours, table, 528
Butyl, nitrate of, action of short waves of ether upon, 750

CALIBRATION of the galvanometer, Melloni's method of, Appendix to Chap. X. p. 330
Calms, regions of, 214
— cause of torrents of rain in region of, 488
Calorescence, experiments on, 614, 617
Caloric proved not to exist by Rumford and Davy, 20, 22
Calorific power of a body, Rumford's estimation of, 179
Calorific conduction, three axes of, in wood, 271
— — of liquids, 291
Candle, combustion of, 50
Capacity for heat, different in different bodies, 18
— — — explained, means of determining, 165
Carbon atoms, collision of, with oxygen, 48
— light of lamps due to solid particles of, 52, 527
— amount of heat generated by its combination with oxygen, 179
— diathermancy of the bisulphide, 580
— experiments with the bisulphide, 599
Carbonic acid, how produced by combustion, 49
— — solid, properties of, 190
— — power of radiation and absorption possessed by, 414, 535
Carbonic oxide, table of absorption of heat by, at different pressures, 411
— — flame, radiation from through carbonic acid gas, 535
— — — — — — olefiant gas, 536
— — — — — — human breath, 538
Celestial dynamics, essay by Mayer on, 83, 602
Chantrey and Blagden, their exposure of themselves in heated ovens, 256
Chemical combination, its effect on radiant heat, 425
Chilling an effect of rarefaction, 26
— when produced, 319

The Numbers (except where p. is prefixed) refer to the Paragraphs, and not to the Pages.

Chilling by radiation, how modified, 492
— — — dew an effect of, 652
Climate, cause of dampness of English, 215
— mildness of European, 215
— effect of aqueous vapour on, 215, 492, 544
Clothes, their philosophy, 282
Clothing, conductivity of materials used in, 283
Clouds, cause of generation of, 489
— composition of, 224
Coal mines, cause of explosions in, 289
Co-efficient of expansion of a gas, 69
— — — linear, superficial and cubic, explained, with table, Appendix to Chap. III. p. 91
Cohesion, force of, lessened by heat, 59
Cohesion of water increased by removal of air, 131
Cold, effect of, on thermo-electric pile, 4
— produced by rarefaction, 28
— produced by the stretching of wire, 109
— of snow and salt, 184
— generated in passing from the solid to the liquid state, 182
— — — — from the liquid to the gaseous state, 186
— — by stream of carbonic acid, 189
— conduction of, 254
— apparent reflection of rays of, 333
Colding, his researches on the equivalence of heat and work, 38 *note*
Collision of atoms, heat and light produced by, 57
Colour, physical cause of, 317, 346
— influence of, on radiation, 341
— of sky, possible cause of, 496, 753*a*
— Dr. Franklin's experiments on the absorption and heat of various colours, 634
Combustion, effect of height on, 55
— Dr. Frankland's memoir on, 56
— theory of, 57
— of gases in tubes, sounds produced by, paper on, Appendix to Chap. VIII. p. 254
Cometary theory, the, p. 547
Compounds good absorbers and radiators, cause of, 424
Compressed air, expansion of, produces cold, 15
Compression, heat generated by, 7
Condensation, congelation, and combination, mechanical value of each in the case of water, 181
— effect of, on specific heat, 171
— of aqueous vapour in tropics, cause of, 488
— — — — by mountains, ditto, 491
— and congelation promoted by water in its different states, 491
Conduction of heat defined and illustrated, 242
— — — not the same in every substance, 244
— — — by metals, 147
— — — experiments of Ingenhausz, 246
— — — Despretz's method of observing, 247
— — — by different metals determined by MM. Wiedemann and Franz, 247
— — — by crystals, 258
Conduction of heat by wood in different directions, table, 259
— — — by bark of various trees, table, 272
— — — importance of knowing specific heat in experiments on, 279
— — — by liquids, 291
— — — by hydrogen gas, 292, 295
— of cold, illustrations of, 254
— power of, not always the same in every direction, 258
Conductivity of metals, table, 247
— — crystals and wood, 258, *et seq.*
— — wood in three directions, table, 269
— — bark of various trees, ditto, 272
— — organic structures, ditto, 275, 277
— — woollen textures, ditto, 283
— — liquids and gases, 291
Conductors, withdrawal of heat by, 287
— good and bad, defined, 244
Conservation of force shown in steam-engine, 9, 140
— — energy, law of, 155
Contraction, generally the result of solidification, 120

The Numbers (except where p. is prefixed) refer to the Paragraphs, and not to the Pages.

Contraction of india-rubber by heat, 112
Convection of heat defined, 221
— — — examples of, 222
— — — by hydrogen, 293
Cooling a loss of motion, 297, 319
— effect of air and hydrogen on heated bodies, 292
— how it may be hastened, 183, 342
Cryophorus, or ice carrier, 187
Crystals, expansion of, 108
— of ice, 127
— of snow, 225
— difference of conductivity in different directions, 258
Cumberland, traces of ancient glaciers in, 238
Currents, aerial, how produced, 206
— upper and lower in atmosphere, 209

DAVY, Sir H., his views of heat, 13, 23, 61
— — discharges a gunlock in vacuo, 11
— — his experiment on the liquefaction of ice by friction, 22
— — — first scientific memoir, Appendix to Chapter III. p. 94
— — — his 'Chemical Philosophy' referred to, 29
— — — investigation of flame, 48
— — — discovery of the safety lamp, 289
— — — experiment on the passage of heat through a vacuum, 297
De la Rive and De Candolle on conduction of wood, 259, 272
Density, point of maximum, in water, 98
— of gas, relation of absorption to, 400
Despretz, his experiments on the conductivity of solids, 247; of liquids, 291
Dew, Dr. Wells' experiments on, and theory of, 648, *et seq.*
— cause of deposition of, 652
— a still night necessary for the formation of, 658
Diamond, Newton's opinion of, 48
— combustion of in oxygen, 48
Diathermancy explained and illustrated, 350
— not a test of transparency, 351, 367
Dilatation of gases effected without chilling, 94
— remarks on Appendix to Chap. III. p. 91
Distillation, locomotive force compared to, 9
Donny, M., his experiments on water purged of air, 132
Dove, Prof., quotation from his work, 209
Drying tubes, difficulties in selecting and obtaining, 390, 471
Dynamic energy defined, 154
— radiation and absorption, discovery of, 447
— — — — of gases, table of, 455
— — — — — — vapours, ditto, 459
— — — — — — boracic ether vapour, table, 461
— — — — — in different lengths of tube, 466
Dynamical theory of heat, 20

EARTH, amount of heat that would be generated by stoppage of its motion, 46; by falling into the sun, 46; by resisting the rotation of, 702
— crust of, thicker than generally supposed, 122
— its rotation and shape, effect of on trade winds, 207, 211
— time required to cool down, 706
— all the energies of, due to the sun, 707
Earthquake at Caraccas, 210
Elective power possessed by bodies with regard to rays of light and heat, 348
Electric light, rise in intensity of the obscure rays of the, 567
— sifting the, 577
Electricity and heat, their relationship shown in the conductivity of various bodies, 250
— current of, increased by cooling conducting wire, 252
Elementary bodies, table of their specific heats, 165, 169
Elements, bad absorbers and radiators, 419, 424, *et seq.*
Emission theory of Newton, 303
Energy, mechanical, converted to heat, 9

Energy, potential or possible, defined, 154
— dynamic or actual, ditto, 154
— potential and dynamic, sum of, constant, 155
— all terrestrial, due to the sun, 707
England, cause of even temperature of, 215, 223
Equatorial ocean, winds from, cause the dampness of England, 215
Equivalent, mechanical, of heat, 38, *et seq.*
— — — — how calculated, 75, *et seq.*
Ether, sulphuric, cause of the cold produced by its evaporation, 186
— — absorption of heat by vapour of, at different pressures, 404
— — — — — by different measures of vapour of, 407
— the luminiferous, mode of transmission of light and heat by, 305
— — — fills all space and penetrates all bodies, 346
— — — the power of imparting motion to and accepting motion from, are proportional, 343, 354, 415
— — — action of ether waves of short period upon gaseous matter, 725–738
— — — — mode of experiment, 741
— — — invisibility of, 743
Euler, his argument for the undulatory theory of light, 303
Europe the condenser of the Atlantic, 224
— cause of mildness of climate of, 215, 223
Evaporation produces cold, 186
— water frozen by, 187
Exchanges, Prevost's theory of, 319
Expansion of volume, 59
— — gases by heat, 67
— — — co-efficient of, 69
— — — without performing work, 94
— — liquids by heat, 97
— — water in freezing, 98
— — — — — use of, in Nature, 101
— — alcohol by heat, 97
— — water by *heat*, 98
— — — — *cold*, 98
— — bismuth in cooling, 102
— — solid bodies by heat, 103
Expansion of lead, curious effect of, 107
— — crystals, 108
Expansive force of heat, 66
Explosions of steam-boilers, 134
— — — — possibly due to the spheroidal state, 201
— in coal mines, cause of, 289
Extra-red and extra-violet rays, 313

FARADAY, his discovery of magneto-electricity, 13
— — experiments on melted ice, 133
— — discovery of the regelation of, ice, 232
— mercury first frozen in a red-hot crucible by, 202
Fibre of wood, power of conduction of heat by, 271
Fire produced by friction, 11
— — by savages, p. 557
— syringe, 25
— balloon, 68
— screens of glass, action of, 363
Flame, constitution of, 49, *et seq.*
— cause of its inability to pass through wire gauze, 288
Flames, singing, paper on, Appendix to Chap. VIII. p. 254
— — Count Schaffgotsch's experiments on, Appendix to Chap. VIII. p. 262
— examination of radiation from, 526, *et seq.*
Florentine experiment, 158, *note*
Fluorescence of sulphate of quinine in the invisible spectrum, 313
— experiments on fluorescence and calorescence, 608, *et seq.*
Fog-signalling, applicability of radiant heat to, 646
Foot-pounds, explanation of, 38
Forbes, Prof. J. D., his viscous theory of ice, 235
Force of heat in expanding bodies, 105
— vital, supposed conservative action of, 257
Forces, molecular, energy of, 99, 158
— polar, heat required to overcome, 168
— convertibility of natural, remarks on, p. 543
Frankland, Dr., his experiments on combustion, 56

The Numbers (except where p. is prefixed) refer to the Paragraphs, and not to the Pages.

Franklin, Dr., his experiment on colours, 634
Fraunhofer's lines, 677
Freezing, effect of, on water pipes, 100
— point lowered by pressure, 123
— of water produced by its own evaporation, 187
— together of pieces of ice, 231
Friction, generation of heat by, 6
— against space, heat developed by, 30
Frost, means of preserving plants from, 654
— cause of their preservation, 654
Fusible alloy liquefied by rotation in magnetic field, 36
Fusion, point of, effect of pressure on, 121

GALVANOMETER described, 3
— note on the construction of, Appendix to Chap. I. p. 19
— peculiarity of, in high deflections, 378
— Melloni's method of calibrating, Appendix to Chap. X. p. 362
Gas, carbonic acid, liberation of from soda-water, consumes heat, 16
— — combustion of, 49
Gas, illuminating power of, 53
— co-efficient of expansion of, 69
— absorbs those rays which it emits, 411
— radiation from a luminous jet of, 526
Gaseous condition of matter, 60
Gases, constitution of, 62, 65
— velocity of particles of, 64
— expansion of, by heat, 67
— specific heat of simple and compound, 169, *et seq.*
— conductivity of, 291
— first experiments on their absorption of heat, 372
— mode of experiment improved, 387
— different powers of accepting motion from the ether, or difference in absorption possessed by, 397, 412
— different powers of imparting motion to the ether or difference in radiation possessed by, 413
— table of dynamic radiation of, 455
— chemical action of short waves of ether upon, 725–738
Gassiot, iron cylinders burst by, 99
Gauze wire, cause of its stopping passage of flame, 288
Geyser, the great, of Iceland, description of, 142
— Bunsen's theory of, 145
— produced in lecture room, 148
— its history, 151
Glaciers, formation of, 228
— motion of, described, 229
— point of swiftest motion shifts, 229
— their daily rate of motion, 229
— viscous theory of, 230
— regelation ditto, 232
— ancient, evidences of in various places, 236, *et seq.*
— — hypothesis to account for, 239, *et seq.*
— cold alone cannot produce, 240
Glaisher, his table of nocturnal radiation, 657
Glass, why cracked by hot water, 105
— broken by a grain of quartz, 106
— opacity of, to heat, 349
— absorption of heat by different thicknesses of, 356
— fire-screens, use and philosophy of, 363
Gmelin, his definition of heat, 17
Gore, his experiments on revolving balls, 118
Gravity, velocity imparted to a body by, 44
Grease, philosophic use of, on wheels and axles, 9
Gulf-Stream, 223
Gypsum, powdered, bad conduction of heat by, 285

HARMONICA, chemical, 301
Heat and cold, opposite effects upon thermo-electric pile, 4
Heat, generated by mechanical processes, 5
— — — friction, 6, 13
— — — compression, 7
— — — percussion, 8
— — — falling of mercury or water, 8
— consumption of, in work, 15
— nature of, 17, 20, 61
— a *motion* of ultimate particles, 20
— considered thus, by Locke, 20
— — — — Bacon, 20

The Numbers (except where p. is prefixed) refer to the Paragraphs, and not to the Pages.

Heat considered a *motion* of ultimate particles by Rumford, 13, 23
— — — — Davy, 23
— developed when air compressed, 25
— — — motion of air stopped, 16, 27
— — by rotation in magnetic field, 36
— mechanical equivalent of, 38, 75, 91
— proportional to height through which a body falls, 40
— relation of, to velocity, 42, 46
— an antagonist to cohesion, 59
— of friction, Rumford's essay on the source of, Appendix to Chap. II. p. 54
— *expansion* of gases by, 67
— — — liquids by, 97
— — — solids by, 103
— imparted to a gas under constant pressure, 69
— — — — — at constant volume, 73
— produced by stretching india-rubber, 110
— direct conversion into mechanical motion, 117
— developed by electricity, 118, 251
— performance of work by, in steam-engine, 140
Heat, M. Hirn's experiments, 140 *b*, *et seq.*
— power of, in expanding bodies, 158
— two kinds of motion produced in bodies by, 159
— interior work performed by, 160
— consumed in forcing atoms asunder, 157, 189
— generated by atoms falling together, 160
— quantity yielded up by different bodies in cooling. 164
— *specific*, 165, *et seq.*
— causes change of state of aggregation in bodies, 176
— *latent*, of water, steam, and aqueous vapour. 176, *et seq.*, 240
— — definition of, 177
— generated in passing from liquid to solid state, 185
— cause of more equal distribution of, 215, 223
— *convection* of, 221
Heat necessary for the production of glaciers, 241
— distinction between it and ordinary motion, 242
— *conduction* of, defined and illustrated, 243
— — — not equal in all substances, 244
— method of determining the conductibility of bodies for, 247
— and electricity, relationship of, 250
— motion of, interferes with the motion of electricity, 252
— conversion of, into potential energy, 256
— difference of conductivity of, in crystals and wood, 258, *et seq.*
— transmission of, through wood, 255, 269
— — — influenced by the mechanical state of the body, 285
— doubtful conduction of, by hydrogen gas, 294
— its passage through a vacuum, 297
— analogy of, to sound, 302
— to what motion of, imparted, 306
— *radiant*, 306
— rays beyond visible spectrum, 311
— obeys the same laws as light, 323
— action of, on oxygen and hydrogen, 330
— laws of inverse squares applied to, 337
— transversal undulation of waves of, 339
— *quality* of, 360
— transmission of, through opaque bodies, 367
— effect of, on ice, 366
— *absorption* of, by gases, first mode of experiment, 372
— — — — — means of detecting minute amount of, 386
— absorption of, by gases, improved apparatus for researches on, described, 387
— free passage of, through dry air, oxygen, hydrogen, and nitrogen, 419
— tables of absorption of, by gases and vapours, 414, 418, 421, 433, 437, 439, 522

The Numbers (except where p. is prefixed) refer to the Paragraphs, and not to the Pages.

Heat, spectrum, detached from luminous, 432
— absorption and radiation of a gas or vapour determined without external heat, 449
— absorption of, by aqueous vapour, 470, *et seq.*
— nocturnal radiation of, the cause of dew, 652, *et seq.*
— amount of, generated by collision of meteors with the sun, 692
— developed by friction of tidal wave, 696
— source of this heat, 697
— Sir William Thomson's statement of the chemical action necessary to produce the sun's heat, 700
— amount of heat equivalent to the rotation of the sun, and the revolution of the planets, 702
— rays, invisible, experiments on, 614
— polarisation of, p. 553
Height, influence of, on combustion, 55
Helmholtz, his calculation of the heat that would be developed by stoppage of earth's motion, 46
— — remarks on the exhaustion of the mechanical force of our system, 706
Herbs, aromatic, action of their odours on radiant heat, 439
Herschel, Sir William, his discovery of the obscure rays of the spectrum, 311, 560
— Sir John, note on rocksalt, 373
— — — measurements of solar radiation, 681
Hirn, M., his experiments on the steam-engine, 140 *b*, *et seq.*
Humboldt, on the coal of Central Asia, 492
Huyghens, his theory of light, 303
Hydrogen, collision of atoms with oxygen, 49
— velocity of its particles, 64
— amount of heat generated by combining with oxygen, to form water, 179
— cooling effect of on heated bodies, 293
— low power of absorption possessed by, 392, 421
Hydrogen, flame, radiation from through liquids and vapours, 530, 542, 547
— — character of its radiation, 547
— — experiments on its radiation, 548, *et seq.*

ICE liquefied by friction, 22, and Appendix to Chap. III. p. 94
— why it swims on water, 101
— liquefied by pressure, 125
— structure and beauty of, 126
— dissected by heat, 127
— flowers, 127, *et seq.*
— carrier, or cryophorus, 187
— viscous theory of, 230
— regelation ditto, 232
— moulded by pressure, 234
— its absorption of heat, 351, 365
— artificial formation of by nocturnal radiation, 655
— this theory supplemented, 655
— amount melted per minute by solar radiation, Herschel and Pouillet's measurements, 681
— amount melted per hour by total emission of sun, 686
Iceland, geysers of, 142
India-rubber, stretching of, produces heat, 110
— — contraction of, by heat, 112
Ingenhausz, his experiments on the conduction of heat, 246
Intensity of light and heat, to what due, 534
Interior work, different kinds of, 168
Interior work performed by heat, 160
Iodine dissolved in bisulphide of carbon, diathermancy of, 430, 582, *et seq.*
— diathermancy of solvents of, 597
Ireland, more rain on west side than on east, 219
— traces of ancient glaciers in, 238
Iron bottle burst by freezing water, 99
— expansion of, by heat, 104
— presence of, in sun, proved, 583
Isothermal line runs north and south in England, 223
Ivory, bad conductivity of, 275

The Numbers (except where p. is prefixed) refer to the Paragraphs, and not to the Pages.

JOULE, Dr., his experiments on the mechanical equivalent of heat, 13, 84, *et seq.*
— — — — — — heat and work, 37
— — — — — — magneto-electricity, 84
— — — — — — the shortening of india-rubber by heat, 111
— — explains heat of meteorites, 12
— — his experiments on the cold produced by stretching wires, 109

KNOBLAUCH, explanation of some of his results, 548
Kopp, Professor, his determination of the cubic co-efficients of expansion, Appendix to Chap. III. p. 92

LAMPBLACK, anomalous deportment of, 427
— radiation of heat through, 367, 428
— — from, 524
Land and sea breezes, how produced, 212
Landscape colours, siftings of light in, 777
Latent heat of water, 26, 177, 240
— — — — — mechanical value of, 181
— — — — liquids, 182
— — — — vapours, 186
Lead ball heated by collision, 40
— curious effect of expansion of, 107
Leidenfrost, first observer of the spheroidal state of liquids, 199
Leslie's cube, radiation from, 340, 524
— æthrioscope, 493
— observations explained, 494
Light produced by friction of quartz, 11
— of lamps, to what due, 49
— of gas destroyed when mixed with air, 52
— law of diminution with distance, 335
— theories of, 303
— analogy of sound to, 303
— propagation and sensation of, 305
— reflection of, 322
— action of on chlorine and hydrogen, 329
— undulations of transversal, 329
Light, capacity to produce, 733
— sifting of a beam of, 744, 777
— substances not hitherto known to be chemically susceptible to, 749
— polarisation of, explained, 735
— chemical value of sky-light, 780
Lime-light, dark rays, experiments on, 626
Liquefaction of ice by friction, 22
— — — — pressure, 125
Liquid condition of matter, 59
— changing to solids produces heat, 185
Liquids, expansion of by heat, 97
— the spheroidal state of, 193
— conductivity of, 291
— calorific transmission of, Melloni's table, 353
— apparatus for determining their absorption of heat at different thicknesses, 500
— table of absorption of heat by, 506
— and their vapours, order of their absorption of heat, 510, 514
Lloyd, Dr., his tables of rainfall in Ireland, 217
Locke, his view of heat, 20
Luminous and obscure radiation, 370

MAGNETIC field, apparent viscosity of, 32
Magnus, Professor, his experiments on gaseous conduction, 291
— — — — — — the conductivity of hydrogen, 294
Material theory of heat, 17
Matter, liquid condition of, 59
— gaseous ditto, 60
Maximum density of water, 98
Mayer, Dr., compares locomotive force to distillation, 9
— — enunciates the relationship between heat and work, 37
— — his calculation of the heat that would be produced by stoppage of earth's motion, 46
— — — mechanical equivalent of heat, 82
— — — essay on celestial dynamics, referred to, 83, 698
— — meteoric theory of sun's heat, 698

The Numbers (except where p. is prefixed) refer to the Paragraphs, and not to the Pages.

Mayer, Dr., elected to the French Academy, p, 233, *note*.
Mechanical processes, generation of heat by, 5
— work, consumption of heat in, 15
— theory of heat, 20
— equivalent of heat, 38
— — — — Mayer's determination, 37, 81
— — — — Joule's determination, 37, 84
Meidinger, M., his experiments on ozone, 444 *note*
Melloni, his mode of proving the diminution of heat as the square of the distance, 338
— his researches on radiant heat, 350
— his table of the transmission of heat through solids, 351
— — table of the transmission of heat through liquids, 353
— source of error in his experiments on transmission of heat through liquids, 499
— his theory of serein, 495
— explanation of some of his results, 551
— his addition to the theory of dew, 660
— — experiments on the warmth of the lunar rays, 661
Mercury, low specific heat of, 163
— frozen by solid carbonic acid, 192
— — in red-hot crucible, 203
Metals, good conductors of heat, 245
— bad radiators, 340
— — absorbers, 345
— effect of their bad radiation, 653
— bands seen in spectra of their vapours, 664
— presence of terrestrial, in sun, proved, 679
Meteorology, absorption of heat by aqueous vapour applied to phenomena of, 487, *et seq.*
Meteors, zodiacal light supposed to be, 47, 691, 699
— number of, seen in Boston, 690
— amount of heat generated by collision of, with sun, 692
— sun's light and heat possibly kept up by, 693
Miller, Dr., on rays of burning hydrogen, 547
Mitscherlich, Professor, his experiments on the expansion of crystals, 108
Molecular motion, heat defined as, 20, 58, 242
— vibration of a body more intense when heated, 58
— forces irresistible, 99
— — power of, 158
— — — — calculated, 180
— action in wood, effect of, 271
Molecules, their motions as wholes, 730
Mongolfier, equivalence of heat and work, maintained by, 37
Moon-blindness, cause of, 654
— -beams, cause of their putrefying power, 654
— experiments on warmth of, 465
— obscure heat of, cut off by our atmosphere, 465
Moraines of ancient glaciers, cedars of Lebanon grow on, 238
Mosely, Rev. Canon, curious effect of expansion noted by, 107
Motion, heat considered to be, by Bacon and Locke, 20; by Rumford, 13; by Davy, 13, 23
— transference of from mass to molecules, 23, 58
— point of maximum in a glacier, 229
Mountains, their action as condensers, explained, 490
Moving force, amount of heat produced by destruction of, 40
— — produced by steam, 140
— — or dynamic energy, defined, 154

NATURAL philosopher, his vocation, 102
Nature, adaptation of means to ends in, 102
Névé, the feeder of the glacier, 228
Newton, his opinion of the diamond, 48
— — theory of light, 303
Nitrogen, velocity of particles of, 64
Nitrous oxide, absorption and radiation of, 414

The Numbers (except where p. is prefixed) refer to the Paragraphs, and not to the Pages.

Nitrous oxide, dynamic radiation of, 455
— acid gas, bands produced by spectrum of, 671
Nocturnal radiation, artificial formation of ice by, 655
— — experiments on, by Wells, Glaisher, and others, 650, 657
Novum Organum, extract from 2nd book of, Appendix to Chap. II. p. 53

OBSCURE HEAT, rays of, obey same laws as light, 323
— — ratio of, to luminous rays from different sources, 370
Ocean, influence of, on temperature, 175
Olefiant gas, athermancy of, 393, 421
— — table of absorption by at different pressures, 397, 536
— — — — — by various measures, 402
— — radiation of, 414
— — dynamic radiation of, 455
— — varnishing metal by, 457
Optic nerve, effect of heat rays on, 612*a*
— — effect of invisible rays on, 643
Organic motion, extracts from a paper by Mayer on, 83
— structures, table of conductivity of, 275
Oxygen, collision of atoms of with carbon, 48 ; with hydrogen, 180
— velocity of particles of, 64
— small absorption of heat by, 392, 421
Ozone, action of, on radiant heat, 441
— increase of, by radiation in size of electrodes, 443
— probable constitution of, 445

PARABOLIC mirrors, reflection of light and heat from, 328, *et seq.*
Particles, impact of, causes sensation of heat, 65
— ultimate, motion of produces heat, 20
Percussion, heat generated by, 8
Perfumes, how propagated, 62
— table of absorption of heat by, 437
Period, heat and light differ only in, 517
— influence of on absorption, 517
— determines the quality of heat emitted by bodies, 518
Periods, vibrating, of formic and sulphuric ether, 524
— — of a hydrogen flame, 547
— — shortening of, 547
Perspiration, use of in hot climates, 256
Photosphere of sun, action of on solar rays, 678
Physical analysis of the human breath, 540
Pile, thermo-electric, construction and use of, 3, and Appendix to Chap. I. p. 17
Pitch of note, upon what dependent, 316
Planets, orbital velocity of the interior, 12
— heat that would be developed by their falling into sun, or by resisting rotation of, 702
Platinum lamp, described, 503
Polar forces, heat required to overcome, 168
Polarisation of heat, p. 553
— of light, explained, 735
Potential or possible energy defined, 154
Pouillet, M., his experiments on the temperature of air and swan's down, 659
— — —measurement of solar radiation, 681 ; of its partial absorption by our atmosphere, 684
— — — pyrheliometer, 681
Pressure, relating to heating of gases, 69, 74
— effect of on point of fusion, 121
— — — — crust of earth, 122
— liquefaction of ice by, 125
Prevost's theory of exchanges, 319
Pyrheliometer, use and description of, 681
Pyrometers, 104

QUARTZ, clear and smoky, transmit equal amounts of heat, 351
Quality of radiant heat, definition of, 360, 518, 534

The Numbers (except where p. is prefixed) refer to the Paragraphs, and not to the Pages.

RADIANT heat, definition of, 306
— — and light, analogy between, 306
— — emitted by all bodies, 319
— — laws the same as those of light, 323
— — reflection and convergence of rays of, 326
— — law of inverse squares applied to, 338, *et seq.*
— — its origin and propagation, 339
— — apparatus for researches on described, 372–391
— — absorption of, by gases, 418
— — — — — vapours, 433, 522
— — action of perfumes on, 437
— — object of researches on, 497
Radiating body and air, difference between constant, 659
Radiation, effect of colour on, 341
Radiation and absorption, reciprocity of, 343, 354, 415, 459
— of metals, 340
— — heat by solids, 340, 354
— — — — gases, 413, 455
— — — — vapours, 459, 465
— and absorption of a gas or vapour determined without external heat, 449
— — — *dynamic*, table of gases, 455
— dew an effect of chilling by, 652
— nocturnal, Glashier's table of chilling by, 657
— — artificial formation of ice by, 655
— through the earth's atmosphere, Appendix to Chap. XI. p. 370
— luminous and obscure, 370
Rain, cause of the torrents of, in the tropics, 214
— fall, greater on west than on east coast of Ireland, 217
— — Dr. Lloyd's table of, in Ireland, 217
— — place where the greatest occurs, 220 *note*
— — upon what dependent, 220
Rarefaction, chilling effect of, 27
—will not by itself lower mean temperature, 94
Rectilinear motion, atoms of gases move with, 65
Reflection of light and heat obey same laws, 323
Refrigeration by expansion of a gas, 95
Regelation, discovery of, by Faraday, 232
Rendu, his plastic theory of ice, 235
Resistance, heat of electric current proportional to, 251
Retina of the eye, question whether the invisible rays emitted by luminous sources reach the, 643
Revolving balls, Gore's experiments on, 118
Rifle ball, amount of heat generated by stoppage of its motion, 45
Rocker used in the Trevelyan instrument, 114
Rocksalt, transparency of, to heat, 349, 361
—use of in experiments on absorption of heat by gases, 373
—hygroscopic character of, 471
—deposition of moisture on, avoided, 472
—cell, described, 500
Roscoe, Professor, his experiments on sky-light, 780
Rumford, Count, his experiments on heat produced by friction, 13, and Appendix to Chap. II. p. 54
— — — overthrow of the material theory of heat, 20
— — abstract of his essay on heat, Appendix to Chap. II. p. 54
— — his estimation of the calorific power of a body, 179
— — — experiments on the conductivity of clothing, 283
— — — — — — — — liquids and gases, 291
— — — — — — transmission of heat through a vacuum, 297
Rupert's drops, 106

SAFETY LAMP, explanation and use of, 290
Salt and sugar, dissolving of, produces cold, 183
— common, yellow bands emitted and absorbed by vapour of, 413
Scents, action of on radiant heat, 435

The Numbers (except where p. is prefixed) refer to the Paragraphs, and not to the Pages.

Schaffgotsch, Count, his paper on acoustic experiments, Appendix to Chap. VIII. p. 262
Schemnitz, machine for compression of air at, 29
Schwartz, his observation of sound produced by cooling silver, 113
Sea, warmer after a storm, 8
— breeze, how produced, 212
Séguin, equivalence of heat and work developed by, 37
Selenite, absorption of heat by different thicknesses of, 358
— colours of, 769
Senarmont, his experiments on the conduction of heat by crystals, 258
Serein, Melloni's theory of, 495
Shooting stars, theory of, 11
Silica, water of Geysers, contains and deposits, 143
— as crystal, high conductive power of, 273
— as powder, low ditto, 285
Singing flames, 301
— — paper on, Appendix to Chap. VIII. pp. 254–262
Sky, colour of, 496
— — causes of, 753*a*, 761, 762
— — natural and artificial azure compared, 767
— chemical value of sky-light, 780
Snow, shower of, produced by issuing of compressed air, 29
— — — by chill in a room, p. 557
— carbonic acid, 189
— crystals, 225
— line, the, 226
— formation of glaciers from, 227
— ball, cause of coherence of, 233
— bridges, how crossed, 233
— squeezed to ice, 234
Sodium, yellow bands emitted and absorbed by vapour of, 673, 675
Solar spectrum, cause of dark lines in, 677
— rays, dark, discovery of, 557, 560
— — visible and invisible, 559
— — ray-filters, 575
— — thermal image rendered luminous, 589
— — combustion and incandescence by dark solar rays, 592
Solidification accompanied by expansion, 98
Solidification accompanied by contraction, 120
Solids, expansion of by heat, 103
— calorific transmission of, Melloni's table, 351
Sound, conversion of heat into, 114
Sound, mode of its transmission through air, 300, 315
— produced by flame, 301
— undulation of waves of, longitudinal, 339
—analogy of heat and light to, 305
Sounds, inaudible, 318
— musical, produced by gas flame in tubes, 301, and Appendix to Chap. VIII. p. 254
Specific heat of bodies, how determined, 165
— — — elementary bodies, 165
— — — simple and compound gases, 169, *et seq.*
— — of water the highest, consequences attending, 174
— — masking the conductive power of a body, 279
Spectra of zinc, copper, &c., 666
Spectrum, heat of non-luminous, proved, 311
— non-luminous, obtained, 432
— solar, Herschel and Müller's experiments on the distribution of heat in the, 561, 562
— — cause of dark lines in, 677
— of incandescent carbon, 664
— of solids, similar to, 664
Spheroid, floating of, on its vapour, 195
— not in contact with vessel, proved, 197
Spheroidal state of liquids, 103, *et seq.*
— condition, first observer of, 199
Spiral of platinum wire heated by electric current, radiation from, 522, 547
Springs, boiling, of Iceland, described, 142
Steam, how produced, 137
— elastic force of, increased by heating, 140
— latent heat of, 181
Storms produced by heated air, 205
Strokkur, the imitation of, 149
Sulphate of soda, cold produced by dissolving, 185

The Numbers (except where p. is prefixed) refer to the Paragraphs, and not to the Pages.

Sulphate of soda, heat produced by crystallising, 185
Sulphuric acid used for drying gases, 390
— ether, absorption of heat by vapour of, 404, 433
Sun, probable cause of continuance of heat and light of, 47, 693
— production of winds by heat of, 206
— does not heat dry air sensibly, 364
— discovery of the dark rays of the, 560
— generation and intensification of rays visible and invisible, 563
— rise of intensity with temperature, 563, 564
— combustion by dark rays, 592
— fluorescence and colorescence of the sun, 608, 617
— Sir W. Thomson's statement of the chemical action necessary to produce the sun's heat, 700
— constitution of, 677
— flame atmosphere surrounding, 678
— and planets, supposed common origin of, 679
— heating power of, measurements by Herschel and Pouillet, 681
— mode of determining the radiation from, 682
— atmospheric absorption of heat of, 684
— total amount of heat emitted by, 685
— all organic and inorganic energy referred to, 707, 721
— small fraction of its heat that produces all terrestrial energy, 724
Switzerland, evidences of ancient glaciers in, 236

TEMPERATURE, absolute zero of, 96
— influence of, on conduction of electricity, 252
— high, how endured, 256
— dew caused by lowering of, 650
— influence of on the quality of heat emitted by a body, 518
— — — on transmission, 522
Temperature, difficulties in ascertaining the true, 653
Teneriffe, Peak of, two currents blow on, 215
Thermal effects produced by stoppage of motion, 24, 39
Thermo-electric pile, 3
— — — pile, note on the construction of, Appendix to Chap. I. p. 17
— — — used in researches on radiant heat, 307
— Ruhmkorff's, 568
Thermograph of coal-points, 604
Thermometer, construction of, Appendix to Chap. III. p. 92
Thickness, influence of on absorption of heat, 357
Thomson, Professor Sir Wm., on earth's crust, 101, 122
— — — his suggestion that india-rubber would shorten by heat, 111
— — — — meteoric theory of the sun, 47
— — — — tables of energy, 702
— Professor James, on the influence of pressure on fusion, 124
Tidal wave, velocity of earth's rotation diminished by, 697
Trade winds, upper and lower, 206
Transmission of heat through solids, Melloni's table, 351
— — — — liquids, ditto, 353
— — — influence of temperature of source on, 522, 524, 547
Transparency of bodies, cause of, 346, 517
— not a test for diathermancy, 351; 367
Trevelyan, Mr. A., his instrument, 113
— — — cause of vibrations of, 115
Tropics, flow of air from and to, 211
— the region of calms or rains, 214
— cause of the torrents of rain in, 488

UNDULATION THEORY, 304

VACUUM in centre of ice flowers, 128
— passage of heat through, 297

The Numbers (except where p. is prefixed) refer to the Paragraphs, and not to the Pages.

Vacuum, dry air similar to, with regard to radiant heat, 392
Vaporous condition of matter, 60
Vapour of water condensed by rarefaction of air, 28
— — — its action on radiant heat, 470, 492
— — — condensation promoted by, 491
— production of, consumes heat, 186, 240
— supporting of spheroid by, 194
— of metals, spectrum of, 664
— absorbs those rays which it emits, 532
Vapours and liquids, their absorption of heat compared, 510, 514
— tables of absorption of heat by, 433, 437, 459, 522, 524
— — — dynamic radiation and absorption of, 459, 466
— matter entering into our conception of a vapour, 730
— chemical absorption by liquid and vapour, 747
Varnishing a metal or feeble gas by a powerful one, 457
Velocity of planets and aërolites, 12
— relation of heat to, 42
Vibration of heated metal, 113
— — sounding disks, 299
Viscous theory of ice, 230
Vision, a differential effect, 772
Vital force, supposed conservative action of, 257
Volcanic eruptions showing upper currents of air, 209
— eruption of Morne Garou, 210
Volume of a gas augmented by heat, 66, *et seq.*
Volumes of vapour proportional to liquid, table, 512

WATER boiled by friction, 13, and Appendix to Chap. II. p. 54
— expanded by heat, 98
— — — cold, 98
— maximum density of, 98
— pipes, why burst, 100
— cohesion of, increased by removing air from, 131, *et seq.*
— hammer, 131
— effects of, when in a highly cohesive condition, 132
Water formerly regarded as incompressible, 157
— Bacon's experiment on the compression of, 157 *note*
— amount of heat yielded by, in cooling 1°, 163
— has the highest specific heat, 165
— amount of work equal to heating of 1°, 167
— effect of high specific heat of, 174
Water, latent heat of, 177
— mechanical value of combination, condensation and congelation of, 181
— evaporation of, produces cold, 186
— frozen by its own evaporation, 187
— spheroidal state of, 193, *et seq.*
— frozen in red-hot crucible, 202
— opacity of to heat, 348, 507
— distilled, colour of, 355
— effects of its energy as a radiant, in all its states, 491
— absorbs same rays when solid, liquid, or vapour, 365, 496, 544
— amount of would be boiled by the total emission of sun, 686
— cause of its hardness, 286
— — — — transparency to light, 542
— — — — opacity to heat, 542
— absorption of heat from hydrogen flame at different thicknesses, 542
Waterston, his meteoric theory of the sun, 47, 699
Waves of sounds, 316
— — light, 317
— — heat and sound, difference between, 339
Wells, Dr., his theory of dew, 648, *et seq.*
— — many curious effects explained by, 654
Wiedemann and Franz, their table of conductivities, 247
Winds, extinction of light of gas by, 54
— produced by sun, 206
— trade, 206
— direction of, influenced by earth's rotation, 207
Winds, lesser, cause of, 212

The Numbers (except where p. is prefixed) refer to the Paragraphs, and not to the Pages.

Wollaston, Dr., his cryophorus, 187
— — lines in solar spectrum observed, by, 687
Wood, bad conductivity of, 255
— difference of conductivity in, 259, 269
— apparatus for ascertaining calorific conductivity of, 260
— three axes of conductive power in, 271
Woollen textures, imperfect conduction of, 283
Work, constant proportion between it and heat, 15, 37, 156
Work, interior, 160, 167

YOUNG, Dr. Thos., establishment of the undulatory theory, 13, 304

ZERO, absolute, of temperature, 96
Zinc, bands seen in spectrum of vapour of, 665
Zodiacal light, probable cause of, 47, 691, 699

LONDON: PRINTED BY
SPOTTISWOODE AND CO., NEW-STREET SQUARE
AND PARLIAMENT STREET

Printed in Dunstable, United Kingdom

91231272R00338